Plasticulture Engineering and Technology

Plasticulture Engineering and Technology

Rohitashw Kumar and Vijay P. Singh

CRC Press
Taylor & Francis Group
Boca Raton London New York

CRC Press is an imprint of the
Taylor & Francis Group, an **informa** business

First edition published 2022
by CRC Press
6000 Broken Sound Parkway NW, Suite 300, Boca Raton, FL 33487-2742

and by CRC Press
2 Park Square, Milton Park, Abingdon, Oxon, OX14 4RN

© 2022 Taylor & Francis Group, LLC

CRC Press is an imprint of Taylor & Francis Group, LLC

Reasonable efforts have been made to publish reliable data and information, but the author and publisher cannot assume responsibility for the validity of all materials or the consequences of their use. The authors and publishers have attempted to trace the copyright holders of all material reproduced in this publication and apologize to copyright holders if permission to publish in this form has not been obtained. If any copyright material has not been acknowledged please write and let us know so we may rectify in any future reprint.

Except as permitted under U.S. Copyright Law, no part of this book may be reprinted, reproduced, transmitted, or utilized in any form by any electronic, mechanical, or other means, now known or hereafter invented, including photocopying, microfilming, and recording, or in any information storage or retrieval system, without written permission from the publishers.

For permission to photocopy or use material electronically from this work, access www.copyright.com or contact the Copyright Clearance Center, Inc. (CCC), 222 Rosewood Drive, Danvers, MA 01923, 978-750-8400. For works that are not available on CCC please contact mpkbookspermissions@tandf.co.uk

Trademark notice: Product or corporate names may be trademarks or registered trademarks and are used only for identification and explanation without intent to infringe.

Library of Congress Cataloging-in-Publication Data

Names: Kumar, Rohitashw, author. | Singh, V. P. (Vijay P.), author.
Title: Plasticulture engineering and technology/Rohitashw Kumar, Vijay P. Singh.
Description: First edition. | Boca Raton, FL: CRC Press, 2022. | Includes index.
Identifiers: LCCN 2021048665 (print) | LCCN 2021048666 (ebook) | ISBN 9781032045696 (hardback) | ISBN 9781032227450 (paperback) | ISBN 9781003273974 (ebook)
Subjects: LCSH: Plastics. | Plastics in agriculture. | Plastics in irrigation.
Classification: LCC S494.5.P5 K86 2022 (print) | LCC S494.5.P5 (ebook) | DDC 631.3--dc23/eng/20211122
LC record available at https://lccn.loc.gov/2021048665
LC ebook record available at https://lccn.loc.gov/2021048666

ISBN: 9781032045696 (hbk)
ISBN: 9781032227450 (pbk)
ISBN: 9781003273974 (ebk)

DOI: 10.1201/9781003273974

Typeset in Times
by Deanta Global Publishing Services, Chennai, India

Dedication

Rohitashw Kumar: My late parents Smt. Rajwan Devi and Sh. Chhalu Ram; Brother Sh. Balbir Singh and Bhabhi Smt Kitabo Devi; my wife Mrs Reshma Devi; daughter Meenu; and son Vineet Kumar.

Vijay P. Singh: My wife Anita who is no more; son Vinay; daughter Arti; daughter-in-law Sonali; son-in-law Vamsi; grandsons Ronin, Kayden, and Davin.

Contents

Preface xxiii
Acknowledgments xxvii
About the Authors xxix

Chapter 1 Introduction 1

    1.1   Definition 1
    1.2   Plasticulture Development 1
    1.3   Film Manufacturing 3
    1.4   Types and Quality of Plastics Used in Soil and Water Conservation 4
          1.4.1   Micro-Irrigation 5
          1.4.2   Protected Cultivation 5
          1.4.3   Mulching 7
          1.4.4   Reducing Water Usage in Agriculture 8
    1.5   Production Agriculture and Postharvest Management 9
          1.5.1   Postharvest and Storage Management 10
          1.5.2   Fruit and Vegetable Packaging 10
          1.5.3   Silage Cover 11
    1.6   Preference of Plasticulture 11
    1.7   Importance of Plastics 12
    1.8   Types of Plastics 12
          1.8.1   Thermoplastic and Thermosetting Plastics 12
          1.8.2   Types of Thermoplastics and Their Uses 13
          1.8.3   Types of Thermosetting Polymers and Their Uses 13
    1.9   Plasticulture in India 14
    1.10   How Plastic Is Used 15
    1.11   Status of Plastics 15
    1.12   Agencies Involved in Promotion of Plasticulture in India 16
          1.12.1   Formation of NCPAH 16
    Questions 16
    Multiple-Choice Questions 16
    Answers 17
    References 17

Chapter 2 Plastic Film Properties 19

    2.1   Introduction 19
    2.2   Polymer 19
    2.3   Classification of Polymers 20
          2.3.1   Classification of Polymers Based on the Source of Availability 20
    2.4   Thermoplastics 20
    2.5   Polyethylene 21
          2.5.1   Classification 21
    2.6   Low-Density Polyethylene 21
          2.6.1   Application of LDPE Areas 23
    2.7   High-Density Polyethylene 23
    2.8   LDPE Application for Film Blowing and Extrusion 23

2.9	Cross-Linked Polyethylene	23
2.10	General Properties and Test Method of LDPE	24
2.11	Polymerization	24
	2.11.1 High-Pressure Polymerization	24
	2.11.2 Natural Polymers	26
	2.11.3 Semisynthetic Polymers	26
	2.11.4 Synthetic Polymers	26
2.12	Classification of Polymers Based on the Structure	26
	2.12.1 Linear-Chain Polymers	26
	2.12.2 Branched-Chain Polymers	26
	2.12.3 Cross-Linked Polymers	26
2.13	Classification of Polymers Based on Mode of Synthesis	27
	2.13.1 Addition Polymers	27
2.14	Classification of Polymers Based on Monomers	27
	2.14.1 Homomer	27
	2.14.2 Heteropolymer or Copolymer	27
2.15	Classification Based on Molecular Forces	27
	2.15.1 Elastomers	27
	2.15.2 Fibers	27
	2.15.3 Condensation Polymers	27
	2.15.4 Thermosetting Polymers	28
2.16	Structure of Polymers	28
2.17	Types of Polymers	28
	2.17.1 Classification Based on the Type of Backbone Chain	28
	Organic Polymers	28
	Inorganic Polymers	28
	2.17.2 Classification on the Basis of Their Synthesis	28
2.18	Biodegradable Polymers	29
	2.18.1 High-Temperature Polymers	29
2.19	Properties of Polymers	29
	2.19.1 Physical Properties	29
	2.19.1.1 Degree of Polymerization and Molecular Weight	29
	2.19.1.2 Molecular Mass of Polymers	29
	2.19.1.3 Number Average Molecular Weight	29
	2.19.1.4 Weight Average Molecular Weight	30
	2.19.1.5 Polydispersity Index or Heterogeneity Index	30
	2.19.2 Mechanical Properties	31
	2.19.2.1 Tensile Strength	31
	2.19.2.2 Young's Modulus of Elasticity	31
	2.19.3 Transport Properties	31
	2.19.4 Chemical Properties	32
	2.19.5 Optical Properties	32
	2.19.6 Electrical Properties	32
2.20	Polymer Extrusion	32
	2.20.1 Variance Between Single- and Twin-Screw Extruders	34
	2.20.2 Distinctive Zones of Single- and Twin-Screw Extruders	35
2.21	Die and Screw Parameters	35
Short Questions		36
Multiple-Choice Questions		37
Answer		37
References		37

Contents ix

Chapter 3 Micro-Irrigation 39
    3.1   Introduction 39
    3.2   History of Micro-Irrigation 40
            3.2.1   General Principles of Micro-Irrigation 41
                    3.2.1.1   Wetting Patterns Under Micro-Irrigation 41
    3.3   Advantages 41
            3.3.1   Water Saving 41
            3.3.2   Lower Water Application Rates 42
            3.3.3   Improved Fertilizer and Chemical Application 43
            3.3.4   Water Sources with High Salinity 44
            3.3.5   Improved Crop Yield 44
            3.3.6   Feasibility for Any Topography 44
            3.3.7   Other Advantages 45
    3.4   Disadvantages 45
            3.4.1   Clogging 45
            3.4.2   High Maintenance Costs 45
            3.4.3   Salt Accumulation at Root Zone Periphery 45
            3.4.4   Moisture Distribution 45
            3.4.5   Restricted Root Development 46
            3.4.6   High Cost of Drip Irrigation Systems 46
    3.5   Need of Micro-Irrigation 46
    3.6   Types of Micro-Irrigation 46
            3.6.1   Online Emitter/Dripper System 47
            3.6.2   Inline Drip System 47
                    3.6.2.1   J-Turbo Line 47
                    3.6.2.2   J-Turbo Aqua 47
                    3.6.2.3   Twin-Wall Drip Tap 47
            3.6.3   Micro Jets 48
            3.6.4   Mini Sprinklers 48
    3.7   Drip Irrigation 48
            3.7.1   Components of Drip Irrigation System 49
    3.8   Sprinkler Irrigation 49
            3.8.1   Components of Sprinkler Irrigation System 50
    3.9   Installation of Micro-Irrigation Systems 50
    3.10 Maintenance and Troubleshooting 51
    3.11 Micro-Irrigation under Protected Cultivation 51
    3.12 World Scenario of Micro-Irrigation System 52
    3.13 Micro-Irrigation Potential in India 54
    3.14 Conclusion 55
    Questions 55
    Multiple-Choice Questions 55
    Answers 56
    References 56

Chapter 4 Design and Components of Micro-Irrigation 59
    4.1   Definition 59
    4.2   General System Design 59
            4.2.1   Initial Assessment 60
    4.3   Objectives of Design 63

4.4	Design Inputs Parameters	63
4.5	Steps to Design Micro-Irrigation System	63
	4.5.1 System Capacity	63
	4.5.2 Selection of Emitting Devices	63
	4.5.3 Selection and Design of Laterals	63
	4.5.4 Design of Submain	64
	4.5.5 Design of Main Line	64
	4.5.6 Selection and Design of Filtration Unit	64
	4.5.7 Selection and Design of Pump	64
4.6	Layout and Components	64
4.7	Design Process	64
	4.7.1 Preparatory steps of the micro-irrigation system design process:	64
	4.7.2 Steps for the general micro-irrigation system design process	66
4.8	Sources of Water	67
4.9	Types of Micro-Irrigation Systems	68
	4.9.1 Bubbler Irrigation	68
	4.9.2 Application and General Suitability	68
4.10	Advantages and Disadvantages	69
4.11	System Design	70
4.12	Drip Irrigation	70
	4.12.1 Advantages	71
	4.12.2 Surface and Subsurface Drip Irrigation Systems	73
	4.12.3 Components of Drip Irrigation Systems	74
	4.12.3.1 Drippers	75
4.13	Design of Drip Irrigation System	75
	4.13.1 Discharge of Drippers	75
	4.13.2 Water Distribution Network	76
	4.13.3 Main and Submain Pipes	76
	4.13.4 Laterals	77
	4.13.5 Manifold	78
	4.13.6 Pipeline Accessories and Fittings	78
	4.13.7 Valves	78
	4.13.7.1 Air Release and Vacuum Relief Valves	79
	4.13.7.2 Pressure Relief Valves	81
	4.13.7.3 Pressure-Regulating Valves	81
	4.13.7.4 Nonreturn Valves	81
	4.13.7.5 Flow Control Valves	81
	4.13.8 Filtration Systems	81
	4.13.8.1 Settling Basins	82
	4.13.8.2 Gravel/Sand Media Filters	83
	4.13.8.3 Screen Filters	84
	4.13.8.4 Hydrocyclones or Centrifugal Sand Separator	85
4.14	Application of Fertilizers and Chemicals Using Drip Irrigation Systems	87
	4.14.1 Equipment and Methods for Fertilizer Injection	87
	4.14.2 Fertilizer Tank	87
	4.14.3 Fertilizer Dissolver	88
	4.14.4 Fertilizer Injection Devices	88
	4.14.4.1 Location of Fertilizer Injection Systems	89
	4.14.5 Computation of the Quantity of Fertilizer to Be Applied	89
4.15	Irrigation Water Requirement	93
4.16	Capacity of Drip Irrigation System	94

	4.17	Sprinkler Irrigation	95
		4.17.1 Adaptability of Sprinkler Irrigation	95
		4.17.2 Advantages	96
		4.17.3 Limitations	97
		4.17.4 Components of Sprinkler Irrigation	97
	4.18	Design of Sprinkler System	102
		4.18.1 Uniformity Coefficient of Sprinklers	102
		4.18.2 Capacity of Sprinkler Irrigation Systems	106
	Questions		107
	Multiple-Choice Questions		108
	Answers		109
	References		109
Chapter 5	Application of Plastic in Water Management		111
	5.1	Introduction	111
	5.2	Micro-irrigation	111
	5.3	Moisture Conservation	112
	5.4	Canal Lining	112
		5.4.1 Seepage Reduction	113
		5.4.2 Prevention of Waterlogging	113
		5.4.3 Increase in Commanded Area	114
		5.4.4 Increase in Channel Capacity	114
		5.4.5 Less Maintenance	114
		5.4.6 Safety Against Floods	114
		5.4.7 Advantages of Canal Lining	114
	5.5	Water Harvesting	115
	5.6	Farm Ponds	116
		5.6.1 Types of Ponds	116
		5.6.1.1 Dugout Farm Ponds	116
		5.6.1.2 Embankment-Type Farm Ponds	117
		5.6.1.3 Spring- or Creek-Fed Ponds	117
		5.6.1.4 Off-Stream Storage Ponds	117
		5.6.2 Design of Farm Ponds	117
		5.6.2.1 Site Selection	118
		5.6.2.2 Capacity of the Pond	118
		5.6.2.3 Design of Embankment	119
		5.6.2.4 Design of Mechanical Spillway	122
		5.6.2.5 Design of Emergency Spillway	122
	5.7	Types of Polythene Sheets	123
	5.8	Procedure of Poly Tank Construction	123
	5.9	Farm Pond and Its Benefits	124
	5.10	Lining of Water Bodies/Farm Ponds	125
		5.10.1 Plastic Film as Lining Material	125
	5.11	Role of Plastic Film (Agrifilm) in Lining	125
	5.12	Future Thrust Area for Agrifilm Lining	126
	5.13	Lined Ponds for Storage of Canal Water	127
	5.14	Underground Pipeline System	127
		5.14.1 Data Required for Underground Pipeline System Planning	127
		5.14.2 Advantages of Underground Pipeline System	128
		5.14.3 Application of Underground Pipeline System	128

5.15	Lysimeter		128
5.16	Application of Plastic in Agriculture Drainage		129
	5.16.1	Design Parameters of the Subsurface Drainage System	131
5.17	Plastic Mulch		131

Questions 131
Multiple Choice Questions 131
Answer 132
References 133

Chapter 6 Soil Conditioning and Solarization Using Plastics 135

- 6.1 Introduction 135
- 6.2 Importance and Functions of Soil Conditioner 135
- 6.3 Types of Soil Conditioners 136
  - 6.3.1 Organic Soil Conditioners 136
    - 6.3.1.1 Composts 136
    - 6.3.1.2 Farm Yard Manure (FYM) 136
    - 6.3.1.3 Green Manure 136
    - 6.3.1.4 Sewage Sludge 136
    - 6.3.1.5 Crop Residues 136
    - 6.3.1.6 Peat Moss 137
    - 6.3.1.7 Biochar 137
  - 6.3.2 Inorganic Soil Conditioners 137
  - 6.3.3 Mineral Soil Conditioners 137
    - 6.3.3.1 Gypsum 137
    - 6.3.3.2 Lime 137
    - 6.3.3.3 Fly Ash 138
  - 6.3.4 Synthetic Binding Agents 138
    - 6.3.4.1 Cationic Polymers 138
    - 6.3.4.2 Anionic Polymers 138
- 6.4 Soil Solarization 138
  - 6.4.1 Mechanisms of Solarization 138
    - 6.4.1.1 Thermal Mechanism 139
    - 6.4.1.2 Chemical Mechanism 139
    - 6.4.1.3 Biological Mechanism 139
  - 6.4.2 Factors Affecting Solarization 139
    - 6.4.2.1 Soil Temperature 139
    - 6.4.2.2 Soil Moisture 140
    - 6.4.2.3 Climate and Weather 140
    - 6.4.2.4 Plastic Film 140
  - 6.4.3 Solarization Results 140
    - 6.4.3.1 Increased Soil Temperature 140
    - 6.4.3.2 Improved Soil Physical and Chemical Features 141
    - 6.4.3.3 Control of Pests 141
    - 6.4.3.4 Fungi and Bacteria 141
    - 6.4.3.5 Nematodes 141
    - 6.4.3.6 Weeds 141
    - 6.4.3.7 Encouragement of Beneficial Soil Organisms 141
    - 6.4.3.8 Increased Plant Growth 142
  - 6.4.4 Improving Solarization Efficacy 142
  - 6.4.5 Current Usage 143

		Questions	143
		Multiple-Choice Questions	144
		Answers	144
		References	145
Chapter 7		Irrigation Scheduling to Enhance Water Use Efficiency	147
	7.1	Irrigation	147
	7.2	Irrigation Scheduling	147
	7.3	Full Irrigation	148
	7.4	Deficit Irrigation	148
	7.5	Irrigation Interval	148
		7.5.1 Factors Affecting Irrigation Interval	149
	7.6	Benefits of Irrigation Scheduling	149
	7.7	Factors Affecting Irrigation Scheduling	149
	7.8	Difficulties in Irrigation Scheduling at Farm Level	150
	7.9	Irrigation Scheduling Methods	151
	7.10	Observation of the Plants and Soils	151
		7.10.1 Methods for Monitoring Soil Moisture	154
		7.10.2 Soil Moisture Measuring Devices	154
		7.10.3 Tensiometer	154
		7.10.4 Soil Moisture Sensors	155
	7.11	Resistance Devices	156
		7.11.1 Gypsum Blocks	156
		7.11.2 Granular Matrix Blocks	157
		7.11.3 Neutron Probe	157
		7.11.4 Gravimetric Method	157
		7.11.5 Thermocouple Psychrometry	158
		7.11.6 Stomata Resistance	158
		7.11.7 Infrared Thermometer	158
		7.11.8 Climatological Approach (IW:Cumulative Pan Evaporation (CPE) Ratio)	158
		7.11.9 Canopy Temperature	159
		7.11.10 Water Budget Method	159
	7.12	Guidelines for Planning Irrigation Schedules	160
	7.13	Nonpeak Irrigation Depth Adjustment	160
	7.14	Calculation of Approximate Irrigation Schedules Using a Simple Process	160
	7.15	Soil Water Constants, Root Depth of Crops and Irrigation Scheduling	160
	7.16	Water Balance Approach	162
	7.17	Criteria for Scheduling Irrigation	163
		7.17.1 Soil Moisture as a Guide	163
		7.17.2 Climate as a Guide	164
		7.17.3 Plant as a Guide	164
	7.18	Assessment of the Scheduling Criteria for Surface, Sprinkler, and Drip Irrigation	164
		7.18.1 Surface Irrigation Scheduling	164
		7.18.2 Sprinkler Irrigation Scheduling	165
		7.18.3 Drip Irrigation Scheduling	165
	7.19	Lysimeter Set-Up	165
	7.20	Modeling Approach for Irrigation Scheduling	166
		7.20.1 Reference Evapotranspiration (ET_0)	166

		7.20.2	Reference Evapotranspiration (ET_0) Estimation Methodologies .. 167
		7.20.3	Notation in Reference Evapotranspiration Determination 168
		7.20.4	Crop Coefficient.. 172
		7.20.5	Modification of the Standard Crop Coefficients.................................. 173
	7.21	Future of Irrigation Scheduling – How to Take It Forward? 174
	7.22	Numerical Problems... 174
	Questions.. 178
	Multiple Choice Questions... 179
	Answer.. 182
	References... 182

Chapter 8 Plastics for Crop Protection.. 183

    8.1 Introduction 183
    8.2 Low Tunnels 183
    8.3 Mulches 184
    8.4 Nethouses 185
        8.4.1 Structure 185
        8.4.2 Advantages of Nethouses.. 185
        8.4.3 Net Types 186
        8.4.4 Types of Materials 186
        8.4.5 Types of Threads and Texture 186
        8.4.6 Mesh Size, Porosity, Solidity, and Weight 187
        8.4.7 Mechanical Properties 187
        8.4.8 Color 187
        8.4.9 Transmissivity, Reflectivity, and Shading Factor 187
        8.4.10 Air Permeability 188
    8.5 Agricultural Application of Nets... 188
        8.5.1 Protection against Meteorological Hazards.................................... 188
        8.5.2 Reduction of Solar Radiation.. 188
        8.5.3 Protection against Insects 188
    8.6 Anti-hail Net.. 189
        8.6.1 Features of Anti-Hail Net 189
        8.6.2 Anti-Hail Net Light Transmittance 189
    8.7 Anti-Insect Net 191
        8.7.1 Advantages.. 191
    Short and Long Questions 193
    Multiple-Choice Questions 193
    Answers 194
    References... 194

Chapter 9 Plastics in Drying and Storage 197

    9.1 Introduction 197
    9.2 Drying of Crops... 197
        9.2.1 Low-Cost Poly House Technology for Drying 197
        9.2.2 Poly House Drying 199
        9.2.3 Refractive Window Drying... 201
        9.2.4 Open Sun Drying 203
    9.3 Unit Operation after Harvesting.. 203
        9.3.1 Field Handling of Crops 203

Contents

		9.3.2	Removal of Field Heat 203
		9.3.3	Field Curing 203
		9.3.4	Grading and Sorting 204
		9.3.5	Conveying 204
		9.3.6	Storage 204
		9.3.7	Transportation of Crops 205
	9.4	Packaging Fresh and Processed crops 205
		9.4.1	Classification of Packaging Systems 205
		9.4.2	Plastic Bags 205
		9.4.3	Shrink-Wrap 205
		9.4.4	Rigid Plastic Packages 206
		9.4.5	Biodegradable Films 206
		9.4.6	Modified Atmospheric Packaging 206
	Questions 208
	Multiple-Choice Questions 208
	Answers 209
	References 209

Chapter 10 Plastics in Aquaculture 211

	10.1	Introduction 211
	10.2	Use of Plastics in Aquaculture 212
		10.2.1	Fishnet 212
			10.2.1.1 Types of Fishnets 212
		10.2.2	Cages 213
		10.2.3	Seed-Rearing Tanks 213
		10.2.4	Pens 213
		10.2.5	Trays Used for Packing 213
		10.2.6	Films Used for Packing 213
		10.2.7	Catfish Hatchery 214
		10.2.8	Fish Feeder/Feed Dispenser 214
	10.3	Poly House Ponds 214
	10.4	Most Commonly Used Plastics in Aquaculture 214
	10.5	How to Select the Plastics 215
		10.5.1	Specific Gravity Test in Water 216
	10.6	Plastic Packaging for Freshwater Fish Processing and Products 216
		10.6.1	Fresh Fish Packaging 216
		10.6.2	Evaluation of Polyethylene, Polypropylene, and Laminated Polypropylene Packaging Material in Fish Retailing 217
		10.6.3	Evaluation of Polyethylene 217
		10.6.4	Evaluation of Polypropylene Containers with Lid for Retail Marketing of Fish Cutup Parts during Chilled and Frozen Storage 217
		10.6.5	Evaluation of PET Bottles for Packaging of Fish Pickle 218
	10.7	Contribution of Aquaculture to Marine Litter 218
		10.7.1	Abandoned, Lost, or Otherwise Discarded Fishing Gears 219
		10.7.2	Plastic Debris from Aquaculture 220
	Questions 220
	Multiple-Choice Questions 221
	Answers 221
	References 221

Chapter 11 Plastics in Animal Husbandry 225
    11.1   Introduction 225
    11.2   Polymers and Their Products during Preharvest and Postharvest 225
    11.3   Application of Plastics in Farms 226
    11.4   Plastics in Animal Production 227
    11.5   Conservation of Fodder 227
    11.6   Silage 228
            11.6.1   Silage Bags 228
            11.6.2   Advantages of "Silage Bags" 228
    11.7   Selection of Crops for Silage Making 229
    11.8   Silo 229
            11.8.1   Site for Construction of Silo 230
    11.9   Kinds of Silos 230
            11.9.1   Stack Silo 230
            11.9.2   Bunker Silo 230
            11.9.3   Pit/Trench Silo 230
            11.9.4   Plastic Bag Silo 231
            11.9.5   Fenced Silo (Framed Silo) 232
            11.9.6   Tower Silo 232
    11.10 Sealing Methods 232
            11.10.1 Unsealed Silos 233
            11.10.2 Lining Bunker Walls with Plastic 234
            11.10.3 Plastic Film to Cover Silage 234
                      11.10.3.1   Plastic Film Color and Thickness 234
                      11.10.3.2   Oxygen Permeability of Plastic Film 235
    11.11 Plastic for Livestock Shelter 237
            11.11.1 Plastic Use in Animal Shelter 238
            11.11.2 Advantages of Poly House as Livestock Shelter 239
    Questions 239
    Multiple-Choice Questions 240
    Answers 240
    References 240

Chapter 12 Plastics as Cladding Material 243
    12.1   Introduction 243
    12.2   Benefits of Protected Cultivation 243
    12.3   Greenhouse 244
            12.3.1   Advantages of Greenhouses 245
    12.4   Types of Greenhouses 245
            12.4.1   Greenhouse Types Based on Shape 245
                      12.4.1.1   Lean-to Greenhouse Type 246
                      12.4.1.2   Even-Span-Type Greenhouse 246
                      12.4.1.3   Uneven-Span-Type Greenhouse 247
                      12.4.1.4   Ridge and Furrow-Type Greenhouse 248
                      12.4.1.5   Sawtooth-Type Greenhouses 248
                      12.4.1.6   Quonset Greenhouse 249
                      12.4.1.7   Greenhouse-Type Based on Utility 249
                      12.4.1.8   Greenhouses Built for Active Heating 249
                      12.4.1.9   Greenhouses for Active Refrigeration 249

12.5	Greenhouse Type Based on Construction	249
	12.5.1 Framed Timber Buildings	250
	12.5.2 Structures Framed by Pipes	250
	12.5.3 Structures Framed by Truss	250
12.6	Greenhouse Type Based on Covering Materials	251
	12.6.1 Glass Greenhouses	252
	12.6.2 Plastic Film Greenhouses	252
	12.6.3 Rigid Panel Greenhouses	252
12.7	Types of Greenhouses Based on Cost of Installation	252
	12.7.1 Low-Cost Poly House/Greenhouse	252
	12.7.2 Medium-Cost Greenhouse	253
	12.7.3 Hi-Tech Greenhouse	253
	12.7.4 Miniature Forms of Greenhouses	254
	12.7.4.1 Plastic Low Tunnels	254
	12.7.4.2 Nethouses	255
	12.7.4.3 Walk-in Tunnels	256
	12.7.4.4 Shading Nets	257
Questions		257
Multiple-Choice Questions		258
Answers		259
References		259

Chapter 13 Plastics in Postharvest Management 261

13.1	Postharvest Management	261
13.2	Field Handling of Crops	262
13.3	Minimizing Field Heat	263
13.4	Packaging of Fresh and Processed Crops	264
	13.4.1 Classification of Packaging Systems	265
	13.4.2 Plastic Bags	266
	13.4.3 Shrink-Wrap	266
	13.4.4 Rigid Plastic Packages	267
	13.4.5 Bulk Bins	267
	13.4.6 Reusable Plastic Containers	269
	13.4.7 Insert Trays	269
	13.4.8 Clamp Shells	271
	13.4.9 Sleeve Packs	272
	13.4.10 Plastic Corrugated Boxes	272
	13.4.11 Plastic Sacks	272
	13.4.12 Plastic Punnets	273
	13.4.13 Plastic Tension Netting	273
	13.4.14 Plastic Pouch	273
	13.4.15 EPS Tray/Stretch Wrapped	276
	13.4.16 Leno Bag – 5 Kg	276
	13.4.17 Consumer Packs for Whole Food Grains	276
	13.4.18 Biodegradable Films	276
	13.4.19 Modified Atmospheric Packaging	277
	13.4.19.1 Major Requirements for Plastics Films for MAP	278
13.5	Storage	278
13.6	Transportation	279
Questions		280

		Multiple-Choice Questions	280
		Answers	281
		References	281

Chapter 14 Plastics in Horticulture 283

- 14.1 Introduction 283
- 14.2 History 283
- 14.3 Nursery Management 285
  - 14.3.1 Advanced Plastic-Growing Pots 285
  - 14.3.2 Growing Media 286
  - 14.3.3 Soilless Peat 288
  - 14.3.4 Nursery Containers 289
  - 14.3.5 Petroleum-Based Plastic Nursery Containers 290
  - 14.3.6 Alternatives to Petroleum-Based Plastic Containers ...290
  - 14.3.7 Physical Properties of Nursery Containers 291
  - 14.3.8 Pot-In-Pot 291
    - 14.3.8.1 Air-Pruning Pots 291
    - 14.3.8.2 Reusable and Recycling 292
    - 14.3.8.3 Eco-Friendly Growing Containers 293
    - 14.3.8.4 Bioplastic – An Alternative to Petroleum-Based Plastics 293
- Questions 294
- Multiple-Choice Questions 294
- Answers 295
- References 295

Chapter 15 Plastic Mulching 297

- 15.1 Introduction 297
- 15.2 Plastic Mulching 298
- 15.3 Classification and Color of Mulches 299
  - 15.3.1 Color of Film 299
- 15.4 Advantages of Plastic Mulch 300
- 15.5 Limitations of Plastic Mulch 300
- 15.6 Areas of Application 300
- 15.7 Effect of Different Color Mulching 301
  - 15.7.1 White Mulch 301
  - 15.7.2 Black Mulch 301
  - 15.7.3 Red Mulch 301
  - 15.7.4 Green Mulch 302
  - 15.7.5 Blue Mulch 302
- 15.8 Specifications 302
- 15.9 Parameters of Plastic Mulch 303
- 15.10 Selection of Mulch 303
- 15.11 Techniques of Mulch Laying 303
  - 15.11.1 Mulching Techniques for Vegetables or Close-Spacing Crops 304
- 15.12 Irrigation Techniques for Mulching 304
- 15.13 Preventions in Mulch Laying 304
- Questions 305
- Multiple-Choice Questions 305

		Answers	306
		References	306

Chapter 16 Hydroponics and Vertical Farming ... 309

    16.1 Introduction ... 309
    16.2 Does Agriculture Need to Change? ... 311
    16.3 Environmental Impacts of Agriculture ... 312
        16.3.1 Water Use Problem ... 312
        16.3.2 Water Use: Solution ... 312
        16.3.3 Land Use: Problem ... 312
        16.3.4 Land Use: Solution ... 312
        16.3.5 Chemical Use: Problem ... 313
        16.3.6 Chemical Use: Solution ... 313
    16.4 Soilless Cultivation ... 313
        16.4.1 History of Soilless Cultivation ... 314
        16.4.2 Advantages of Soilless Cultivation ... 315
        16.4.3 World Scenario of Soilless Cultivation ... 315
    16.5 Hydroponics ... 315
        16.5.1 Types of Hydroponic Setups ... 316
    16.6 Advantages of Hydroponics ... 319
    16.7 Disadvantages of Hydroponics ... 319
    16.8 Aeroponics ... 320
        16.8.1 Equipment Considerations ... 320
        16.8.2 How Does Aeroponics System Work? ... 321
        16.8.3 Types of Aeroponic Systems ... 322
        16.8.4 What Can Be Grown with Aeroponics? ... 322
        16.8.5 Advantages of Aeroponics ... 323
        16.8.6 Drawbacks of Aeroponics ... 323
    16.9 Feasibility and Suitability of These Technologies in Indian Background ... 324
    Long Answer-Type Questions ... 324
    Short Answer-Type Questions ... 324
    Multiple-Choice Questions ... 324
    Answers ... 325
    References ... 326

Chapter 17 Design of Protected Structures ... 327

    17.1 Introduction ... 327
    17.2 World Scenario ... 328
    17.3 Principles of Protective Cultivation ... 329
    17.4 Selection of Film for Poly House ... 330
    17.5 Benefits of Greenhouse Technology ... 330
    17.6 Effect of Wind on Structural Design of Poly House ... 330
    17.7 Constraints of Climate in Hilly and Mountainous Region ... 331
    17.8 Protected Vegetable Cultivation at High Altitudes ... 331
    17.9 Low-Cost Poly House Technology for Vegetable Production ... 332
    17.10 Site Selection ... 332
    17.11 Prospects for Protected Cultivation ... 332
    17.12 Principles of Greenhouse Design ... 333
    17.13 Site Characteristics that Affect the Design ... 333

		17.13.1	Wind	333
		17.13.2	Climate (Altitude)	333
		17.13.3	Snow	333
	17.14	Different Types of Poly Houses		333
	17.15	Greenhouse Technology for Cold Arid Regions of Ladakh		335
	17.16	Vegetable Production		341
	17.17	Collection of Solar Radiation		341
		17.17.1	Thermal Storage and Insulation	342
			17.17.1.1 Double Wall	342
			17.17.1.2 Color	342
			17.17.1.3 Roof	342
			17.17.1.4 Ground	342
			17.17.1.5 Door	342
			17.17.1.6 Ventilation	342
	17.18	Low-Cost Poly House Technology for Drying		342
	17.19	Low-Cost Poly Tunnel Drier		345
	17.20	Walnut Propagation under Poly House		345
	17.21	Design Parameter of Greenhouse		347
		17.21.1	Dead Loads	347
			17.21.1.1 Weight of Roof Covering	347
			17.21.1.2 Weight of Purlin	348
			17.21.1.3 Weight of Truss	348
			17.21.1.4 Live Loads	348
			17.21.1.5 Wind Loads	348
			17.21.1.6 Design Wind Pressure	348
			17.21.1.7 Wind Load on Individual Members	349
			17.21.1.8 Snow Loads	349
			17.21.1.9 Method of Joint	349
			17.21.1.10 Procedure for Analysis	350
			17.21.1.11 Load Combinations	350
			17.21.1.12 Design Strength of Tension Member	350
			17.21.1.13 Design Strength due to Rupture of Critical Section	350
			17.21.1.14 Design Strength due to Block Shear	351
			17.21.1.15 Design Strength of Compression Member	351
			17.21.1.16 Slenderness Ratio	351
			17.21.1.17 Greenhouse Microclimate	351
			17.21.1.18 Functional Design of Greenhouse	351
			17.21.1.19 Greenhouse Orientation	352
	Short-Answer Type			352
	Long-Answer Type			353
	Multiple-Choice Questions			353
	Answers			355
	References			356
Chapter 18	**Application of Plastic in Farm Machinery**			**357**
	18.1	Introduction		357
	18.2	Materials Used		359
		18.2.1	Advantages	359
	18.3	Plastic Bearings Withstand High Forces		361
	18.4	Plastic Types		361

	18.5	Application	361
		18.5.1 Agriculture Injection Molding	361
	18.6	Other Plastic Applications in Farm Machinery	365
		18.6.1 Plastics in Mulching Machines	365
		18.6.2 Sprayer Tanks	366
		18.6.3 Knapsack Sprayer	367
		18.6.4 Poultry Feeders	367
		18.6.5 Fertilizers Spreader	367
		18.6.6 Hand Tools	368
	18.7	Bicycle Sprayer	369
	18.8	Saffron Corm Grader	370
	18.9	Solar-Operated Knapsack Sprayers	370
	18.10	Improved Sickle	370
	18.11	Zenoah Reciprocator	371
	18.12	Tree Planting Auger	372
	18.13	Budding Cum Grafting Knife	372
	18.14	Pruning Knife	372
	18.15	Grafting Tools (Omega Cut)	372
	18.16	Pruning Saw	373
	18.17	Tree Climber	374
	18.18	Trowels (Planting/Digging)	375
	18.19	Rake (Leveling)	375
	18.20	Watering Can	377
	18.21	Aerator	377
	18.22	Wheelbarrow	378
	Questions		378
	References		378
Chapter 19	Smart Farming Using Internet of Things		379
	19.1	Introduction	379
	19.2	IoT in Agriculture	380
	19.3	IoT Sensors for Agriculture	382
	19.4	IoT Software for Agriculture	385
	19.5	Some Applications of IoT in Agriculture	385
		19.5.1 Irrigation	386
		19.5.2 Water Quality Monitoring	387
		19.5.3 Soil Monitoring	388
		19.5.4 Greenhouse Condition Monitoring	388
		19.5.5 Pest and Disease Control	389
	19.6	Benefits of IoT in Agriculture	389
	19.7	Automation in Water Management	390
	19.8	Conclusion	390
	Short Answer-Type-Questions		391
	Long Answer-Type Questions		392
	Multiple-Choice Questions		392
	Answers		395
	References		395
Sample Question Papers			401
Index			403

Preface

Agriculture is the economic backbone of many countries. The growth in agricultural productivity is vital for food and nutritional security, which is being threatened by global warming and climate change. These days climate change is receiving much attention. The impact of climate change is felt almost everywhere. To ensure food security, governments and development agencies are promoting policies to increase agricultural productivity and meet associated challenges. To meet increasing food demand and growing challenges, we have to pursue new technologies for revolutionizing agricultural productivity for sustainable food and nutritional security.

The use of plastics in agriculture can increase crop output, improve food quality, and enhance sustainability. Thus, plasticulture can play a major role in multiplying farmers' income. It can also play an important role in the efficient use of water in agriculture by growing more crops per drop of water. The water table and water quality are gradually decreasing in many parts of the world, and steps have to be taken to save water through the use of micro-irrigation. The productivity, production, and farmers' income need to be increased for the growth of the agricultural sector. This may be enabled by plasticulture which encompasses plant production using plastics. Plasticulture covers such topics as plastic mulch, row covers, drip irrigation, and high/low tunnels. It also covers the many uses of plastics in all aspects of agriculture, including plastic greenhouses, disposal of plastics, and plastics in animal production.

Precision agriculture uses recent developments in micro-irrigation, sensor networks, and greenhouse and protected agriculture structures. It is being increasingly experienced that the availability of labor for agricultural activities is in short supply. The time has now come to exploit modern technologies in tandem with agricultural sciences for improved economic and environmentally sustainable crop production. Improving food security and environmental conservation should be the main goals of innovative farming systems. Improper agricultural practices can reduce the ability of ecosystems to provide food and other services. On the other hand, efforts to promote food security and environmental sustainability can often reinforce each other and enable farmers to adapt to and mitigate the impact of climate change and other stresses. Plasticulture engineering offers a promise for improving soil health, increasing productivity, reducing pollution, and enhancing the sustainability and resilience of agriculture.

This book is designed to provide a discussion of important aspects of plasticulture engineering for crop production, such as the use of plastic in agriculture, animal husbandry, water management, soil solarization, micro-irrigation, automation, vertical farming, sensor-based automation, and farm machinery. The book provides information to deal with these aspects. The book also discusses plasticulture, the basis of plant production using plastics, including plastic mulch, row covers, drip irrigation, and high/low tunnels. It covers the process of producing polyethylene and polypropylene plastics that are used in plant and animal production agriculture, and the many uses of plastics in agriculture, including plastic greenhouses, rigid mold plastics, disposal of plastics, and plastics in animal production.

The subject matter of the book (*Plasticulture Engineering and Technology*) spans 19 chapters. Chapter 1 presents the development, history, application, and status of plasticulture in the world. It also describes applications of plastics in agriculture in different ways, such as plastic mulch films, micro-irrigation, protected cultivation and low tunnels, crop covers, water management, canal lining, silage bags, hay bale wraps, plastic trays, and pots in the production of transplant and bedding, as well as in hi-tech horticulture.

Chapter 2 provides a broader picture of "plastic film" which is basically a thin continuous polymeric material. The thicker plastic material is often called a "sheet." Plastic films also can be clear or colored, printed or plain, and single- or multi-layered and can be combined with alternate materials, such as aluminum metal or paper. Plastic films have an advantage over other materials because they resist corrosion and provide electrical insulation. Compared with a plastic sheet, plastic films are thinner and more flexible and are often transparent or translucent. Chapter 3 describes micro-irrigation, which is a modern method of irrigation used to enhance water use efficiency and allows for a high level of water control application. The losses due to evaporation from soil are significantly reduced compared with other irrigation systems since only a small surface area under the plant is wetted and it is usually well shaded by the foliage.

The design steps of the micro-irrigation system are highlighted in Chapter 4, along with the supply of irrigation water to crops uniformly and efficiently. Focusing on design criteria and examples related to drip and sprinkler irrigation, the chapter also highlights the need, advantages, and design criteria of micro-irrigation technology and estimation of water requirement. Chapter 5 describes the application of plastics in water management which is essential for conservation and regulation to obtain maximum benefit. The chapter also describes the construction of poly-lined farm ponds, canal lining, plastic mulch, micro-irrigation, lysimeter, and plastics in waterlogged areas.

Chapter 6 discusses soil conditioning and solarization using plastics. It also treats the process that increases soil's ability to boost crop yield or improves soil's efficiency. Soil conditioning involves the creation and stabilization of soil aggregates that are conducive to seed germination and seedling emergence. The solarization treatments are primarily related to the thermal action of solarization and the resulting chemical and biological changes in the soil. The main cause for the decrease in soil production potential may be its physical constraints, such as surface crusting and hardening, hardpan, compactness of the subsurface, and high or slow permeability. This suggests that soil must be maintained in such a physical condition as to increase crop production in order to allow sufficient crop growth.

Irrigation scheduling to enhance water use efficiency is discussed in Chapter 7. It describes crop water requirement and root water uptake pattern of different crops, irrigation scheduling methods, parameters, and crop water requirements. Water is vital for plant growth. Deficiency of water in the root zone of soil results in reduced plant growth and affects crop yield, and thus the objective of irrigation is to maintain the adequate moisture content in the root zone, so that crop yield is not affected adversely.

Chapter 8 describes applications of plastics for crop protection from hail, wind, snow, and rains in fruit-farming and ornamental crops, greenhouses shade nets, and nets used for micro-environment management of crops. Plastics are also used to shield insects and birds from viruses as well as for harvesting and postharvest operations.

Chapter 9 discusses plastics in drying and storage. Drying is a good preservation and cost-effective method for the enhancement of storage life. Plastics used in packaging solutions help in increasing the shelf life and during collection, storage, and transportation of fruits and vegetables. Plastics can play a major role in energy conservation. Plastics are used in each and every step of unit operation of postharvest management from packaging to transportation.

Chapter 10 describes the use of plastics in aquaculture and fishing commonly used for the packaging of raw fish and the manufacturing of crafts and gear. It also describes the application of poly house-mounted ponds to accelerate fish productivity. The water temperature is one of the most important parameters for fish growth. Therefore, fish culture inside the poly house is used for the production potential and growth in the greenhouse and open-pond environment during low temperate periods.

The application of plastics in animal husbandry is discussed in Chapter 11. The use of plastics to store animal grains and straws during winter. Plastic films used for storing silage are resistant and are easy to transport and store material for long periods. Plastics can be used in animal production and preproduction, such as mulching, fertilizer, pesticide insecticide to produce food and postproduction for the storage of animal food.

Chapter 12 describes plastics as a cladding material, an important aspect of greenhouse design. Roofing materials used in this way to cover greenhouses can have a significant impact on the amount and quality of radiation exposed to plants, ultimately affecting crop yields. The layer of the

Preface

greenhouse is actually one of the most important components of the growing environment, so it is important to choose the right cladding material for the greenhouse.

Plastics in postharvest management are described in Chapter 13. A major part of the postharvest losses can be reduced by following better packaging practices. This in turn leads to an increase in the availability of produce and a reduction in market prices. Various types of packaging materials are available and are used for packaging diverse food products; however, plastic packaging materials are preferred. This is attributed to their durability, safety owing to their shatter-proofness, hygienic property, and security.

Application of plastics in horticulture is described in Chapter 14, which offers large payouts and is considered a major indirect horticultural resource, resulting in moisture retention, reduced fertilizer use, more appropriate water and nutrient application, and a controlled horticultural environment. Plastic nets and innovative packaging help protect crops and increase shelf life, as well as collection, storage, and transport of garden products, especially fruits, vegetables, and flowers.

Chapter 15 describes plastic mulching, classification of mulches, advantages and limitations, areas of application, effect of different color mulches, specifications of plastic mulching, parameters of plastic mulches, and selection of mulches under different conditions. Plastic mulching is a widely used agricultural practice due to the immediate economic benefits it provides, such as increased yields, better harvests, enhanced fruit quality, and reduced water use. Mulching has the potential to reduce soil evaporation, conserve moisture, control soil temperature, reduce weed development, and improve microbial activity. Mulches could also benefit agriculture and landscaping in terms of cost, aesthetics, and the environment. Mulches are used to manage a variety of stress conditions in both agricultural and landscape settings.

Hydroponics and vertical farming are described in Chapter 16. The combination of vertical farming with hydroponics has contributed a new innovative chapter to agricultural engineering, helping alleviate future food scarcity. Vertical farming is the process of producing agricultural products in vertically stacked structures to produce more food crops in a compact area, particularly in urban and peri-urban areas where agricultural land is scarce. It can be one of the most effective ways to produce crops to meet the food demands of the urban population, whereas hydroponics is a method of cultivating plants using a liquid nutrient solution as growing medium and other necessary minerals to maintain plant growth. Thus, the chapter highlights the need for adopting vertical farming and hydroponics, types of hydroponic and aeroponic systems, their advantages and disadvantages, and their feasibility under Indian conditions.

Chapter 17 describes the design of protected structures constructed in different agroclimatic zones, feasibility, and benefit of technology. Greenhouse technology has a lot of potential in terms of improving productivity, mitigating the effects of climate change, and increasing agricultural growth periods under prevailing climatic conditions. The technology has a direct bearing on ensuring food security, ensuring yields and productivity much higher than those achieved under open conditions.

Chapter 18 describes the application of plastics in farm machinery which helps farmers to improve the quality of products, increase crop production, and reduce the ecological footprint. Different plastic materials are used in agriculture, including polypropylene polyolefins, ethylene-vinyl acetate, polyvinyl chloride, polycarbonate, and poly-methyl methacrylate. Plastics are used to wrap silage, to cover crops, in tubing for irrigation, and to transport feed and fertilizer. Plastic improves the performance, efficiency, and longevity of all heavy equipment types.

In order to increase production, competitiveness, and world economy, and reduce human interference, time and expense, there is a need to shift toward digital technologies called the Internet of Things (IoT). The society around the globe is industrialized by replacing manual operations with technical advances, as they are energy-consuming and engross limited human resources. IoT technologies facilitate the manufacturing of systems that support various agricultural processes. From farmers to consumers, IoT technologies could change the industry, contribute to food safety, and reduce agricultural inputs and food waste. Chapter 19 describes smart farming using the Internet of

Things (IoT) and main components, IoT sensors and software that are available, and the advantages of IoT in the agricultural sector.

It is emphasized that we have just begun to scratch the surface with some of the recent advances in plasticulture engineering and technology. This book discusses real-world examples of plasticulture applications in agriculture and is based on experiences in different agroclimatic regions in the world. It is hoped that this book will be useful to increase productivity, profitability, and suitability in agriculture and precision farming for climate change mitigation.

Rohitashw Kumar
SKUAST-Kashmir, Srinagar
Vijay P. Singh
Texas A&M University, College Station, Texas

Acknowledgments

I consider it a proud privilege to express my heartfelt gratitude to ICAR – All India Coordinated Research Project on Plastic Engineering in Agriculture Structures & Environment Management, for providing financial assistance to carry out this project at SKUAST-Kashmir, Srinagar. The practical utility data that have been presented in this book are taken from AICRP on the PEASEM project. I am greatly thankful to the Project Coordinated unit, CIPHET, Ludhiana, for providing all necessary facilities. Their timely help, constructive criticism, and painstaking efforts made it possible to present the work contained in this book in its present form.

I sincerely acknowledge the efforts of Professor Vijay Pal Singh, Distinguished Professor, Department of Biological and Agricultural Engineering, Texas A&M University, Texas, US, for his support and guidance to bring this manuscript in final refined shape, constructive criticism, and utmost cooperation at every stage of this work.

I am highly obliged to Hon'ble Vice Chancellor Prof. Nazir Ahmad Ganai for his support and encouragement. I am highly thankful to him provide support under NAHEP project. I wish to extend my sincere thanks to the College of Agricultural Engineering and Technology, SKUAST-Kashmir for their support and encouragement.

I am grateful to Dr. Rishi Richa, Dr. Showkat Rasool, and Dr. Kalay Khan and my students Dr. Saba Parvez, Er. Sakeel Ahmad Bhat, Er. Munjid Maryam, Er. Zeenat Farooq, Er. Tanzeel Khan, Er. Dinesh Vishkarma, Er. Faizan Masoodi, Er, Noureen, Er Iqra, and Er. Mahrukh for providing all the necessary help to write this book.

I express my regards and reverence to my late parents as their contribution to whatever I have achieved till date is beyond expression. It was their love, affection, and blessed care that has helped me to move ahead in my difficult times and complete my work successfully. I thank my family members, especially my brother Balbir, who has been a source of inspiration since my schooldays. I thankfully acknowledge the contribution of all my teachers since schooldays, for showing me the right path at every step of life.

I sincerely acknowledge with love the patience and support of my wonderful wife Reshma. She has loved and cared for me without ever asking anything in return and I am thankful to God for blessing me with her. She has spent the best and the worst times with me but her faith in my decisions and abilities have never wavered. I would also like to thank my beloved daughter Meenu and son Vineet for making my home lively with their sweet activities.

Finally, I bow my head before the almighty God, whose divine grace gave me the required courage, strength, and perseverance to overcome various obstacles that stood in my way.

(Rohitashw Kumar)
Associate Dean,
College of Agricultural Engineering and Technology,
SKUAST – Kashmir, Srinagar

About the Authors

Dr. Rohitashw Kumar (BE, ME, PhD) is Associate Dean and Professor in College of Agricultural Engineering and Technology, Sher-e-Kashmir University of Agricultural Sciences and Technology of Kashmir, Srinagar, India. He is also Professor Water Chair (Sheikkul Alam Shiekh Nuruddin Water Chair), Ministry of Jal Shakti, Government of India, at the National Institute of Technology, Srinagar (J&K). He is also Professor and Head, Division of Irrigation and Drainage Engineering. He obtained his PhD in Water Resources Engineering from NIT, Hamirpur, and Master of Engineering in Irrigation Water Management Engineering from MPUAT, Udaipur. He got a leadership award in 2020, a Special Research Award in 2017, and a Student Incentive Award-2015 (PhD Research) from the Soil Conservation Society of India, New Delhi. He also got the first prize in India for best M.Tech thesis in Agricultural Engineering in 2001. He has published over 110 papers in peer-reviewed journals, more than 25 popular articles, 4 books, 2 practical manuals, and 25 book chapters. He has guided 1 PhD and 14 MTech students in soil and water engineering. He has handled more than 12 research projects as a principal or co-principal investigator. Since 2011, he has been Principal Investigator of ICAR – All India Coordinated Research Project on Plastic Engineering in Agriculture Structural and Environment Management.

Prof. Vijay P. Singh is Distinguished Professor, Regents Professor, and the inaugural holder of the Caroline and William N. Lehrer Distinguished Chair in Water Engineering at the Texas A&M University. His research interests include surface-water hydrology, groundwater hydrology, hydraulics, irrigation engineering, environmental quality, water resources, water-food-energy nexus, climate change impacts, entropy theory, copula theory, and mathematical modeling. He graduated with a BSc in Engineering and Technology with emphasis on Soil and Water Conservation Engineering in 1967 from the U.P. Agricultural University, India. He earned an MS in Engineering with specialization in Hydrology in 1970 from the University of Guelph, Canada; a PhD in Civil Engineering with specialization in Hydrology and Water Resources in 1974 from the Colorado State University, Fort Collins, US; and a DSc in Environmental and Water Resources Engineering in 1998 from the University of the Witwatersrand, Johannesburg, South Africa. He has published extensively on a wide range of topics. His publications include more than 1,365 journal articles, 32 books, 80 edited books, 305 book chapters, and 315 conference proceedings papers. For his seminar contributions, he has received more than 100 national and international awards, including 3 honorary doctorates. Currently, he serves as Past President of the American Academy of Water Resources Engineers, the American Society of Civil Engineers (ASCE), and previously he served as President of the American Institute of Hydrology and Chair, Watershed Council, ASCE. He is Editor-in-Chief of two book series, three journals, and serves on the editorial boards of more than 25 journals. He has served as Editor-in-Chief of three other journals. He is Distinguished Member of the American Society of Civil Engineers, Honorary Member of the American Water Resources Association, Honorary Member of the International Water Resource Association, and Distinguished Fellow of the Association of Global Groundwater Scientists. He is Fellow of five professional societies. He is also Fellow or Member of 11 national or international engineering or science academies.

1 Introduction

1.1 DEFINITION

Plasticulture is defined as the practice of using plastic materials in agricultural applications, and according to the American Society for Plasticulture, it means "use of plastics in agriculture." Plastic materials are broadly referred to as "ag plastics." Plastics are applied in agriculture in different ways, including plastic mulch films, micro-irrigation (sprinkler, drip irrigation, rain gun, foggers, etc.), crop covers, greenhouses and low tunnels, silage bags, hay bale wraps, plastic trays, and pots used in the production of transplant and bedding plants (Lamont and Orzolek, 2004). Plastic materials gave way to different types of polyethylene films, which have revolutionized the protection of crops (Emmert, 1957; Agrawal, and Agrawal, 2005; Singh and Asrey, 2007). The history of plastic cultivation dates back to 1948, when Professor Emmert from the University of Kentucky first used polyethylene as a greenhouse film to replace glass (Anderson and Emmert, 1994; Jensen, 2004). In the United States, Emmert is known as the father of agricultural plastic development. With his contributions to greenhouses, mulches, and row covers, he described the principles of plastic technology (Emmert, 1957).

Although plasticulture includes micro-irrigation (drip and sprinkler), soil fumigation film, mulching, water storage, silage bags, and nursery pots are most frequently used, denoting the use of all kinds of plastics. Such variety ranges from plastic mulch film, row coverings, and high and low tunnels (polytunnels) to plastic greenhouses. Because of its affordability, flexibility, and easy manufacturing, polyethylene (PE) is used by a majority of growers. Low-density polyethylene (LDPE) and linear low-density (LLDPE) forms have different thicknesses. It can be modified by adding some elements of plastic that give properties beneficial to plant growth, such as reduced water loss, UV stabilization to cool soil and prevent insects, elimination of photosynthetically active radiation to prevent weed growth, and antidrip/antifog.

1.2 PLASTICULTURE DEVELOPMENT

Baekel, who coined the term "plastic," invented the world's first fully synthetic plastic (Bakelite) in New York in 1907. Many chemists have contributed to the materials science of plastics including Nobel laureate Hermann Staudinger, who is called "the father of polymer chemistry," and Herman Mark, who is known as "the father of polymer physics."

The first use of plastic film in agriculture was an effort to make a cheaper version of a glasshouse. Professor E.M. Emmert built the first plastic greenhouse in 1948, a wooden structure covered with cellulose acetate film and further used a more effective polyethylene film. After the introduction of plastic film in agriculture, by the early 1950s, larger-scale mulching was used in vegetables. By 1999, almost 30 million acres worldwide were covered by plastic mulch, but only a small percentage of this acreage (185,000 acres) was in the United States, and the majority of the plastic growth was occurring in economically poor areas of the world and unproductive desert regions, such as the Thar Desert.

In the agricultural sector, plastic farming is a technology that entails the use of plastics. In the late 1900s, polyethylene was available, and tar-coated paper mulches began to be used. When Warp (1971) developed the first glass replacement for widespread agricultural use, the science of plastic cultivation began. In 1938, British scientists first produced polyethylene as a film sheet. The earliest method of modifying the microclimate of crops using organic and inorganic materials was mulching (Jaworski *et al.*, 1974).

The greenhouse technology around the world was used in two areas, with 80% throughout the Far East (Japan, Korea, China) and 15% in the Mediterranean basin. The area of greenhouse cover is increasing at a fast rate; during the last decade, it was estimated to grow every year by 20%. In the Middle East and Africa, the use of plastic greenhouses increased by 15–20% per year. In Europe, there was a weak growth area, i.e., covering greenhouses. In China, its growth is about 30% per year and, further, China translating into a volume of plastic film reaching 1,000,000 tons/year. In 2006, 80% of the area covered by plastic mulch was found in China with a growth rate of 25% per year. Plastic film is designed and developed to increase the yield of agricultural produce, shorten growth time, and increase produce size. Developments in plastic film encompass durability, optical (ultraviolet, visible, near-infrared, and middle infrared) properties, and the antidrip or antifog effect.

In the 1980s, plasticulture involved more than 2 million tons of plastic consumption per year in the world. A great number of polymers were used, from the expanded polystyrene of seedling trays to the polypropylene strings for plant knitting. The plastics in agriculture in the United States (polyethylene (PE) or polypropylene (PP) resin) were first used in greenhouses that produced vegetables for commercial sale. Emmert evaluated in 1960 plastic film that covered his greenhouse at the University of Kentucky. Interest in this new technology led to the formation of the National Agricultural Plastics Association (NAPA) that was later changed to the American Society for Plasticulture (ASP) to include all the facets of plastic use in agriculture. This professional organization included academic professionals, manufacturers in the production of plastics, government personnel, and students. The first meeting of NAPA was held in 1960 at Lexington, Kentucky, US, to discuss the needs and challenges of using plastics to improve agriculture. Several researchers and leaders promoted plasticulture for horticultural crop production in the late 1960s and the 1970s in the United States.

Bernard J. Hall from the University of California introduced plastic tunnels, row covers, and drip irrigation to Southern California growers on a large scale. The first large-scale use of PE film worldwide was made in the construction of plastic-covered greenhouses, and the first plastic-covered greenhouse was erected in 1955 in England. In the late 1960s and the early 1970s, plastic-covered greenhouses were constructed in Russia, southern Europe, and Asia. Meanwhile, by 1971, Israel had extensive areas of high-value horticultural crops under plastic tunnels, row covers, and plastic soil mulches, paralleling the development and use of drip or trickle irrigation to maximize crop production. China began using plastic film covers to protect rice seedbeds from inclement weather, such as cold and wind, as experienced in the spring of 1958 in the central and southern provinces. There was widespread use in both China and Japan by 1965 and rapid expansion through the 1970s in Korea, China, and Japan. By 1980, there were 8,600 ha of plastic greenhouses in Japan and by 1984, 16,000 ha in China used plastics. After the oil crisis in 1973, the European crop diversification programs expanded the use of plasticulture for crop production. Population increases and better life standards spread into the cultivated areas and the production of fresh vegetables, which led to the expansion of the use of plastics for crop production. New techniques, like hydroponics, soilless culture, have further increased the use of plastics.

In European countries (France, Italy, and Spain) the use of plastic materials for agricultural production in the 1980s was about 550,000 tons per year, without including plastic materials for packaging of agricultural products. The most common plastic material was low-density polyethylene (LDPE), which in Spain reached around 60% of the total plasticulture use, followed by high-density polyethylene (HDPE), polyvinylchloride (PVC), ethylene-vinyl-acetate copolymer (EVA), and polyester. Plastic films accounted for around 65% of the total thermoplastic compounds used in Spain's agricultural sector. The plastic material consumption for agriculture in Spain reached 9% of the total national consumption in the 1980s. Presently, millions of pounds of plastic mulches are used extensively around the world, including China that has to feed 1.4 billion people every day.

The past decades have seen a rapid increase in the production and consumption of plastics and these numbers are still increasing. The estimated amounts of plastic used for different applications in agriculture are summarized in Table 1.1. Plastic cultivation films are used in agriculture the most,

TABLE 1.1
Quantity of Plastics Used

S No	Use	Estimated Quantity Worldwide/Year (million tons)
1.	Total	348
2.	Total for agriculture purpose	6.96
3	Applications	
i	Cultivation film (2017)	6.5
ii	Mulching	2.75
iii	Greenhouse and high tunnels	1 6.5
iv	Low tunnels	0.17

TABLE 1.2
Plastic Use in Agriculture

S. No	Protected Cultivation (Films)	Packaging	Nets	Irrigation/Drainage and Piping	Others
1.	Low tunnel	Anti-hail	Containers	Water reservoir	Silage films
2.	Greenhouse and tunnel	Shading	Fertilizer sacks	Channel lining	Fumigation film
3.	Mulching	Anti-bird	Agrochemical cans	Irrigation taps and pipes	Bale twins
4	Nursery films	Wind breaking	Tanks for liquid storage	Drainage pipe	Bale wraps
5	Direct covering	Nets for nut picking		Micro-irrigation Lateral Drippers main lines	Nursery pots, strings, and ropes

accounting for more than 90% of the total plastics used. Relatively few plastics are used in agriculture for silage, irrigation, and nets. Polyethylene (PE) is mainly used for plastic cultivation films and, therefore, the type of plastic that is used most in agriculture.

A brief summary of the application of plastic in agriculture for different purposes is given in Table 1.2.

1.3 FILM MANUFACTURING

Plastic is the main material that comes from natural gases and petroleum that constitute the building blocks of plastics and forms chemically linked subunits called monomers. Long chains of monomers are called polymers. All plastics are made up of polymers. Different mixtures of polymers are what make up resins. Virgin or raw resin material is processed for a variety of products and uses. Plastics can be formed into objects or films of fibers. The name "plastic" is derived from the fact that it has the property of plasticity. Depending on their physical properties, plastics may be classified as thermo-softening plastics and thermosetting materials. Thermo-softening plastic materials can be formed into desired shapes under heat and pressure and become solids on cooling.

Plastics have found tremendous applications, such as fibers, textiles, plasticulture, service ware, and packaging containers via established processing technologies. Poly products are finding use in many applications, including packaging, paper coating, fibers, films, and a host of molded articles. The first products were aimed at packaging film and fibers for textiles and nonwoven products. For packaging, clear films with good barrier properties but low heat-seal properties are recommended.

The process for making polyethylene film and bags is called extrusion. Plastic films are used in a wide variety of applications. These include packaging, plastic bags, labels, building construction, landscaping, electrical fabrication, photographic film, film stock for movies, videotape, etc. Almost all plastics can be formed into a thin film. Some of the primary ones are:

- Polyester – Biaxially-oriented polyethylene terephthalate (BoPET) is a biaxially oriented polyester film
- Polyethylene
  The most common plastic film is made up of one of the varieties of polyethylene: low-density polyethylene, medium-density polyethylene, high-density polyethylene, or linear low-density polyethylene.
- Polypropylene
  Polypropylene can be made a cast film, biaxially oriented film (BOPP), or as a uniaxially oriented film.
- Nylon
- Nylon is synthetic polymers composed of polyamides. Nylon is a silk-like thermoplastic, made from petroleum further processed into films, fibers, or shapes.
- Polyvinyl chloride film (with or without a plasticizer)
- Cellophane (made up of regenerated cellulose)
- Cellulose acetate (an early bio-plastic)
- Bio-plastics and bio-degradable plastics
- Semiembossed film (to prevent dust and other foreign matters from sticking to the rubber while calendaring and during storage)

Plastic films are usually thermoplastics and are formed by melting. The following are top plastic film and sheet manufacturers:

- Bemis Company
- Berry Plastics
- Sigma Plastics
- Sealed Air
- Inteplast
- Dupont Teijin Films
- Novolex
- Printpack
- Winpak
- Glad Products
- PolyOne
- Reynolds

Among the plastic manufacturers, Polyplex is a renowned name (polyester film), which produces a different range of plastic films across several substrates: both standard plain films and a range of value-added films with offline coating and metalizing capabilities. The details of different plastic manufacturers are shown in Table 1.3.

1.4 TYPES AND QUALITY OF PLASTICS USED IN SOIL AND WATER CONSERVATION

HDPE (high-density polyethylene), PVC (polyvinyl chloride), and polyethylene are mainly used as the lining material. Drip and sprinkler irrigation components are made up of plastics, i.e., plastic pipes, valves, filters, and emitters. The different applications of plastic are given below.

TABLE 1.3
Plastics Manufacturers

Name	Year Founded
Jain Irrigation System Ltd	1963
Netafim	1965
Premier Irrigation India Pvt Ltd	2007
Harvel Irrigation India Pvt Ltd	1984
Mahindra EPC Irrigation Ltd	1981
A & S Mold & Die Corp.	1969
Adapt Plastics, Inc.	1973
Applied Plastics Co., Inc.	1955
CDS Plastics, Inc.	1986
Hoehn Plastics, Inc.	1996
Mar-Bal, Inc.	1970
Plastic Design International, Inc.	1977
Plastics Services Network (PSN)	1990
PlastiFab/Leed Plastics	1977
Roscom	1980
The Rodon Group	1956

1.4.1 Micro-Irrigation

Around the world, irrigated land has increased for increasing crop production in arid and subhumid zones. The agriculture sector utilizes more water than industries, municipalities, and other sectors. Judiciously, water for crops requires an understanding of evapotranspiration and the use of efficient irrigation methods. The micro-irrigation method plays an important role in increasing water use efficiency. Micro-irrigation mainly consists of sprinkler and drip irrigation systems. Drip irrigation is also known as trickle irrigation, localized irrigation, or high-frequency or pressurized irrigation. Through this irrigation method, farmers can save water and fertilizer by allowing water to flow very slowly to the roots of plants, either onto the soil surface or directly onto the root zone, using a network of pipes, valve tubing, and emitters.

In India, perhaps more than 80% of the available water is used for irrigation. Irrigation is the controlled application of water through man-made systems to meet the water requirements of agriculture. India has the largest irrigated area in the world and faces acute water scarcity. We need to adopt irrigation methods that help in not only saving freshwater but also providing sufficient water to plants for growth. Micro-irrigation gained prevalence when the Parliament was rocked with the issue of farmer suicides. Sensing the significance and probable benefits of the process to double the farmers' income along with agricultural sustainability and environmental quality, the Union government launched a comprehensive flagship program called Pradhan Mantri Krishi Sinchai Yojana or "more crop per drop." Drip irrigation is mainly used for vegetable and fruit crops. Figure 1.1 shows the drip irrigation system in high-density apple crops.

1.4.2 Protected Cultivation

The films for greenhouse covers and low tunnels represent the most important application quantitatively. Greenhouse covers have been evolving since their launch in the 1950s. Greenhouse technology in different agroclimatic regions holds a lot of potential in terms of improving productivity, mitigating the effects of climate change, and increasing the agricultural growth periods under

FIGURE 1.1 Drip irrigation system in high-density apple crop

FIGURE 1.2 Naturally ventilated poly house for vegetable production

prevailing climatic conditions. The agricultural growth period is only 8–9 months under temperate climatic conditions and even less i.e., 5–6 months under cold arid climatic conditions and erratic precipitation patterns have further worsened the scenario. Greenhouse technology that has shown exponential growth in other countries is still moving at a very sluggish pace in India. A naturally ventilated polyhouse is shown in Figure 1.2. This technology has a direct bearing in ensuring food security, with yields and productivity much higher than those achieved under open conditions. The design and development of suitable greenhouses hold the key to the large-scale popularization of this technology. The climatic conditions, orientation, shape, availability of various resources, and cultivation technologies of greenhouses are very important parameters to be well established before

Introduction

large-scale adoption of this technology is achieved in farmers' fields. However, due to the sheer magic associated with the greenhouse effect that is being utilized in growing crops during the winter months, the technology is showing adaptability in farmer's fields. A low tunnel widely used for vegetable and nursery production is illustrated in Figure 1.3.

The market share of monolayer and three-layer films is different depending on the technological level of the country. More than 80% of the worldwide market comprises films made of LDPE, ethylene-vinyl acetate (EVA), and ethylene-butyl acrylate (EBA) copolymers. Other polymers used include plasticized PVC in Japan and linear low-density polyethylene (LLDPE) in the rest of the world. The area covered by greenhouses has been steadily increasing at a rate of 20% per year during the past decade. Development in Europe is very weak but Africa and the Middle East are growing at 15–20% annually. Of special interest is the case of China, which has grown from 4,200 ha in 1981 to 1,250,000 ha in 2002 (30% per year). The volume of plastic films used for this application would thus be about 1,000,000 tons/year (Espí et al., 2006). Small tunnels are different from the high ones, which are considered to be greenhouses in their smaller size, the smaller thickness of the films used (generally below 80 mm). The area covered with low tunnels has been very stable during the past decade, which has had an annual growth rate of 15% during the past decade (Figure 1.4).

An anti-hail net is used for protecting the crop from adverse conditions like rains, hailstorms, sunburn, etc. It offers several advantages: production risk is comparatively less than open-field condition (Figure 1.5). In the case of apples, it has been found that covering apple orchards with hail nets is an effective strategy in reducing the surface temperature of apples and the amount of UV radiation they receive. In areas that will incur above optimal temperatures and much solar radiation with climate change, the use of hail nets can reduce sunburn losses, improve flavor, and improve consumer desirability of apples.

1.4.3 Mulching

Mulches are applied to the soil surface, as opposed to materials that are incorporated into the soil profile. It is an effective way of manipulating the crop growth environment to increase crop yield and improve product quality by controlling soil temperature, retaining soil moisture, and reducing soil evaporation. Mulch is a layer of material that covers the soil surface, and mulching is a water conservation technique that increases water infiltration into the soil, retards soil erosion, and reduces surface runoff (Sica et al., 2015). Plastic mulches are primarily used to protect seedlings

FIGURE 1.3 Low tunnels for vegetable and nursery production

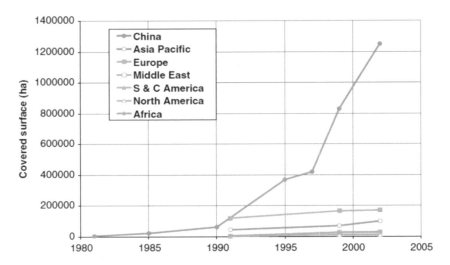

FIGURE 1.4 Agricultural surface covered with plastic film (greenhouses and walk-in high tunnels) (Espí et al., 2006)

FIGURE 1.5 Anti-hail net used for apple protection

and shoots through insulation and evaporation prevention, thus, maintaining or slightly increasing soil temperature and humidity. Plastic mulch films consist mostly of low-density and linear low-density polyethylene, which do not readily biodegrade (Figure 1.6). As a result, these polyethylene-based mulches must be retrieved and disposed of after usage. Agricultural plastic mulch films are often contaminated with soil and, therefore, are not accepted by many recycling facilities. The amount of mulch used is 1.25 million tons in China alone in 2011.

1.4.4 Reducing Water Usage in Agriculture

Plasticulture, the use of plastics in agricultural applications, can help save water up to 45–70%. Plasticulture can play an important role in water scarcity areas. Monsoons are becoming erratic

Introduction

FIGURE 1.6 Different-color plastic mulches used in vegetable crop

FIGURE 1.7 Drip irrigation in tomato crop to increase water use efficiency

and the fall in groundwater levels is becoming alarming. In these situations, the use of plastics plays an important role in water conservation. It has been estimated that appropriate applications of micro-irrigation technologies can result in water saving up to 45–65%. Thus, plasticulture can play an important role in facilitating the judicious usage of water application through micro-irrigation which can help increase productivity by 40–100% with significant savings of water along with enhanced efficiency of fertilizer use (Figure 1.7).

1.5 PRODUCTION AGRICULTURE AND POSTHARVEST MANAGEMENT

Plastic has become a popular material in our daily life due to its chemical property, structural integrity, and versatile nature. It has many valuable applications in high-tech horticulture, including drip irrigation, plastic mulches, packaging, and storage for the superior quality of produce and postharvest management. The use of plastics in horticulture crop production has increased dramatically in the past ten years, even though the number of agricultural plastic manufactures has reduced by 40%

over the same period. The use of plasticulture in the production of horticultural crops helps mitigate the, sometimes, extreme fluctuations in weather, especially temperature, rainfall, and wind, which occur in many parts of the world. The controlled environment of horticulture using plastic plays an important role in enhancing crop productivity. Plastics offer an option for increased both vertical and horizontal expansion in horticultural crop production and needed improvements in the quality of products grown. To ensure that the product quality is maintained throughout the distribution environment, guidelines regarding short-term postharvest storage, packaging, and handling are necessary. The postharvest handling activities that maintain fruit quality as fresh produce travels throughout the supply chain are essential.

1.5.1 Postharvest and Storage Management

Fruit and vegetable growers usually possess the necessary skills to improve the value of their crops during the growing season. However, once the harvest begins, good postharvest handling practices must be implemented to safeguard the product throughout the distribution environment. The primary objective of postharvest handling is to maintain quality by

- Minimizing the bruising, friction damage, and other mechanical injuries
- Reducing the metabolic rates that result in undesirable changes in composition, color, texture, flavor, and nutritional status and undesirable growth such as sprouting or rooting
- Reducing the water loss that results in wilting, softening, and loss of salable weight and crispness
- Reducing the spoilage caused by decay, especially of damage
- Preventing the development of freezing injury or physiological disorders

A fruit is the mature ovary of a plant, whereas a vegetable is considered the edible portion of an herbaceous plant, such as leaf, stem, root, tuber, bulb, immature fruit, or flower. Normally, fruits are regarded as fleshy, sweet structures with seeds, whereas vegetables are other edible parts of the plant. Fruits are classified in terms of two categories: (1) climacteric and (2) non-climacteric, based on differences in their patterns of respiration and ethylene production that impact their optimum postharvest handling practices. Because of their ripening behavior, when climacteric fruits can be harvested mature but unripe, the onset of ethylene production and the ripening rate are delayed, and the fruit can be stored for extended periods of time or transported over relatively long distances. The storage life of commodities varies inversely with the respiration rate, which itself increases with an increase in temperature. In other words, the higher the temperature, the greater the respiration rate, and the more rapid the deterioration of the commodity, which reduces shelf life.

1.5.2 Fruit and Vegetable Packaging

There are many types of packages available in the market. Packaging systems are available in a variety of materials, such as plastic, wood, corrugated fiberboard, and even sustainable materials, such as bio-plastics and fibers that decompose. Packaging plays an important role in the fruit and vegetable distribution chain and serves four main functions: (a) containment, (b) protection and preservation, (c) convenience, and (d) communication.

Bulk bins allow producers to store, handle, and transport large amounts of product with relative ease. They are manufactured from plastic polymers, wood, or corrugated fiberboard. Plastic and wood bulk bins are often only used on the farm because they are quite heavy and bulky. Corrugated fiberboard cartons comprised a paper-based material manufactured from the Kraft paper pulping process. Corrugated fiberboard cartons are commonly made of single-wall design with an inside

Introduction

and outside liner surrounding a fluted medium. Reusable plastic containers are becoming more prevalent in the fresh produce distribution systems because producers and companies are trying to minimize their carbon footprint while simultaneously reducing costs.

Farming is the largest supplier of livelihoods in India. A significant part of the Indian economy is made up of agriculture. Current agricultural practices have to be reformed and modernized. One of the recent approaches followed in several farming activities is the use of plastics in the form of polyethylene, polypropylene, polyvinyl chloride, polycarbonate, and polyester films. Plastic materials are used in agriculture, horticulture, water management, storage of food grains, and packaging. Plastic applications mainly for water conservation enhance irrigation efficiency, crop, and environmental protection, as well as end-product storage and transportation.

1.5.3 Silage Cover

Green and nutritious "forage" (maize, sorghum, etc.) can be conserved through a natural "pickling" process in a sealed airtight container; this is known as "ensiled forage" or "silage." Silage can be prepared when surplus green fodder is available and can be used during lean periods.

There are numerous advantages of converting green fodder to silage:

- Silage is a substitute for green fodder during the lean period
- Silage ensures improved quality and digestibility of fodder to the livestock
- Increases milk production and helps sustain higher milk production during the lean period

1.6 PREFERENCE OF PLASTICULTURE

The unique advantages of plastics over other conventional materials are:

- Higher strength/weight ratio
- Superior thermal insulation properties
- Excellent corrosion resistance
- Superior flexibility
- Resistance to most of the chemicals
- Excellent moisture barrier properties
- Smoother surface—resulting in the reduction in friction losses
- Excellent light transmissibility
- Plastics are easily processed with low energy consumption (consider that a temperature of about 1,000°C is required to make a glass bottle, whereas 250°C is enough for a plastic bottle)
- Plastics are relatively easily recovered and recycled

The plastic material is preferred due to the following reasons:

- Conservation of energy
- Higher proportion of strength/weight
- Superior electrical characteristics
- Superior properties of thermal isolation
- Excellent resistance to corrosion
- Superior versatility
- Impermeability to water, gas, etc.
- Resistance to chemical substances
- Less friction because of a smoother surface

1.7 IMPORTANCE OF PLASTICS

Plastics are important for the following reasons:

- Plasticulture technologies have been shown to enhance productivity and profitability worldwide
- In India, polymer utilization is just 2% as compared to the world average of 8%
- Agricultural output can be increased by INR 74,000 crores through plasticulture
- Production improvement (50–60%)
- Water saving (60–70%)
- Fertilizer saving (30–40%)
- Labor saving (7–18%)
- Early fruiting (10–25 days)
- Excellent quality of seedlings
- Enhances quality of produce

1.8 TYPES OF PLASTICS

Due to its affordability, versatility, and easy production, polyethylene (PE) is the plastic film used by the majority of farmers. It comes in a variety of thicknesses, such as a form of low density (LDPE) and that of linear low density (LLDPE). These can be changed by adding certain components to the plastic, such as decreased water loss, UV stabilization to cool soil and prevent insects, removal of photosynthetically active radiation to prevent weed growth, antidrip/antifog, and fluorescence, to provide beneficial plant growth properties. Polypropylene is also used for packaging cords for agricultural plants.

1.8.1 THERMOPLASTIC AND THERMOSETTING PLASTICS

Thermoplastic materials have low melting points, making it further possible to re-mold or recycle them. On the other hand, plastic thermosetting is quite the contrary. Thermosetting plastics are able to withstand high temperatures and cannot be reformed or recycled once hardened, even with the application of heat. In any case, let's look at some of the essential differences between these two compounds, which are illustrated in Tables 1.4 to 1.6.

TABLE 1.4
Comparison between Thermoplastic and Thermosetting Plastics

Thermoplastic	Thermosetting Plastic
The process called addition polymerization will synthesize thermoplastics	By condensation polymerization, thermosetting plastics are synthesized
Injection molding, extrusion, blast molding, rotational molding, and thermoforming are used to process thermoplastics	Thermosetting plastic is manufactured by compression molding and injection molding of reactions
There are secondary bonds between molecular chains in thermoplastics	There are primary bonds between molecular chains in thermosetting plastics and strong cross-links hold them together
They have low melting points and low tensile strength	They have high melting points and low tensile strength
Thermoplastic has a low molecular weight	The molecular weight of thermosetting plastic is high
Examples of thermoplastics are • Polystyrene • Teflon • Nylon	Examples include: • Vulcanized rubber • Bakelite • Polyurethane

TABLE 1.5
Types of Thermoplastics and Their Uses

Thermoplastic	Properties and Uses
Polyamide (nylon)	Used for casing power tools, curtain rails, bearings, gear parts, clothing, and durable and comparatively hard materials
Polymethyl methacrylate (PMMA, acrylic)	Used for signage, the fuselage of aircraft, curtains, sinks, and bathtubs for bathrooms
Polyvinyl chloride (PVC)	Widely used for tubing, flooring, toys, industrial fittings, strong, and durable materials
Polypropylene	With excellent chemical resistance, light, but hard material that scratches reasonably easily, used for medical and laboratory equipment, strings, ropes, and utensils
Polystyrene (PS)	Light, rigid, durable, fragile, waterproof material, used primarily for rigid packaging
Polytetrafluoroethylene (PTFE, Teflon)	The material used is very strong and flexible for non-stick cooking utensils, machine components, gears, and gaskets
Low-density polythene (LDPE)	For packaging, toys, plastic bags, and film wrap, it is a durable, relatively soft, chemical-resistant material
High-density polythene (HDPE)	Rigid, heavy, chemical-resistant material used in plastic bottles and household goods casings

TABLE 1.6
Types of Thermosetting Polymers and Their Uses

Thermoset	Properties and Uses
Epoxy resin	Strong plastic without additional reinforcement, which is brittle. Used for adhesive and material bonding
Melamine formaldehyde	Used for strong, rigid, and sturdy working-surface laminates, tableware, and electric insulation, with good chemical and water resistance
Polyester resin	When unlaminated, it is hard, rigid, and brittle. Used to encapsulate, bond, and cast
Urea formaldehyde	Hard, rigid, sturdy, and fragile, primarily used in electrical devices due to their strong electrical insulation characteristics
Polyurethane	Used in paint, foam insulation, shoes, car parts, adhesives and sealants, and materials which are rough, sturdy, and durable
Phenol formaldehyde (PF) resin	Strong, thermal, and electrical-resistant materials used in electrical appliances, sockets and plugs, vehicle parts, and industrial components

1.8.2 Types of Thermoplastics and Their Uses

The types of thermoplastics and their uses are summarized in Table 1.5.

1.8.3 Types of Thermosetting Polymers and Their Uses

Different types of thermosetting polymers and their uses are summarized in Table 1.6.

Plastics are used for the following purposes with various fields in plasticulture applications in agriculture:

- In situ moisture conservation and weed control by plastic mulching
- Plastic pipes for water distribution network
- Controlling seepage losses in pond or reservoir by plastic lining

- In pressurized irrigation (drip/sprinklers, micro-sprinklers, bubbler irrigation, and rain guns)
- Controlled environment agriculture (poly houses/shade net houses, polytunnels, and low tunnels)
- Surface-covered cultivation (crop covers by nonwoven covers)
- Protection nets (anti-bird net, insect proof net, net for protection from hail)
- Nursery raising (nursery bags, root trainers, and pots)
- Subsurface drainage
- Use of plastics in postharvest operations (plastic crates)
- Use of plastics in farm machinery
- Silage preservation
- Packaging of agricultural products

Other applications of plasticulture include:

- Covers for outdoor food grain storage
- Partly underground storage bins
- Packing of milk
- Unit packaging of fruits and vegetables
- Anti-hail nets/bird protection nets
- Nutrient film technique
- Soil fumigation/solarization
- Modified atmosphere packaging
- Silage preservation
- Plastics pots and containers
- Biogas plant holders/conveyance piping system
- Leno bags for packaging of fruits and vegetables

1.9 PLASTICULTURE IN INDIA

A modern approach to promote the development of agricultural products is plastic cultivation or the use of plastics in agriculture. In the mid-20th century, plastic farming emerged. Some countries have been skeptical about embracing this viable way of supporting agricultural development, perhaps, because of a lack of awareness of its advantages or the fear of elimination of their current strategies. Using plastics in cultivation is one of the alternative approaches involved. Other fields, such as food preservation, water conservation, and horticulture, have also found their application. For the sole purpose of improving the existing output of food manifold, the use of plastic culture has been suggested. Plastics have made their way into almost all professions, but a massive boon is its use in agriculture.

The elasticity of individual material growth with regard to GDP has been developed in the past and is projected for India for the next three decades, thus, assuming a development comparable to that of Western Europe. On this basis, the global use of plastics is expected to rise by a factor of 6 between 2000 and 2030. In the 1990s, plastic use in India rose exponentially. Over the last decade, overall plastic consumption has risen twice as quickly (12% per year) as the growth rate of the gross domestic product based on buying-power parities (6% per year). The current growth trend in the use of Indian polymers (16% per annum) is clearly higher than that in China (10% per year) and many other important Asian nations. In 2000/2001, the average Indian per capita intake of virgin plastics reached 3.2 kg (5 kg if recycled content is included) from a mere 0.8 kg in 1990/1991. This is, however, barely one-fourth of China's intake (12 kg/capita, 1998) and one-sixth of the world average (18 kg/capita).

Introduction

1.10 HOW PLASTIC IS USED

For growing plants and other crops, the way farmers use plastics is quite different, including micro-irrigation mulching, nets for plant protection, shade, among several others. Using plastics for food processing is not a modern concept. Proper usage, along with government-assisted services, structures, and organizations, provides producers with the best advice. It also opens up a range of opportunities for crop conservation and irrigation purposes. Mulching is required for the coverage of roots between plants. As it keeps the roots and soil warm, this method has several advantages. In fact, the warmth makes it unpleasant for mosquitoes and other destructive pests to survive. Plasticulture is expected to have a strong future in order to further improve food production in the region. Farmers need to be equipped to allow for the adequate use of plastics in agriculture. If plastic cultivation takes form, then germicides, toxins, and germicides can be minimized. It would also help and improve organic farming in this way.

India has around 140 Mha of net sown area, of which 86% of landholdings are less than 2 ha. Irrigation networks cover around 66 Mha of land under cultivation. Depletion of water resources, falling groundwater reserves, and deteriorating water quality are among serious concerns faced by the agriculture sector in India. The Central Government had planned the Accelerated Irrigation Benefit Program (AIBP) during the 2016–2017 budget at a cost of INR 86,500 crores to fast-track irrigation projects.

In India, plastic production, the use of plastics in agriculture, horticulture, water conservation, the storing of food grains, and related areas are at a nascent level. The Government has set up a National Committee on Plasticulture Applications in Horticulture (NCPAH) to popularize plasticulture adoption in India as a solution for improving resource use and crop yields. Precision Farming Production Centers have been set up in several states to facilitate high-tech horticulture precision farming and plasticulture applications.

1.11 STATUS OF PLASTICS

The use of plastics in cultivation has improved both production and productivity by minimizing the population of rodents, diseases, and weeds. The focus must be placed on developing methods to make plastic effective and inexpensive for farmers to use in their sectors. In agriculture, plastics are gaining significance, because, in one way or the other, their introduction has supported agriculture. The usage has ranged from soil planning to crop processing: sterilization of fields, sowing seeds, propagation methods, covering soils, drainage, growth of crops in managed conditions, management of insect pests, and harvesting mature crops from the field including film, net, or crates of plastics. With the emerging value of plastic with its flexible nature, the shape of farming in India seems to be getting modified for the following reasons:

- Plastics demand is growing rapidly at a rate of 10% annually
- Present consumption is 14 million tons per annum (MnTPA)
- India is the net importer of polyethylene (PE)
- Significant regional diversity in consumption: Western India (47%), Northern India (23%), Southern India (21%), and Eastern India (9%)
- Per capita consumption: 9.7 kg (US: 109 kg)

Key government schemes that promote plasticulture include National Mission on Micro Irrigation (NMMI), National Horticulture Mission (NHM), Horticulture Mission for North East and the Himalayan States (HMNH), Pradhan Mantri Krishi Sichai Yojana (PMKSY), NCPAH, and ICAR–AICRP on Plastic Engineering in Agriculture Structure and Environment Management.

1.12 AGENCIES INVOLVED IN PROMOTION OF PLASTICULTURE IN INDIA

Precision farming and plasticulture applications are promoted by the Ministry of Agriculture & Farmers' Welfare and the Department of Agriculture, Cooperation & Farmers' Welfare, Government of India (GoI), through various central-funded schemes/mission programs in different regions. The mandate of the National Committee on Plastic Applications in Horticulture (NCPAH) is to encourage and improve the use of plastics in agriculture, horticulture, water, and other related fields. During its tenure, the committee submitted several reports and policy documents to the Government of India that paved the way for the production of plastic cultivation in the region. The Government of India has also set up 22 Precision Farming Development Centres (PFDCs) across the country, on the recommendation of NCPAH, to facilitate research and extension activities in the promotion of such applications under the GoI flagship programs.

1.12.1 Formation of NCPAH

In 1981, the Department of Chemicals & Petro-Chemicals, Government of India, considered promoting and expanding plastic use in agriculture and irrigation as a significant step to increasing agricultural yield. In 1996, the commission was reconstituted. NCPAH was reconstituted in 2001 as the National Committee on Plasticulture Applications in Horticulture to make this committee more functional and concentrate its efforts in an organized way to encourage applications in horticulture. The NCPAH has been reconstituted at periodic intervals, recognizing the significance of precision farming & plasticulture interventions in livestock, horticulture, and allied sectors. The Government of India's Ministry of Agriculture is the nodal agency for the country's promotion of plastic cultivation and precision farming applications. Implementation projects are introduced via relevant state agencies. The National Committee on Plastic Applications in Horticulture (NCPAH) of the Ministry of Agriculture has a mandate to encourage and improve the use of plastics in agriculture, horticulture, water management, and other related fields.

QUESTIONS

1. What is plasticulture?
2. Enlist different types of plastics.
3. Describe the applications of plastics in agriculture.
4. Why have we used plastic material?
5. Differentiate between thermoplastic and thermosetting plastics.
6. Describe the types of thermoplastics and their uses.
7. Discuss the history and present status of plasticulture in the world.
8. Enlist different companies that are promoting plasticulture applications.
9. Describe the status of plasticulture in India.
10. How plasticulture is important in water management.
11. Describe the role of plastics in nursery production.
12. Enlist use of plastic in postharvest management of the crop.

MULTIPLE-CHOICE QUESTIONS

1. Plastic can be used for water management, such as
   a) Farm pond lining b) Canal lining, c) Mulching d) All above
2. At what rate is plastics demand growing in India annually?
   a) 10%, b) 20%, c) 30% d) 50 %?
3. The present consumption of plastics in India is
   a) 14 MnTPA b) 20 MnTPA c) 30 MnTPA d) 35MnTPA

4 Which agency is involved in the promotion of plastics in India?
   a) Reliance b) NCPAH c) DRDO d) None
5 In India, per capita consumption of plastic is:
   a) 9.7 kg b) 7.0 kg c) 15.0 kg d) 50 kg

ANSWERS

1	2	3	4	5
d	a	a	b	a

REFERENCES

Agrawal, N. and Agrawal, S. 2005. Effect of drip irrigation and mulches on the growth and yield of banana cv. Dwarf Cavendish. *Indian Journal of Horticulture*, 62(3), 238–240.

Anderson, R. G. and Emmert, E. M. 1994. The father of plastic greenhouses. The 25th National Agricultural Plastics Congress.

Emmert, E. M. 1957. Black polyethylene for mulching vegetables. *Proceedings of the American Society for Horticultural Science*, 69, 464–469.

Espí, A., Salmerón, A., Fontecha, Y., García, Y. and Real, A. I. 2006. Plastic films for agricultural applications. *Journal of Plastic Film and Sheeting*, 22(2), 85. doi: 10.1177/8756087906064220

Jaworski, C. A., Johnson, A. W., Chalfant, R. B. and Sumner, D. R. 1974. A system approach for production of high value vegetables on southeastern coastal plain soils. *Georgia Agricultural Research*, 16(2), 12–15.

Jensen, M. H. 2004. *Plasticulture in the Global Community—View of the Past and Future*. American Society for Plasticulture, Bellefonte.

Lamont, W. and Orzolek, M. 2004. *Plasticulture Glossary of Terms*. The American Society for Plasticulture, Bellefonte.

Sica, C., Dimitrijevic, A., Mugnozza, G. S. and Picuno, P. 2015. Technical properties of regenerated plastic material bars produced from recycled agricultural plastic film. *Polymer-Plastics Technology and Engineering*, 54(12), 1207–1214.

Singh, R. and Asrey, R. 2007. Cultivating strawberry the plasticulture way. *Indian Horticulture*, 52(2), 6–7.

Warp, H. 1971. Historical development of plastics for agriculture. *Proceeding Natural Agricultre Plastics Congress*, 10, 1–7.

2 Plastic Film Properties

2.1 INTRODUCTION

Polymers are chemical compounds that consist of very long chains composed of small repeating units, monomers. Today is a world of a competitive marketplace where various manufacturers of products and packages are under immense pressure to satisfy the needs and conflicting demands, like lowering cost, improving the performance of products, and enhancing attributes that are environmentally friendly. Within this arena, the material that a manufacturer chooses for processing products and packaging can be one of the greatest factors affecting the competition in the market. The varying marketplace demand has been met with the usage of plastic film and thus plastic has proved to be beneficial for the manufacturers to meet their increasing demands by enabling them to do more with less. Unfortunately, this information on the broad categories of materials is not well disseminated. Oftentimes, people talk about plastic films which technically are defined as plastic and thus they get treated and grouped together with flexible plastic packaging into a single category. However, there is a broader picture to see, as 'plastic film' is not just ordinary plastic but is composed of materials that may vary from comparatively simple to complicated, depending upon the need for a specific product or package. Now, it is understood that "plastic film" is basically a thin continuous polymeric material. The thicker plastic material is often called a "sheet." The materials that are used for separating areas or volumes or to hold items, even to act as barriers, or as printable surfaces are all examples of the usage of plastic films (Singh *et al.*, 2012).

Plastic films can be made of various resins which have a unique combination of properties that make them ideal for various applications like plastic bottles and containers. The low-density polyethylene (LDPE) film has the unique property of acting as a gas barrier and thus is highly used in the markets for packaging things. On the other hand, polyvinyl chloride (PVC) film which is gas permeable is used for packaging things, such as red meat, which needs a small amount of oxygen inside the package in order to remain fresh. Plastic films also can be clear or colored, printed or plain, single- or multilayered, and combined with alternate materials, such as aluminum metal or paper. Thus, if the plastic films are analyzed, then it is realized that they have only one thing in common, i.e., they are flexible in nature, as seen in grocery bags, or rigid, as seen in soft drink bottles and butter tubs. Plastic films have an advantage over other materials because they resist corrosion and provide electrical insulation. Plastic films are lighter than plastic sheets and often transparent or translucent.

2.2 POLYMER

The word polymer has been derived from two Greek words: 'poly' meaning many and 'meros' meaning parts. The compounds of very high molecular mass, formed by the combination of a very large number of simple molecular or repeating units, are called polymers. The simple combining molecules are called monomers and the process of combining is called polymerization. Polymers are present all around us (Figure 2.1). From the strand of DNA to polypropylene, which is used throughout the world as plastic, are examples of polymers. Polymers occur naturally in plants and animals and are thus called natural polymers or may be man-made, called synthetic polymers. All of these polymers whether natural or synthetic have a number of unique physical and chemical properties which make them ideal for usage in everyday life.

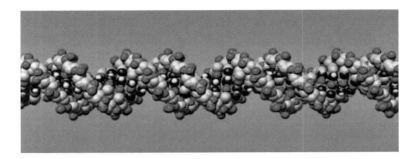

FIGURE 2.1 Polymer chain

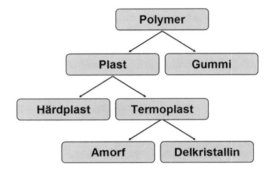

FIGURE 2.2 Polymer description

2.3 CLASSIFICATION OF POLYMERS

Polymers have a very complex structure, show a different behavior, and have a very wide application and thus cannot be classified under one category. Thus, the classification of polymers is described in Figure 2.2.

2.3.1 CLASSIFICATION OF POLYMERS BASED ON THE SOURCE OF AVAILABILITY

It is extremely rare to work with a pure polymer. As a rule, different additives (modifiers) are used to modify a material's properties. On the basis of source, polymers can be broadly divided into:

1. Surface lubricants
2. Heat stabilizers (improve the process window)
3. Color pigments
4. Reinforcement additives such as glass or carbon fiber (increase stiffness and strength)
5. Impact or toughness modifiers
6. UV modifiers (e.g., to protect against UV light)
7. Fire retardants
8. Antistatic agents
9. Foaming agents

2.4 THERMOPLASTICS

The intermolecular forces which act on thermoplastics lay in-between elastomers and fibers. These polymers have a unique property. They become soft when heated and become hard when cooled.

Plastic Film Properties

Polyvinyl chloride, Teflon, and polythene are examples of such type plastic. They are easy to process with a variety of methods, such as:

1. Injection molding (the most common process method for thermoplastics)
2. Blow molding (for making bottles and hollow products)
3. Extrusion (for pipes, tubes, profiles, and cables)
4. Film blowing
5. Rotational molding (for large hollow products such as containers, buoys, and traffic cones)
6. Vacuum forming (for packaging, panels, and roof boxes)

2.5 POLYETHYLENE

Polyethylene or polyethene is a semicrystalline commodity, denoted as PE. It is the most common plastic, and more than 60 million tons are manufactured each year worldwide. "Low-density" polyethylene was launched on the market by the British chemicals group ICI in 1939.

2.5.1 CLASSIFICATION

Polyethylene can be classified into different groups depending on its density and the lateral branches on the polymer chains:

1. UHMWPE – Ultrahigh molecular weight
2. HDPE – High-density
3. MDPE – Medium-density
4. LLDPE – Linear low-density
5. LDPE – Low-density
6. PEX – Cross-linked

Polyethylene types are described in Table 2.1.

2.6 LOW-DENSITY POLYETHYLENE

LDPE is more flexible than HDPE, which makes it a good choice for prosthetic devices, most of which are either drape-formed or vacuum-formed. Its impact resistance makes it natural for impact pads. It is a good choice for fabricated parts where chemical and corrosion resistance is demanded.

It is manufactured by heating ethylene at high temperature and pressure in the presence of traces of oxygen, which acts as a free radical. This polymer has a branched structure which results in loose packing and hence low density and low melting point. It is mainly used as packing material, as an insulator for wires and cables, and in the manufacture of toys, pipes, and bottles.

The general properties of LDPE are given below:

- Light-weight
- Good impact resistance
- Extremely flexible
- Easily cleaned
- Thermoforming performance
- Meets food-handling guidelines
- No moisture absorption
- Chemical- and corrosion-resistant

TABLE 2.1
Polyethylene Type

PE Type	Molecular Structure (schematic)	Degree of Branching per 1,000 C-atoms	Degree of Crystallinity (%)	Melting (Crystallization) Range (°C)	Density (g/cm³)	Strength	Elongation Toughness	Max. Use Temperature (°C)
PE-LD		High, 20–40 long- and short-chain branches	40–55	100–110 (80–95)	0.915–0.935	Low	High	ca. 90
PE-LLD		Middle, 15–30 short-chain branches up to 6 C-atoms	55–65	120–130 (105–115)	0.92–0.94	Middle	Medium, locally very high	ca. 95
PE-HD		Low, 1–10 short-chain branches from 1 to 2 C-atoms	70–75	125–135 (115–120)	0.93–0.97	High	Low	ca. 100

Plastic Film Properties 23

2.6.1 APPLICATION OF LDPE AREAS

UHMWPE is processed mainly by extrusion into pipes, film, or sheets (Figure 2.3).

2.7 HIGH-DENSITY POLYETHYLENE

It is manufactured by heating ethylene at high temperature and pressure in the presence of a Zeigler-Natta catalyst (i.e., triethylaluminum titanium tetrachloride). It is used in the manufacture of buckets, tubes, housewares, pipes, etc. HDPE is used for injection molding, blow molding, extrusion, film blowing, and rotational molding (Figure 2.4).

2.8 LDPE APPLICATION FOR FILM BLOWING AND EXTRUSION

A large part of all the polyethylene produced is used for film blowing. If the film is soft and flexible, it is either made of LDPE or LLDPE. If it has the rustle of the free bags at the grocery store, it is probably made of HDPE. LLDPE is also used to improve the strength of LDPE film (Figure 2.5).

2.9 CROSS-LINKED POLYETHYLENE

Cross-linked polyethylene is mainly used in the extrusion of tubes. The cross-linking provides improved creep resistance and better high-temperature properties (Figure 2.6).

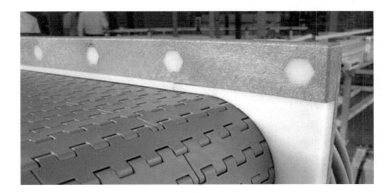

FIGURE 2.3 Ultrahigh molecular weight polyethylene application

FIGURE 2.4 High-density polyethylene application

FIGURE 2.5 LDPE application for film blowing and extrusion

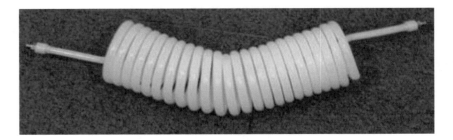

FIGURE 2.6 Cross-linked polyethylene application in the extrusion of tubes

2.10 GENERAL PROPERTIES AND TEST METHOD OF LDPE

The general properties and test methods of LDPE are illustrated in Table 2.2.

Commercially, polyethylene is produced from ethylene; the polymer was produced by this method in March 1933 and reported verbally by Fawcett in 1935. Until the mid-1950s, all commercial polyethylene was produced by high-pressure processes developed from those described in the basic patent. These were somewhat branched materials and the average molecular weight, generally less than 50,000. However, around 1954, two other routes were developed, one using metal-oxide catalysts (e.g., the Phillips process) and the other aluminum alkyl or similar materials (the Ziegler process). By these processes, polymers could be prepared at lower temperatures and pressures and with a modified structure. Because of these modifications, polymers had a higher density, were harder, and had high softening points. These materials are known as high-density polyethylene, while the earlier materials are known as low-density polyethylene.

2.11 POLYMERIZATION

There are five quite distinct routes to the preparation of high polymers of ethylene:

(1) High-pressure processes
(2) *Ziegler processes*
(3) The Phillips process
(4) The Standard Oil (Indiana) process
(5) Metallocene processes

2.11.1 HIGH-PRESSURE POLYMERIZATION

With regard to ethylene polymerization under high pressure, little information has been made publicly available concerning details of current commercial processes. It may, however, be said that

Plastic Film Properties

TABLE 2.2
General Properties of LDPE

Property	Test Method	Units	LDPE
Physical			
Density	ASTM D-792	lbs/ft^3	57.39
Water absorption	ASTM D-570	%	Slight
Mechanical			
Yield point	ASTM D-638	psi	1,363
Tensile break	ASTM D-638	psi	1,943
Elongation at break	ASTM D-638	%	515
Tensile modulus	ASTM D-638	psi	41,615
Flexural modulus	ASTM D-790	psi	28,565
Flexural strength	ASTM D-790	psi	1,175
Izod impact	ASTM D-4020	ft-lbs/in	No break
Tensile impact	DIN 53448	ft-lbs/in^2	401
Hardness	ASTM D-2240	Shore D	55
Thermal			
Melt point	ASTM D-3417	°F	230
Heat deflection	ASTM D-668		
264 psi		°F	98
66 psi			110
Electrical			
Volume resistivity	ASTM D-257	ohm-cm	>10^{15}
Surface resistivity	ASTM D-257	ohm/square	>10^{15}

commercial high polymers are generally produced under conditions of high pressure (1,000–3,000 atm) and at temperatures of 80–300°C. A free-radical initiator such as benzoyl peroxide, azodiisobutyronitrile, or oxygen is commonly used. Because of the high heat involved in polymerization, care must be taken to prevent runaway reactions. This can be done by having a high cooling surface–volume ratio in the appropriate part of a continuous reactor and in addition by running water or a somewhat inert liquid such as benzene (which also helps to prevent tube blockage) through the tubes to dilute the exotherm. Local runaway reactions may be prevented by operating at a high flow velocity. In a typical process, 10–30% of the monomer is converted to polymer. After polymer–gas separation, the polymer is extruded into a ribbon and then granulated. Film grades are subjected to the homogenization process in an internal mixer or a continuous compounder machine to break up high molecular weight species present. Although in principle the high-pressure polymerization of ethylene follows the free-radical-type mechanism discussed in particular characteristics, the high exothermic reaction and a critical dependence on the monomer concentration. It is particularly important to realize that at the elevated temperatures employed, other reactions can occur leading to the formation of hydrogen, methane, and graphite. These reactions are also exothermic and it is not at all difficult for the reaction to get out of hand. It is necessary to select conditions favorable to polymer formation, which allow a controlled reaction. Most vinyl monomers will polymerize by free-radical initiation over a wide range of monomer concentrations. Methyl methacrylate can even be polymerized using the homogenization process in an internal mixer or a continuous compounder machine to break up high molecular weight species present. Unless they have reacted within a given interval, they undergo changes that terminate their growth. Since the rate of reaction of radicals with monomers is much greater with higher monomer concentration (higher pressure), it will be appreciated that the probability of obtaining high molecular weights is greater at high pressures than at low pressures.

At high reaction temperatures (e.g., 200°C) much higher pressures are required to obtain a given concentration or density of monomer than at temperatures say 25°C, and it might appear that better results would be obtained at lower reaction temperatures. This is, in fact, the case where a sufficiently active initiator is employed. This approach has an additional virtue in that side reactions leading to branching can be suppressed. For a given system, the higher the temperature the faster the reaction and the lower the molecular weight. By varying temperature, pressure, initiator type, and composition, by incorporating chain transfer agents and by injecting the initiator into the reaction mixture at various points in the reactor, it is possible to vary independently of each other polymer characteristics such as branching, molecular weight, and molecular weight distribution over a wide range without needing unduly long reaction times. In spite of the flexibility, most high-pressure polymers are of the lower density range for polyethylene (0.915–0.94 g/cm^3) and usually also of the lower range in terms of molecular weights.

2.11.2 Natural Polymers

Polymers occurring naturally, i.e., through plants and animals, are called bio-polymers. Proteins, starch, cellulose, nucleic acids, and natural rubber are examples of natural polymers. There is also a class of biodegradable polymers, which are called biopolymers.

2.11.3 Semisynthetic Polymers

When naturally occurring, polymers undergo further chemical modification, and they form semisynthetic polymers. Cellulose nitrate and cellulose acetate are examples of semisynthetic polymers.

2.11.4 Synthetic Polymers

These polymers are artificially made in the laboratory, called man-made polymers. Polythene plastic (which is the most common and widely used synthetic polymer), nylon, Teflon, Bakelite, synthetic rubber, etc. are examples of synthetic polymers.

2.12 CLASSIFICATION OF POLYMERS BASED ON THE STRUCTURE

On the basis of structure, polymers can be divided into three categories.

2.12.1 Linear-Chain Polymers

In these polymers, monomers are linked together to form linear chains. These are well packed and they have high density, high melting points, and high tensile strength. Polythene, nylon, and PVC, i.e., polyvinyl chloride, are examples of a linear polymer.

2.12.2 Branched-Chain Polymers

When the monomers are linked to form long chains with side chains or branches of different lengths, they form branched-chain polymers. The branched-chain polymers are irregularly packed and, therefore, have low tensile strength and low melting points as compared to linear polymers. Examples are low-density polythene, glycogen, starch, etc.

2.12.3 Cross-Linked Polymers

When the monomer units are cross-linked together to form a three-dimensional network, they form a cross-linked polymer. The presence of cross-links makes these polymers hard, rigid, and brittle. Bakelite and melamine-formaldehyde are examples of this category.

2.13 CLASSIFICATION OF POLYMERS BASED ON MODE OF SYNTHESIS

On the mode of synthesis, polymers can be classified as given below.

2.13.1 ADDITION POLYMERS

Addition polymer is defined as a polymer formed by the direct addition of the same or completely different monomers without the elimination of any simple molecule as a by-product. In this case, the monomers are unsaturated compounds, generally derivatives of ethane. Examples are polythene, PVC, polypropylene, Teflon, rubber, etc.

2.14 CLASSIFICATION OF POLYMERS BASED ON MONOMERS

On the basis of monomer units, polymers can be classified as given below.

2.14.1 HOMOMER

Homomer is a type of polymer in which a single type of monomer unit is present. An example is polythene.

2.14.2 HETEROPOLYMER OR COPOLYMER

Heteropolymer is a type of polymer that consists of different types of monomer units. An example is nylon 6,6.

2.15 CLASSIFICATION BASED ON MOLECULAR FORCES

Depending upon intermolecular forces, polymers can be classified as follows.

2.15.1 ELASTOMERS

These polymers have an elastic property and polymeric chains are joined together by the weakest intermolecular forces (van der Waal's forces). Because of weak forces, these polymers can be easily stretched by applying small stress and they regain their original shape when the stress is removed. An example is rubber.

2.15.2 FIBERS

These polymers have the strongest intermolecular forces between chains. These forces are generally hydrogen bonding or dipole–dipole interaction. Due to strong intermolecular forces, fibers have high tensile strength and least elasticity. These polymers are long, thin, and thread-like with high melting points and low solubility. Examples are nylon 66, Dacron, silk, etc.

2.15.3 CONDENSATION POLYMERS

A polymer formed by the condensation of two or more than two of the same or different monomers with the elimination of simple molecules like water, ammonia, hydrogen chloride, alcohol, etc. is called condensation polymer. Examples are nylon 610.

2.15.4 Thermosetting Polymers

These polymers once heated undergo permanent change. They become hard, infusible, and insoluble because they get highly cross-linked and become permanently rigid. Examples are Bakelite, epoxies, and silicones.

2.16 STRUCTURE OF POLYMERS

The polymers present around us are made up of a hydrocarbon backbone defined as a long chain of linked carbon and hydrogen atoms, which is possible due to the tetravalent nature of carbon. The various types of hydrocarbon backbone polymer are polypropylene, polybutylene, and polystyrene. However, there are certain polymers which, instead of carbon, have other elements in their backbone. An example is nylon that contains nitrogen atoms in the repeated unit backbone.

2.17 TYPES OF POLYMERS

2.17.1 Classification Based on the Type of Backbone Chain

Polymers can be classified on the basis of the type of backbone chain (Figure 2.7) as follows.

Organic Polymers

Organic polymers are the type of polymers that have a carbon backbone present in their structure.

Inorganic Polymers

Inorganic polymers are the type of polymers that have a backbone constituted by elements other than carbon.

2.17.2 Classification on the Basis of Their Synthesis

On the basis of their synthesis, polymers can be classified as:

- Natural polymers
- Synthetic polymers

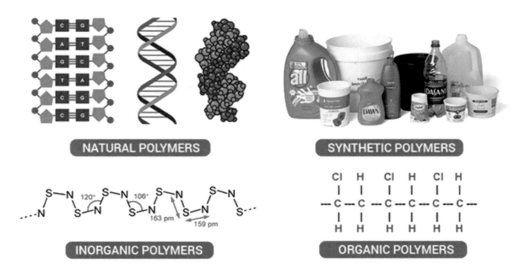

FIGURE 2.7 Types of polymers

Plastic Film Properties

- Biodegradable polymers
- High-temperature polymers

2.18 BIODEGRADABLE POLYMERS

These are polymers that can be broken into small segments by enzyme-catalyzed reactions. The required enzymes are produced by microorganisms. These polymers can be easily degraded by the microorganisms present in the soil when they are buried. They find applications in surgical bandages, capsule coatings, and surgery. An example is polyhydroxybutyrate co vel (PHBV).

2.18.1 HIGH-TEMPERATURE POLYMERS

The polymers that are stable at high temperatures are called high-temperature polymers. These polymers do not get destroyed even at very high temperatures because of their high molecular weight. They find applications in healthcare industries, for making sterilization equipment, and in the manufacturing of heat- and shock-resistant objects. An important polymer is polythene or polyethylene.

This is the most common type of plastic found around us. It is mostly used in packaging from plastic bags to plastic bottles. There are different types of polythene but their common formula is $(C_2H_4)_n$.

Polythene is essentially of two types: low-density polythene and high-density polythene.

2.19 PROPERTIES OF POLYMERS

Polymers have different properties that are briefly discussed in what follows.

2.19.1 PHYSICAL PROPERTIES

The physical properties of polymers include molecular weight, molar volume, density, degree of polymerization, crystallinity of material, and so on. Some of these physical properties are described below.

2.19.1.1 Degree of Polymerization and Molecular Weight

The degree of polymerization in any polymer molecule is simply defined as the number of repeating units in the polymer chain. For example,

$$-(-CH-CH-)-_n$$

The product of the degree of polymerization and the molecular weight of the repeating unit gives the molecular weight of a polymer molecule.

2.19.1.2 Molecular Mass of Polymers

When polymers are formed, different macromolecules have different degrees of polymerization, i.e., they have a different and unique chain length. Thus, if one sees any particular sample of a polymer, the molecular masses of individual macromolecules are different. Hence, one takes an average molecular mass. The polymers have two types of average molecular masses.

2.19.1.3 Number Average Molecular Weight

When the total mass of all the molecules of a sample is divided by the total number of molecules, the result obtained is number average molecular weight. Let N_i be the number of polymer molecules having the molecular weight M_i, then the "number average" probability of the given mass is given below:

$$P_i = \frac{N_i}{\sum_{j=0}^{\infty} N_j}.$$

The number average molecular weight is given by

$$M_n = \frac{\sum_{i=0}^{\infty} M_i N_i}{\sum_{j=0}^{\infty} N_j}.$$

2.19.1.4 Weight Average Molecular Weight

The weight average molecular weight is the result obtained when the total mass of groups of molecules having different molecular masses is multiplied with their respective molecular masses, and the products are added and the sum is divided by the total mass of all the molecules. The weight average probability is given by:

$$P_i = \frac{N_i M_i}{\sum_{j=0}^{\infty} N_j M_j}.$$

The weight average molecular weight is given by

$$M_i = \frac{\sum_{i=0}^{\infty} M_i N_i^2}{\sum_{j=0}^{\infty} N_j M_j}.$$

The weight average molecular weight is more than the number average molecular weight.

The degree of polymerization can, thus, be calculated using the number average molecular weight

$$\text{Degree of polymerization} = \frac{\text{Number average molecular weight}}{\text{Molecular weight of repeating unit}}.$$

2.19.1.5 Polydispersity Index or Heterogeneity Index

The homogeneity of a polymer is defined by the term called polydispersity index or heterogeneity index. It is the ratio of the weight average molecular weight to the number average molecular weight:

$$\text{PDI} = \frac{M_w}{M_n}.$$

The heterogeneity of sizes of molecules or particles in the mixture is measured through dispersity. If the molecules have *the same size*, *shape*, or *mass*, the mixture is called *monodisperse*. The mixture is called *polydisperse* if the molecules in the mixture have *inconsistent sizes, shapes*, and *mass distribution*. All the naturally occurring polymers are generally monodispersed, while the synthetic polymers are polydisperse with some exceptions. The polydispersity index is equal to or greater than one but when the polymer chains approach uniform chain length, the PDI tends to unity.

1. **Polymer Crystallinity**

The polymers that have simple structural chains as linear chains and have a slow cooling rate will generally give good crystallinity. When there is slow cooling, enough time is available for crystallization to take place. If the degree of crystallinity is high in polymers, they become rigid and have a high melting point while their impact resistance is low. However, the polymers that are soft and have lower melting points are categorized as amorphous polymers. A solvent can penetrate the amorphous part more easily than the crystalline part.

Some of the amorphous polymers are polystyrene and poly (methyl methacrylate).

Some of the crystalline polymers are polyethylene and PET polyester.

The range of crystallinity is defined as amorphous which is (0%) to highly crystalline which is > 90%.

The percentage of crystallinity is given by:

$$\text{Percentage Crystallinity} = \frac{\rho_c(\rho_s-\rho_a)}{\rho_s(\rho_c-\rho_a)} \times 100$$

ρ_c = Density of the completely crystalline polymer
ρ_a = Density of the completely amorphous polymer
ρ_s = Density of the sample.

2.19.2 MECHANICAL PROPERTIES

The mechanical properties of polymers are as given below.

2.19.2.1 Tensile Strength

The tensile strength of a material is a measure of how much elongating stress the material will endure before failure. Tensile strength is, therefore, important in applications that solely rely upon the polymer's physical strength or durability. A rubber band, for example, having a higher tensile strength will hold a greater weight before snapping. In general, tensile strength is directly linked to polymer chain length and cross-linking of polymer chains and hence increases when the polymer chain length increases.

2.19.2.2 Young's Modulus of Elasticity

To quantify the elasticity of a polymer, Young's modulus of elasticity is used. Young's modulus for small strains is defined as the ratio of the rate of change of stress to strain. This is highly relevant in polymer applications like tensile strength because it involves the physical properties of polymers, such as rubber bands. Young's modulus is strongly dependent on temperature. Viscoelasticity is used to describe a complex time-dependent elastic response, which exhibits hysteresis in the stress–strain curve as soon as the load is removed. Dynamic mechanical analysis is used to measure this complex modulus by oscillating the load and, thus, measuring the resulting strain as a function of time.

2.19.3 TRANSPORT PROPERTIES

Transport properties such as diffusivity depict how quickly molecules move through the polymer framework. These are exceptionally critical in numerous applications of polymers for films and membranes. The development of individual macromolecules happens by a process called reputation in which each chain molecule is compelled by traps with neighboring chains to move inside a virtual tube. The hypothesis of reputation can clarify polymer particle flow and viscoelasticity.

2.19.4 CHEMICAL PROPERTIES

The polymer's properties can largely be determined by the attractive forces between polymer chains. As the polymer chains are very long, they have many such interchain interactions per molecule and hence amplify the effect of these interactions on the polymer properties in comparison to attractions between conventional molecules. Various side groups present in the polymer can cause the polymer to form ionic bonding or hydrogen bonding between its own chains. These stronger forces usually lead to higher tensile strength and higher crystalline-melting points.

The dipoles in the monomer units affect the intermolecular forces in the polymers. The polymers which contain amide or carbonyl groups form hydrogen bonds between adjacent chains; the partially negatively charged oxygen atoms in the C=O groups on one chain are highly attracted to the positively charged hydrogen atoms present in the N–H groups of another chain. These strong hydrogen bonds are responsible for the high tensile strength and melting point of polymers containing urethane or urea linkages. Polyesters existing between the oxygen atoms in the C=O groups and between the hydrogen atoms in the H–C groups generally undergo dipole–dipole bonding. Dipole bonding is, however, not as strong as hydrogen bonding; that is why polyesters have a low melting point and strength as compared to Kevlar's (Twaron), but polyesters have generally greater flexibility. Polymers that have a nonpolar unit like polyethylene interact only through weak van der Waals forces. Thus, they usually have lower melting temperatures than any other polymers. In commercial products like paints and glues, when a polymer is dispersed or dissolved in a liquid, the chemical properties and molecular interactions are the drivers of the solution flows and also lead to the self-assembly of the polymer into complex structures. However, when a polymer is applied as a coating, the chemical properties strongly influence the adhesion of the coating and also its interaction with external materials, like superhydrophobic polymer coatings, which lead to water resistance. In a nutshell, for designing new polymeric material products, the chemical properties of a polymer are important elements.

2.19.5 OPTICAL PROPERTIES

Polymers, like poly (methyl methacrylate), conjointly referred to as acrylic, abbreviated as PMMA, and hydroxyl ethyl methacrylate or HEMA; methyl methacrylate or MMA are used as matrices within the gain medium of solid-state dye lasers, which are referred to as solid-state dye-doped polymer lasers. These polymers have a high surface quality in addition to being tremendously transparent in order that the laser properties are influenced by the laser dye used to dope the polymer matrix. These kinds of lasers which also belong to the class of organic lasers usually yield very narrow linewidths, which can be used for spectroscopy and analytical applications. A crucial optical parameter in the polymer which is employed in laser applications is the change in refractive index with temperature which is also known as dn/dT. The polymers mentioned here have (dn/dT) ~ -1.4×10^{-4} in units of K^{-1} in the $292 \leq T \leq 332$ K range.

2.19.6 ELECTRICAL PROPERTIES

Typical polymers, for example, polyethylene, are electrical insulators; however, the development of polymers that contain π-conjugated bonds has led to the development of polymer-based semiconductors, like polythiophenes. This has led to many applications within the field of organic electronics.

2.20 POLYMER EXTRUSION

Extrusion may be defined as a process accustomed to creating objects of a fixed cross-sectional profile. The two essential benefits of this process over disparate manufacturing processes are its ability to form very composite cross-sections and to use materials that are fragile, as the material only encounters compressive and shear stresses. It conjointly forms parts with an excellent surface finish.

This limits the number of modifications that can be performed in one step; therefore, it is restricted to less complicated shapes, and multiple stages are usually required. The process of extrusion can be continuous (theoretically manufacturing indefinitely long material) or semicontinuous (producing several pieces). It can be done with the material either hot or cold. Typical examples of extruded materials embody metals, polymers, ceramics, concrete, modeling clay, and foodstuff. The products of extrusion are usually called "extrudates."

Extrusion can be a primary shaping process. Polymer is pumped into the die and forms the shape. Shapes of products vary from profile, plate, and film tube to any type of geometric shape. The types of a process during the extrusion can be mixed, densified, plasticized, homogenized, degassed, or reactive extrusion (chemically change). The types of processes after the extrusion are solidification by pressured air process. Polymer melts throughout extrusion and conveys a new shape, thus, justifying that extrusion is a primary shaping process. Extrusion products have a constant cross section.

There are a couple of requirements for extrusion:

a) Homogeneous transport of the material
b) Production of thermally and mechanically homogeneous parts
c) Avoidance of the thermal, chemical, or mechanical degradation
d) Computation of the suitable and economic operation

A single-screw extruder is mostly used for pumps with very high mass outputs at a very high pressure that is needed for large components like pipes, plates, or profiles (Figure 2.8). Twin-screw extruders are used for mixing and compounding.

The different parts of an extruder are hopper, barrel, screw, heating-cooling, and drive gear. The screw is the main part of the extruder, as it provides transporting, solid feedstock, compression, melting, and homogenizing, and it meters the polymer to provide the right amount of pressure to pump the melt in the die (Figure 2.9). Usually, three-zone screws are used for most thermoplastics. There are different sizes of extruders according to the internal diameter D, which is in the range of 20–150 mm. Also, there is the L/D magnitude relation which can change within 5 to 34. Shorter machines are generally used for elastomers and longer machines are most commonly used for thermoplastics. There are grooved barrels that can be used to increase transport and compression. Electric heater and fans control the barrel temperatures. There are thermocouple junctions and pressure gauges to monitor temperatures and pressures at a certain position of the barrel. Hopper is employed for providing the pellets or powder to the extrusion. For larger systems, there is also an agitating system or a conveyor system. There can also be a dryer for moisture-sensitive materials.

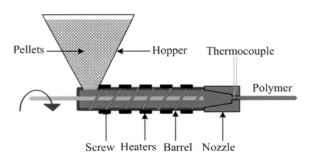

FIGURE 2.8 Parts of extruder (Kocak, 2018)

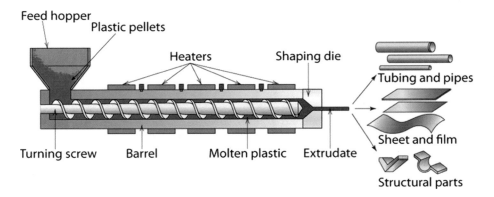

FIGURE 2.9 Different types and shapes of products (Kocak, 2018)

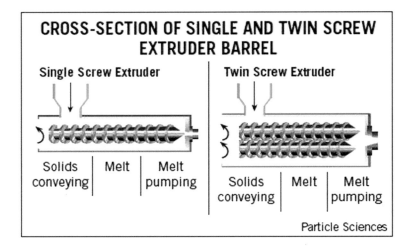

FIGURE 2.10 Illustration of single-screw and twin-screw extruders (Kocak, 2018)

2.20.1 Variance Between Single- and Twin-Screw Extruders

Single-screw extruders are employed for pumping high mass outputs at high pressures required for parts like pipes, plates, or profiles. Twin-screw extruders are used for mixing and compounding and are often used for continuous mixing when compared to single-screw extruders. The different types of twin-screw extruders are counter-rotating and corotating, intermeshing, non-intermeshing, and close intermeshing extruders. These different types can be employed for different mixing and compounding purposes. Twin-screw extruders offer various advantages of mixing efficiency, heat generation, etc. They additionally have other advantages like the addition of polymer additives without mechanical and thermal stresses. They can also be used for the preparation of polymer blends by mixing various kinds of polymers with a large divergence in the melting points or glass transition temperatures or viscosities. Single-screw extruders are generally used for plastic materials or recycled plastic materials or the type of extrusion that doesn't require the compounding ability. Twin-screw extruders are desirable for less heat friction, uniform shearing, larger capacity, stable extrusion, etc., and are generally used for fine material, or temperature-sensitive PVC. Twin-screw extruders have additionally short retention time for temperature-sensitive materials (Figure 2.10). When compared to single-screw extruders, twin-screw extruders have many advantages which are as follows:

Plastic Film Properties

1) Twin extruders have lower energy consumption
2) Twin extruders have perfect exhaustion ability and, thus, have easily degassing operation
3) Materials can be fed easily through a twin-screw extruder
4) Process time is low for a twin extruder owing to less heat friction and, thus, is used for heat-sensitive materials
5) They have higher compounding and plasticizing ability

Despite many advantages, the twin-screw extruders cannot be approached for high-pressure values. Thus, single-screw extruders are generally better for higher pressure and hard-extruder materials.

2.20.2 Distinctive Zones of Single- and Twin-Screw Extruders

The solid conveyance zone which is present in the feed section provides a way for the movement of pellets or powders from the hopper to the screw channel. Feeding on the hopper, generally, is due to gravity (flood feed). The screw is used to permanently remove the resin from the hopper (Figure 2.11). The materials are compacted and then transported down the channel in the screw channel. Friction provides the compaction and movement at the screw surface. The high coefficient of friction is important for compaction and movement, and because of this, the feed section of the barrel should be cooled often by water. Friction forces cause the pressure to increase within the feed section; its pressures compress the solid bed that continues traveling down the channel and thus melts in the compression or melting zone. This situation is controlled by a reduction of the channel depth.

The metering zone provides the pressure for ample pumping and, therefore, the final melt temperatures are formed here. This section essentially has a constant screw depth which is less than the feeding zone. The metering zone homogenizes the melt to offer a constant rate, the material of uniform pressure, and temperature to the die. It is connected to the viscous melt flowing along a uniform channel. The granules are taken and melted, compressed, and homogenized in the first part of the screw. The pressure of melt is then reduced by the atmospheric pressure in the decompression zone. This can provide the advantage; otherwise, the volatile parts can go outside of melt because of low pressure in the (vent port) barrel. The melt is then transported along the barrel to a second compression zone that stops air pockets from being trapped. The venting port works at a temperature of 250°C and thus the water changes into vapor at a pressure of about 4 MPa. This high temperature and pressure allow the vapor to easily pass out of the melt end through the vent. Application of a vacuum effect is employed for the removal of moisture. The standard diameter of the barrel changes according to the screw.

2.21 DIE AND SCREW PARAMETERS

The extrusion process has different die and screw parameters that change directly the properties of products and extrusion process conditions. One of the important components of the extrusion

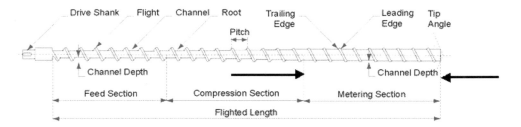

FIGURE 2.11 Different zones for extruder (Kocak, 2018)

process is the screw diameter. Temperature is also an important parameter for the extrusion process because as temperature changes, the viscosity of melts also changes. Geometry, length, and temperature of the die can cause different pressures. The viscosity of the die and the temperature of the barrel are important because as viscosity increases, the pressure also increases. For screw length and extruder diameter, there is time for the longer extruder to melt the polymer. The *L/D* ratio is higher for higher melting temperature thermoplastics and elastomers. Higher rotational speeds mean higher shearing rates and higher shearing rate means lower viscosity. The screw length and extruder diameter ratio determine the type of materials that should be processed. Thick sheets, generally, have low pressure change (Δp bar) and, thus, less restrictive die, whereas thin sheets use high pressure change (Δp bar) and thus higher restrictive die. Deep channel screws have high throughput and low pressure change (Δp bar) and shallow screw channels have low throughput and high pressure change (Δp bar). The die and screw characteristic pressure-throughput curves are illustrated in Figure 2.12.

There are die-screw characteristic pressure-throughput curves; die 1 represents the low-resistance die such as for a thick plate and die 4 represents a restrictive die, which can be used for restrictive die. Different characteristic curves represent the different screw rotational speeds.

SHORT QUESTIONS

1. What is a plastic film?
2. Describe different plastic film properties.
3. What is a polymer? Describe the classification of polymers.
4. Describe biodegradable polymers.
5. Give details of properties of good quality plastic films.
6. Describe the classification of polymers.
7. Describe different types of polyethylene.
8. Describe LDPE and its properties.
9. What is HDPE? Describe its application area.
10. Give details about natural polymers.
11. Describe the classification of polymers based on monomers.
12. What is thermoplastic?
13. Describe degradable plastic.

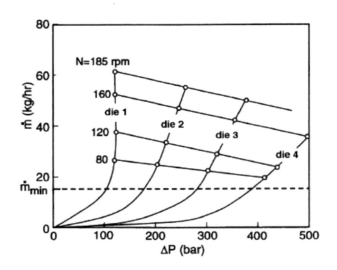

FIGURE 2.12 Die and screw characteristic pressure-throughput curves (Kocak, 2018)

14. Explain high-temperature polymers.
15. What is polymer extrusion?
16. Describe single-screw and twin-screw extruders.

MULTIPLE-CHOICE QUESTIONS

Q1. Which has a unique property of acting as a gas barrier?
   a) LDPE   b) PVC
   c) HDPE   d) None of these
Q2. Which of the following is a biodegradable polymer?
   a) PHBV   b) Polyethylene
   c) Both a and b   d) None of the above
Q3. When the monomers are linked to form long chains or branches of different lengths, they form
   a) Branched-chain polymers   b) Linear-chain polymers
   c) Cross-linked polymers   d) Condensation polymers
Q4 Polythene is an example of _____.
   a) Copolymer   b) Homomer
   c) Heteropolymer   d) None of these
Q5. Which of the following is a semisynthetic polymer?
   a) Cellulose nitrate   b) Cellulose acetate
   c) Both a and b   d) None of these
Q6. _____ polymers have a carbon backbone present in their structure.
   a) Inorganic polymers   b) Organic polymers
   c) Both a and b   d) None of these
Q7. As the polymer chain length increases, the tensile strength
   a) Decreases   b) Remains the same
   c) Increases   d) None of these
Q8. Twin extruders have _____ energy consumption.
   a) Low   b) High
   c) Zero   d) None of these
Q9. The presence of cross-links makes cross-linked polymers
   a) Hard   b) Rigid
   c) Brittle   d) All of the above
Q10. _____ are examples of linear-chain polymer
   a) Polythene   b) Nylon
   c) Both a and b   d) None of these

ANSWER

1	2	3	4	5	6	7	8	9	10
a	a	a	b	c	b	c	a	d	c

REFERENCES

Kocak, A. 2018. *Material Processing: Polymer Extrusion.* Technical report.
Singh, R., Singh, S. and Hashmi, M. 2012. Polymer Twin Screw Extrusion with Filler Powder Reinforcement. *Reference module in material science and material engineering.* Elsevier.

3 Micro-Irrigation

3.1 INTRODUCTION

Water is the most precious gift of nature, is essential for human and animal life, and plays an important role in plant growth. On the macro level, India is a water-stressed country with annual water availability of 1,000–1,700 m^3 per capita per year. Micro-irrigation can play a pivotal role in doubling the farmer's income, as at least a 40–50% increase in income levels due to micro-irrigation is already proven (Kumar, 2019). Micro-irrigation, a modern method of irrigation, is an important irrigation method used in agriculture. Micro-irrigation is the process of slow and frequent application of water and involves the system of drip and sprinkler that saves water, improves the quality of water usage, reduces energy requirements, reduces weed growth, and controls soil erosion (Narayanaswamy and Anandakumar, 2016). The water is applied through a network of economically built plastic pipes and low discharge emitters, directly to the root zone of plants. While there are different methods of irrigating crops, irrigation systems for drip and sprinkler are considered to be the best in achieving the quality of water and fertilizer usage along with improved productivity of crops. In the drip system, water is supplied, via a tubing network, directly to the root zone of each plant. As per plan and convenience, the tubing can be shifted around distinct places, topography, and slopes to deliver water to the plants at the desired pressure through emitters/microtubes. There is enormous space for water conservation, distribution, and farm use and for achieving greater productivity in water usage. The crop yield can be greatly maximized with a small amount of water through micro-irrigation systems.

Irrigation water requirements can be smaller with micro-irrigation when compared with other irrigation methods. This is due to the irrigation of a smaller portion of the soil volume, decreased evaporation from the soil surface, and the reduction or elimination of the runoff. The losses due to the evaporation from the soil are significantly reduced compared with other irrigation systems since only a small surface area under the plant is wetted and it is usually well shaded by the foliage. Since the micro-irrigation system allows for a high level of water control application, water can be applied only when needed and deep percolation can be minimized or avoided.

Smaller flow rates: Since the rate of water application in micro-irrigation systems is significantly lower than in other systems, smaller sources of water can be used for irrigation of the same acreage. The delivery of plastic pipes, the pump, and other components of the system can be smaller and, therefore, more economical. The systems operate under low pressure (5–30 psi) and require less energy for pumping than high-pressure systems.

Application of chemicals (chemigation): Micro-irrigation systems allow for a high level of control of chemical applications. The plants can be supplied with the exact amount of fertilizer required at a given time. Since they are applied directly to the root zone a reduction in the total amount of fertilizer used is possible. There is also an advantage to the frequent application of fertilizers through the system in humid climates. In case of rain, only a small portion of recently applied fertilizer will be washed out and it can be easily replaced through the irrigation system. This application method is more economical, provides better distribution of nutrients throughout the season, and decreases groundwater pollution due to the high concentration of chemicals that could ordinarily move with deep percolated water. Other chemicals, such as herbicides, insecticides, fungicides, nematicides, and growth regulators, can be efficiently applied through micro-irrigation systems to improve crop production.

Water sources with high salt content: A significant advantage of micro-irrigation is that water with relatively high salt content can be used by the system. For optimum plant growth, a certain range of total water potential in the root zone must be maintained. The potential defines how difficult it is for a plant to extract water from the soil. Large negative numbers are characteristic of very dry soils with low total water potentials while potentials near zero reflect soils near saturation. The total water potential in the root zone is a sum of the matric potential and osmotic potential. Since matric potential is close to zero under micro-irrigation (high moisture content) the osmotic potential component can be a relatively large negative value, indicating high salt content, without harmful effect on plant growth. This is not true for other irrigation systems.

Improved quality of the crop: Micro-irrigated plants are supplied very frequently with small amounts of water and the stress due to the moisture fluctuation in the root zone is reduced to the minimum, often resulting in larger and better-quality yield. In arid climates, or during dry seasons, the harvest timing can be controlled by proper water management.

Adaptation to any topography: Micro-irrigation systems can operate efficiently on hilly terrain if appropriately designed and managed. Well-managed micro-irrigation system will not create runoff even on hilly terrain.

Additional advantages of micro-irrigation systems: During dry seasons or in arid climates disease and insect damage can be reduced under the micro-irrigation system since the foliage of the plant is not wetted. With a small portion of soil surface being watered, field operations can be continued during irrigation.

3.2 HISTORY OF MICRO-IRRIGATION

In Germany, earlier attempts were made during 1860 by simply pumping the irrigation water through the underground drainage system into the clay pipes. In 1913, in Colorado, the first work on micro-irrigation systems was conducted and it was concluded that the drip system was too costly. Later, in 1920, a major development was made in Germany when perforated pipes were used to irrigate crops. The peach growers in Australia pumped water through 5 cm GI pipes laid down along the tree rows in 1930, with water emitting points made as small triangular holes on the pipe. In early 1940, Simcha Blass noticed that compared to other trees in the region, a tree near a water leakage point showed vigorous growth. This led to the idea of micro-irrigation, in which water is applied drop by drop in very small quantities. Later, when polyethylene, a crack-resistant and cheaper substitute, was developed in a British laboratory, a remarkable breakthrough was made in material science. Later, high-density polyethylene was introduced by low-density polyethylene and linear low-density polyethylene was introduced in 1977. With advances in the plastic industry, micro-irrigation systems never got off the ground. The orifice emitters were later designed to improve the consistency of "holes drilled into the pipes" and small diameter plastic tubes and microtubes were eventually developed with sophisticated water emissions. Also, turbulent flow emitters were developed which are currently being used.

Irrigation has played a very significant role in the development of pre-historic civilizations. Archaeological records show that the oldest civilizations based on irrigation were along the Nile River in Egypt, Tigris and Euphrates in Western Asia, the Indus River in India, and the Yellow River in China. Conventional methods of irrigation did not witness any change until the twentieth century when pressurized sprinkler irrigation systems were introduced. The novel and innovative approach of micro-irrigation evolved from the concept of subirrigation, where the water table is raised to apply irrigation water to the root zone of the plant.

Despite being based on a simple concept, micro-irrigation could not find large-scale application until recent times because of the unavailability of economic construction materials (Javed et al., 2015). Combinations of irrigation and drainage clay pipes were experimented in 1860. The water in these pipes was pumped into the subsurface drainage systems that lasted for about 20 years. In 1874, Nehemiah Clark (Clark, 1874) was granted one of the first patents for micro-irrigation in the United States. Micro-irrigation has been a global development. The introduction of perforated pipe in Germany in 1920 was considered a significant breakthrough in the development of micro-irrigation systems. Around 1934, Michigan State University started using porous pipes or canvas for subsurface irrigation (Robey, 1934). Thereafter, research and development concentrated on the use of perforated and porous pipes made of several elements. The studies mostly aimed to ascertain whether the flow of water from these pipes to the soil could be regulated by soil water pressure in place of water pressure in the system. Concurrently, experiments were conducted in Germany, USSR, and France using closely spaced channels for raising the groundwater level close to the root zone. Many other countries, including the United Kingdom, the United States, and the Netherlands, were also using several different forms of subirrigation.

During the Second World War, there was considerable improvement in the design and application of plastics. The use of plastic pipelines in agriculture was introduced and found to be feasible. Around the 1940s in the United Kingdom, irrigation in greenhouses was done using plastic pipes. At the same time, the work on the use of plastic pipes for subsurface irrigation (known as underground irrigation at that time) started in Germany (Dorter, 1962). Israel and the United States started work on drip irrigation published in 1963 and 1964, respectively, although work had commenced earlier in both these countries. Ludwig Blass of the United States patented a low-pressure device resembling the present-day micro-sprinkler in 1956. These micro-sprinklers (Figures 3.1–3.2) were made from machined aluminum and were exorbitant for wide-scale application (Blass, 1956). Ischajahu Blass and Simcha Blass in 1969 (Blass and Blass, 1969) were awarded the first US patent of a surface drip irrigation emitter (Figure 3.2). This patent was based on an earlier Israeli patent granted to Simcha Blass in 1966 for developing an emitter. During the 1960s, the concept of drip irrigation spread from Israel to North America, Australia, and South Africa, and finally everywhere in the world.

3.2.1 General Principles of Micro-Irrigation

3.2.1.1 Wetting Patterns Under Micro-Irrigation

How a micro-irrigation system applies water to the soil, only part of the soil surface and root zone of the entire field is wetted. Water flowing from the emitter is distributed in the soil by gravity and capillary forces. The wetting pattern is in the form of contour lines, often known as "onion" patterns shown in Figure 3.3. The texture and initial moisture content of soil along with the rate of water application determine the exact shape of the wetted area. In fields having closely spaced emitters along line-source-type micro-irrigation tubes, individual "onion patterns" connect to create a continuous moisture zone along each row.

3.3 ADVANTAGES

Micro-irrigation systems have several potential benefits in comparison to other irrigation methods. Most of these are associated with low water application rates. Some of these benefits are not peculiar to micro-irrigation systems, but specific combinations of these advantages qualify for the uniqueness of micro-irrigation compared to other systems. The more significant advantages are described in the following sections.

3.3.1 Water Saving

The water requirement for micro-irrigation systems is small as compared to other irrigation methods. This is because, in micro-irrigation, a small part of the soil is irrigated, resulting in reduced soil evaporation and low or no runoff. Soil evaporation is reduced significantly since the small area wetted is mostly under the plant and is well shaded by foliage. Also, there is very low or no deep percolation in micro-irrigation systems. Water is applied when needed and is held in the root zone, which reduces deep percolation.

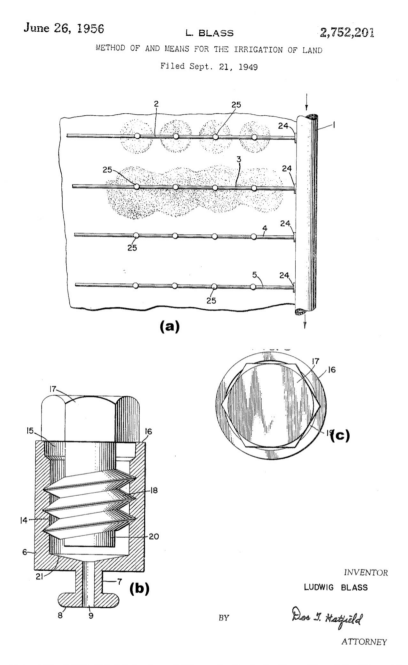

FIGURE 3.1 Patent No. 2,752,201. Methods and Means for the Irrigation of Land, L. Blass; (a) plan view of part of an irrigation system in action (b) and (c) sectional elevation and a plan view, respectively, of an irrigation nozzle

3.3.2 Lower Water Application Rates

The water application rate in micro-irrigation systems is considerably lower than in other systems. Thus, minor water sources can be used for irrigating the same acreage. The pump, the delivery pipes, and various other system components can be smaller and, thus, economical. The systems also

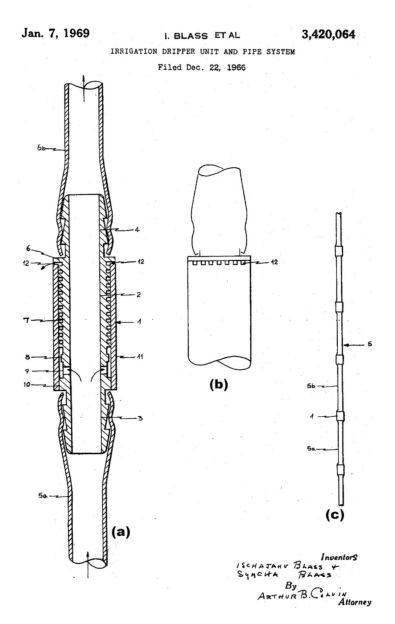

FIGURE 3.2 Patent No. 3,420,064. Irrigation dripper unit and pipe system (a) longitudinal sectional elevation of the unit shown connected to two opposite ends of irrigation supply pipe portions, (b) plan view, on an enlarged scale, of a detail of the unit shown in (a), and (c) side elevation of an irrigation supply pipe including irrigation dripper units as shown in (a)

require low pressure (5–30 psi) for operation and have lower energy consumption than high-pressure systems.

3.3.3 Improved Fertilizer and Chemical Application

Micro-irrigation systems enable effective control of fertilizer and chemical applications. Precise amounts of fertilizer can be provided to the plants at the necessary time. A reduction in the quantity

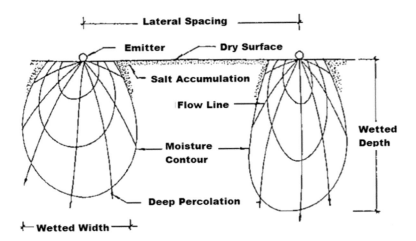

FIGURE 3.3 Profiles of wetting front advance in micro-irrigation in medium-textured soils

of fertilizers to be applied is possible, since they are applied directly to the root zone. Another major advantage is that only a small portion of fertilizers will be washed out in the case of rain, which is replaced by the irrigation system. This method of applying fertilizers is more economical and enables more convenient nutrient distribution throughout the growing season. Because of the reduction in deep percolation and lower amounts of chemicals leaching, there is a significant decrease in groundwater pollution. In addition to fertilizers, other substances, like fungicides, herbicides, nematicides, insecticides, growth regulators, and carbon dioxide can be applied efficiently to enhance crop production.

3.3.4 Water Sources with High Salinity

A significant advantage of micro-irrigation is that the system can use water with relatively high salt content. For optimal plant growth, the total water potential must lie within a specific range. The ease with which the plants extract water from the soil depends on this potential. If the soils are very dry, the total water potential has large negative values. For saturated soils, this value is zero. The total water potential is the sum of the metric potential and osmotic potential. For micro-irrigation systems, the metric potential is nearly zero, and the osmotic potential can be a considerable negative value, which indicates high salt content. This is not possible with other irrigation systems.

3.3.5 Improved Crop Yield

Plants having micro-irrigation receive small amounts of water with a high frequency. This reduces stress due to moisture fluctuation in the root zone and produces a higher and superior quality yield. The harvest timings can also be controlled in arid regions or during dry seasons by proper water management.

3.3.6 Feasibility for Any Topography

Micro-irrigation systems operate well on hilly terrains if adequately designed and maintained. A well-managed micro-irrigation system will not generate runoff even on sloping grounds or hilly terrains.

3.3.7 OTHER ADVANTAGES

  i. Damages due to insects and diseases are reduced in crops irrigated by micro-irrigation systems since the plant foliage is not wetted
  ii. As only a small portion of the field is watered, the field operations are not hampered during irrigation
  iii. For drip and bubbler irrigation systems, the wind does not affect water distribution
  iv. The smaller wet regions in the soils reduce the water uptake by weeds, thereby reducing their growth
  v. Automated micro-irrigation systems lessen the labor and operating costs for irrigation
  vi. Intercultural operations, like spraying, weeding, thinning, and harvesting, are not interrupted during micro-irrigation
  vii. Minimum tillage practice can be used with ease along with micro-irrigation, as dry areas are available in the field
  viii. In some areas, wastewater effluents which serve as a source of nutrients to the plants are also supplied through micro-irrigation systems

3.4 DISADVANTAGES

Despite marked successes and potential advantages, various problems are experienced with the mechanics and economics of applying water with micro-irrigation systems for specific soils, water quality, and environmental conditions. The disadvantages of micro-irrigation are presented in the following sections.

3.4.1 CLOGGING

Clogging of the emitter is one of the major problems encountered in micro-irrigation systems. The minute openings are readily clogged by organic matter, soil particles, algae, bacterial slime, or even chemical precipitates. Good filtration equipment is required for micro-irrigation even with very good quality water. Clogging will not only influence the uniformity of water and fertilizer applications but also increase maintenance costs. This can lead to crop damage and reduced yield, except if the clogging is discovered early and corrected.

3.4.2 HIGH MAINTENANCE COSTS

Besides the servicing cost of emitters, other significant expenses are due to the maintenance problems of pipelines and component damage such as leaks or flow restrictions. Rodent infestations and problems caused due to poor installation further add to the maintenance cost.

3.4.3 SALT ACCUMULATION AT ROOT ZONE PERIPHERY

Saline water can be used in micro-irrigation systems. However, this might lead to salt accumulation at the wetting zone periphery during extended dry periods. These salts can be washed into the root zone by light rain and can cause damage to the plants. A supplementary irrigation system (surface or sprinkler) might be necessary for regions with less than 250 mm of annual rainfall to leach accumulated salts from the root zone. In areas with high rainfall, salts are washed out of the root zone before significant accumulation.

3.4.4 MOISTURE DISTRIBUTION

Moisture distribution depends mostly on the type of soil being irrigated by the irrigation system. For example, in deep sandy soils, minimal lateral water movement (low capillary forces) can generate many problems. The gravity forces dominate in these soils, and the wetted soil volume assumes a cylindrical shape rather than a hemispherical one. Thus, it becomes challenging to wet a considerable portion of the root zone. The problem can be overcome by increasing the number of emitters per plant, which improves the water distribution in the soil. It is necessary to understand that the micro-irrigation systems wet only a restricted region of the potential soil-root volume.

3.4.5 Restricted Root Development

Micro-irrigation ordinarily wets only a part of the root zone. Consequently, the root distribution is almost restricted to the moist zone. Many factors concerning soil and plant characteristics, management practices (volume and frequency of water application), and the design of the irrigation system (number of emitters per plant, placement, and discharge rate of emitters) affect the development of crops.

The concentrated dispersion of roots may decrease the ability of plants to resist strong winds. Further, the capacity of plants to withstand drought, rising from any disruption in the irrigation system, is significantly reduced as the water of the wetted zone gets depleted quickly and the neighboring region becomes dry. Meticulous planning and operation of the irrigation system and prevention of malfunctions will reduce the extent of the problem.

3.4.6 High Cost of Drip Irrigation Systems

The cost of micro-irrigation systems is high as compared to other irrigation systems. Micro-irrigation systems are costly because of the large number of piping and filtration equipment to clean the water. Nevertheless, the cost of these systems differs considerably, depending on the crop and topography. Several pressure regulators might be required for the steep region. Under many conditions, the micro-irrigation system's advantages will generally outweigh the cost of the system compared to other irrigation methods.

3.5 NEED OF MICRO-IRRIGATION

Irrigation has played a very significant role in the development of any country. Following are important points necessitating micro-irrigation:

- Make agriculture productive
- Environmentally sensitive and capable of preserving the social fabric of rural communities
- Help produce more from the available land, water, and labor resources without affecting either ecological or social harmony
- Generate higher farm income, on-farm and off-farm employment
- Use water efficiently
- No water runoff or evaporation and soil erosion
- Reduces water contact with crop leaves, stems, and fruits
- Agricultural chemicals can be applied more efficiently

The following are the major advantages of micro-irrigation (Figure 3.4):

3.6 TYPES OF MICRO-IRRIGATION

Micro-irrigation can be used to increase water use efficiency. The following are different categories of micro-irrigation:

- Online emitter/dripper system
- Emitting pipe system/inline drip system

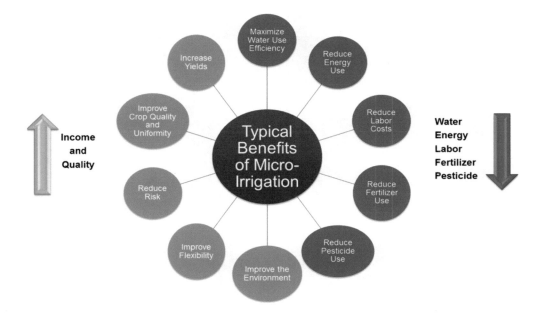

FIGURE 3.4 Major advantages of micro-irrigation

- Micro jets
- Mini sprinklers

3.6.1 Online Emitter/Dripper System

In this system, drippers are fixed externally on laterals at designed spacing. Online dripper system is widely used in orchards, vineyards, and nurseries. It is used for plants having large spacing like mango, grapes, etc. In this system, openable drippers such as J-Turbo key plus, J-Loc, J Vary Flow drippers are used. There are different drippers discharging 2 l/h, 4 l/h, 8 l/h, 14 l/h, and 16 l/h capacity are available.

3.6.2 Inline Drip System

3.6.2.1 J-Turbo Line

J-Turbo inline dripper can be used for high precise water application in the crop. In this system, round drippers are fixed in a lateral tube at the time of manufacturing at different spacings to suit the requirement of different crops. It is available in 12-mm, 16-mm, and 20-mm diameter with standard dripper spacing of 30 cm, 40 cm, 60 cm, 75 cm, and 90 cm. J-Turbo line PC type is available for hilly, undulating terrain, and steep slopes. It is effective for row crops like vegetables and floriculture.

3.6.2.2 J-Turbo Aqua

It is an extruded seamless tube with flat drippers permanently fixed inside. Dripper spacing and diameter are the same as the J-Turbo line (Figure 3.5). It is easy to roll back and relay. It is used for widely spaced horticultural crops like mango, apple, etc.

3.6.2.3 Twin-Wall Drip Tap

It is thin tubing made of special grade virgin plastic. It has a continuous inlet filter which runs for a full length of the tap. It has a slit-type outlet and is available in 16 mm and 22 mm diameter tubes

FIGURE 3.5 Different types of drippers

FIGURE 3.6 Different types of sprinklers

having a wall thickness of 4 mm to 25 mm. The twin-wall drip operating pressure is 0.4 kg/cm^2, 0.7 kg/cm^2, and 1.0 kg/cm^2. It is suitable for surface as well as subsurface installations.

3.6.3 Micro Jets

Micro jets operate at lower pressures than do sprinklers. They apply water at higher rates than emitters. There is a limit to their distance of flow. It is suitable for greenhouses, nurseries, and delicate plants.

3.6.4 Mini Sprinklers

These are like jets operating at low pressure. They discharge water over a larger area which is useful in the landscape, shrubs, etc. The different types of sprinklers are shown in Figure 3.6.

3.7 DRIP IRRIGATION

Drip irrigation is defined as the frequent application of small quantities of water directly above and below the soil surface using a pipe network: usually as discrete drops, continuous drops or tiny streams through emitters placed along a water delivery line. This method applies sufficient moisture to the roots of crops and prevents water stress.

Micro-Irrigation 49

3.7.1 COMPONENTS OF DRIP IRRIGATION SYSTEM

The drip irrigation components are shown in Figures 3.7–3.8. The following are the drip irrigation components.

- Pump
- Header assembly (bypass arrangement, return valve, and air release valve)
- Filters: Hydro cyclone filter
  i. Media filter/sand filter
  ii. Screen filter
- Pressure gauge
- Fertigation equipment: Venturi injector
    Fertilizer tank
- Flow control and safety valves
  i. Ball valve
  ii. Flush valve
  iii. Gate valve
  iv. Air release valve
- Main line
- Submain line
- Laterals
- Lateral drain valve
- Poly fittings and accessories
- Drippers/emitters

3.8 SPRINKLER IRRIGATION

Sprinkler irrigation is defined as applying irrigation water similar to the fashion of rainfall. Water is distributed through a system of pipes usually by pumping. It is then sprayed into the air and irrigates

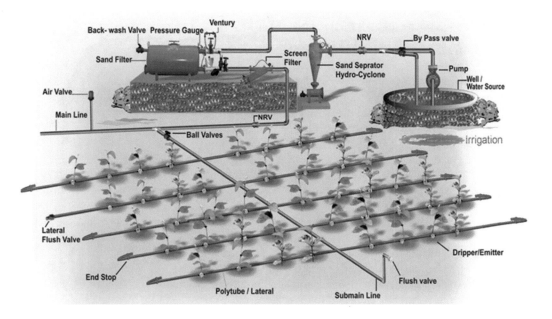

FIGURE 3.7 Drip irrigation component

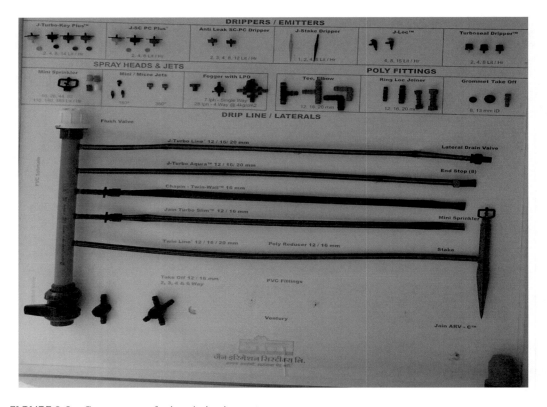

FIGURE 3.8 Components of micro-irrigation system

the entire soil surface through spray heads so that it breaks up into small water drops, which fall on the ground.

3.8.1 Components of Sprinkler Irrigation System

The following are different components of sprinkler irrigation.

- Pipe networks: mains, submains, and laterals
- Couplers
- Sprinkler head
- Other accessories such as valves, bends, plugs, risers, and fittings
- Pumping unit
- Pressure gauge

3.9 INSTALLATION OF MICRO-IRRIGATION SYSTEMS

The following points should be considered while installing micro-irrigation in any field.

- Water supply installation (bucket, barrel, tank, pump, etc.)
- Laying of pipes and emitters/micro-tubes/setting up sprinklers
- If there is no overhead tank, then a water source must be created
- The system then can be connected to the water source
- Micro-sprinkler and overhead sprinkler kits can be directly connected with the equivalent discharge outlet of a pump or water supply system

- Ensure that via the main line, the control valve and filter are connected to the system
- For drip systems, in a straight line or along the plant rows, lateral pipes are laid on the ground. The lateral emitters/microtubes are pre-fixed
- They are placed at equal spacing so that plants receive a uniform amount of water
- For sprinklers, stakes are used to place them properly
- Endcaps at the end of laterals and submain are released before running the machine so that it is washed away and the air is also forced out if there is dirt in the pipes
- Open the control valve and allow the water to flow through the pipes freely for some time (flush the system). Then, close the endcaps and ensure that each emitter has water coming out

In general, the following activities are involved in the installation of micro-irrigation systems:

- Study the installation sketch
- Provide design for the water tank/filter platform and pipe trenches if needed
- Check the kit/material components on-site according to the list of materials in the user manual
- Install water storage tank and filter on the platform
- Connect the filter to the source/pump of water and the main line
- Lay out the primary thread, submain, and lateral pipes
- Cover the pipe trenches, if necessary
- Place/repair the sprinklers/emitters (if microtubes require inflated lateral pipes, then fill the pipes with water and then punch holes and fix microtubes)
- Start the pump/open the valve and fill the water pipes
- Release all endcaps/flush valves to clean the system of dirt
- Pressure and discharge and review to ensure that all emitters work
- Operate according to the plan

3.10 MAINTENANCE AND TROUBLESHOOTING

The biggest problem of any micro-irrigation system is the clogging of emitters. Micro-irrigation system kits use very simple emitters that are less prone to clogging due to a wider flow path. Therefore, it requires less maintenance than other drippers. However, periodic and preventive maintenance is essential for smooth functioning. The following general checks can be carried out periodically depending on the local condition and water quality:

- Clogging of emitters/micro-sprinklers and wetting pattern
- Placement of emitters/micro-tubes/micro-sprinklers
- Leakages in pipes, valves, filters, fittings, etc.
- Flushing and cleaning of filters by opening and cleaning the screen
- Flushing of submain and laterals by releasing the endcaps
- There are dissolved chemical (mainly salts) impurities and even biological impurities, such as algae, bacteria, etc. present in some water sources, apart from physical impurities that can be isolated by using a screen filter.
- They will accumulate and clog the emitters if the dissolved salts are more concentrated. For flushing the salts, hydrochloric acid may be added to the emitters.
- If the system is obstructed by bacteria or algae, chlorine treatment may be applied in the form of bleaching powder (2 mg/l) to clean the emitters and prevent slime development.

3.11 MICRO-IRRIGATION UNDER PROTECTED CULTIVATION

Micro-irrigation plays a vital role in getting maximum production from high-value vegetable crops under protected cultivation. The objective of the micro-irrigation system inside the protected

FIGURE 3.9 Drip-irrigated vegetable crop under protected cultivation

FIGURE 3.10 Layout of drip irrigation system

cultivation is to deliver optimum water and nutrient levels directly into the root zone and reduce wastage and evaporation. In soilless culture, drip irrigation is used to supply the water to the crops, and the amount of irrigation supplied and its timing throughout the crop cycle influence the yield and fruit quality. Fertigation with drip irrigation in soilless culture ensures a good yield and higher fertilizer-use efficiency and avoids nutrient wastage which may decrease the production cost and reduce the risk of water pollution in the surrounding environment (Figures 3.9–3.12).

3.12 WORLD SCENARIO OF MICRO-IRRIGATION SYSTEM

While there are different methods of irrigating crops, the quality of water and fertilizer usage and the improved field production are on the top of the list for drip and sprinkler irrigation systems. In

FIGURE 3.11 Drip irrigation system under soilless medium

FIGURE 3.12 High-value tomato crop under drip irrigation

this method, water is distributed directly through a network of different places, topography, and slopes to the root zone of individual plants, as per plan and comfort, to minimize the low-pressure distribution through emitters and microtubes. In Israel, Australia, Mexico, New Zealand, South Africa, and the United States, effective drip irrigation techniques have been developed and are commonly used effectively for growing different crops. When we look at the global micro-irrigation situation, the United States, led by China, France, and Spain, stands first in terms of the region

FIGURE 3.13 Drip irrigation layout in a high-density apple field

FIGURE 3.14 Lettuce crop under drip irrigation system

under drip and sprinkler irrigation. India is in fifth place, with 1.6% of the total area protected. The micro-irrigation practices used in different crops are shown in Figures 3.13–3.14.

3.13 MICRO-IRRIGATION POTENTIAL IN INDIA

Before the planners and policy makers decide on the various techno-economic aspects of the introduction of micro-irrigation in a state, they should have basic knowledge of its potentiality in the state. It has been assessed that there is a potentiality of bringing around 70 Mha under micro-irrigation in the country. Out of this 70 Mha, about 45 Mha are suitable for sprinkler irrigation for crops like cereals, pulses, and oilseeds in addition to fodder. Out of this drip with a potentiality of around 20 Mha and the major crops suitable for this system are cotton, sugarcane, fruits and vegetables, spices and condiments, and some pulse crops like red gram, etc. In addition, there is a potentiality for bringing an area of about 2.8 Mha under mini sprinklers for crops like potato, onion, garlic, groundnut, and short-stature vegetable crops like cabbage, cauliflower, etc. (Kumar, 2019).

3.14 CONCLUSION

Micro-irrigation is the slow and frequent application of water through a pipe network directly to the root zone of plants. Micro-irrigation will increase the irrigation potential and water use efficiency and increase the productivity of the crop. It will charge the current land use to more remunerative crop production. Assessment of the impact of climate change on water resources also provides information for future strategies for planning irrigation scheduling and water budgeting of different crops. It is used to evaluate the effectiveness of better irrigation water management strategies to reduce the impacts of water scarcity for different crops.

QUESTIONS

1. Define micro-irrigation.
2. Discuss components of micro-irrigation.
3. What is drip irrigation? Describe components of drip irrigation.
4. What is sprinkler irrigation? What are its components?
5. Enlist different techniques used for soil and water conservation using plastics.
6. Discuss the application of plastics in agriculture and the future prospects of plasticulture in India.
7. Explain the role of the irrigation engineer. Describe the necessity and sources of irrigation.
8. Describe different parameters used for the planning of any irrigation system.
9. How we can increase water use efficiency through micro-irrigation.
10. What is irrigation efficiency? How we can improve irrigation efficiency. Discuss high-efficiency irrigation methods.
11. Discuss design and components of drip and sprinkler irrigation systems.
12. Determine the required capacity of a sprinkler system to apply water at the rate of 1.25 cm/h. Two 180 m long sprinkler lines are required. Sixteen sprinklers are spaced at 12 m intervals on each line. The spacing between the lines is 18 m. Allowing 1 h for moving each 180 m sprinkler line, how many hours would be required to apply a 5 cm irrigation to a square 16 ha field? How many days are required, assuming 10-h days?
13. Write a short note on
    i) Use of plastic in moisture conservation
    ii) Plastics for losses of water from canals, ponds, and reservoirs

MULTIPLE-CHOICE QUESTIONS

1. Which one is a method of overhead irrigation?
   a) Drip irrigation
   b) Sprinkler irrigation
   c) Center pivot irrigation
   d) Terraced irrigation

2. Micro-irrigation is also called
    a) Localized irrigation
    b) Flood irrigation
    c) Nano-irrigation
    d) Petite irrigation
3. The ratio of the quantity of water stored in the root zone of crops to the quantity of water actually delivered in the field is known as
    a) Water application efficiency
    b) Water use efficiency
    c) Water conveyance efficiency
    d) None of the above
4. The amount of irrigation water required to meet the evapotranspiration needs of the crop during its full growth is called
    a) Consumptive irrigation requirement
    b) Net irrigation requirement
    c) Effective rainfall
    d) Consumptive use
5. Fertigation is a process in _____ irrigation
    a) Drip
    b) Center pivot
    c) Sprinkler
    d) Surface
6. The field water efficiency of trickle irrigation is
    a) 80–90%
    b) 60–70%
    c) 50–55%
    d) 55–85%
7. Micro-irrigation is needed
    a) To make agriculture productive
    b) Environmentally sensitive
    c) Enhance water use efficiently
    d) All of the above
8. What is the pressure required in drip irrigation
    a) 1.5–2.0 kg/cm^2
    b) 0.5–1.0 kg/cm^2
    c) 1.0–5.0 kg/cm^2
    d) None

ANSWERS

1	2	3	4	5	6	7	8
b	a	a	a	a	a	d	a

REFERENCES

Blass, I. and Blass, S. 1969. *Irrigation Dripper Unit and Pipe System*. U.S. Patent Office. Patent No. 3,420,064, issued January 7. Washington, DC.

Clark, N. 1874. *Improvements in Irrigation Pipe*. U.S. Patent Office. Patent No. 146,572, issued January 20. Washington, DC.

Dorter, K. 1962. *Study on Solution of Existing Problems with Underground Irrigation*. W. Germany.

Javed, Q., Arshad, M., Bakhsh, A., Shakoor, A., Chatha, Z. A. and Ahmad, I. 2015. Redesigning of drip irrigation system using locally manufactured material to control pipe losses for orchard. *Pakistan Journal of Life and Social Sciences*, 13.

Kumar, R. 2019. Doubling farmers income in India through micro irrigation. *Fertilizer Focus*, January/Febuary, 66–69.

Narayanaswamy, T. and Anandakumar, B. M. 2016. Scope of drip irrigation for vegetable production in India. *Indian Journal of Applied Research*, 6(5), 165–165.

Robey, O. E. 1934. *Porous Hose Irrigation*. Extension Bulletin, East Lansing, MI.

4 Design and Components of Micro-Irrigation

4.1 DEFINITION

Micro-irrigation refers to the slow water application on, above, or below the soil employing a surface or subsurface drip, bubbler, or micro-sprinkler system. Water is applied in discrete or continuous small streams or minute sprays through applicators or emitters placed along a water delivery line adjacent to the plant row (ASAE, 2001). Micro-irrigation is also termed localized irrigation since only a specific volume of soil is wetted. The localized aspect of micro-irrigation has implications for evapotranspiration, deep percolation, and soil water. This in turn affects the distribution of roots in the soil. An emitter discharges a small amount of water at the emission point and may reduce the supply line's pressure. Water passes from the emission points through the soil by capillary action and gravity. The characteristic features of micro-irrigation are:

1. The application of water takes place at a low rate
2. The duration of water application is longer than conventional irrigation methods
3. The water is applied at frequent intervals
4. A low-pressure delivery system is used to apply water
5. Chemicals and fertilizers are often transported with irrigation water

Unlike other irrigation methods, micro-irrigation cannot be employed for all crops or land conditions. But for some cropping systems and regions, it offers various exceptional agronomic, agrotechnical, and economic advantages compared to other irrigation practices.

4.2 GENERAL SYSTEM DESIGN

Micro-irrigation systems are designed for transporting water from a source to the crop through a network of delivery pipes and water emission devices. The overall goal of micro-irrigation is to supply irrigation water to the crop uniformly and efficiently. Adequate water is necessary to meet the evapotranspiration demand and maintain a favorable water balance in the root zone. Related goals are to increase crop production and maintain the visual quality of the plant. Other goals may include protection of plants, fruits, or flowers from extreme temperatures, application of soil chemicals (pesticides, fertilizers, etc.), or disposal of wastewater. Therefore, the eventual design will vary from one system to another. System goals are established with well-defined objectives, system constraints, and a desirable set of outcomes to be applied as a common theme during the design process. This will assist with not only the selection and placement of certain components and sizing and layout of the distribution system but also the development of suitable operational strategies and procedures.

The following sections will focus on the design methods related explicitly to micro-irrigation systems concerning water sources, system hydraulics, and micro-irrigation components.

4.2.1 INITIAL ASSESSMENT

The primary purpose of the water delivery system is to supply water to the crop in a timely, effective, and economical way. Various limitations and system characteristics affect this process and can be arranged into the following sections and related characteristics:

i. **Land/Field**—The total irrigated area; field shape(s) and dimensions; slope(s); and soil profile characteristics of texture, porosity, water holding capacity, and depth
ii. **Crop(s)**—Type of crop(s), evapotranspiration requirements, growth, and development characteristics; and cultural characteristics (spacing/orientation, tillage, and structural field characteristics such as bedding, plastic mulch, trellises, etc.)
iii. **Water Source**—Available quantity (flow and volume) and timeliness, quality (pH, EC, Fe, S, Mg, Ca, Na, algae, or other suspended solids, etc.), and physical location concerning the irrigated area

System properties in the above classes are evaluated and then used to define crop water requirements, the ability of the soil to retain and release water for the crop, and the number and extent of the irrigated zones. One or more constraints might exist in the above characteristics that may restrict the "size" of the overall system. Thus, a preliminary study is conducted using a mass balance approach given by Equation (1):

$$Q_s \times T = 2.778 \times A \times I_g \tag{1}$$

The product of the design irrigation system volumetric flow rate (Q_s, l/s) and the system operating time per irrigation cycle (T, h) is balanced with the irrigated area (A, ha) and the gross irrigation depth per irrigation cycle (I_g, mm).

Gross irrigation depth needs to incorporate the net irrigation depth, added water for leaching (if required), and water to recompense for transmission and application efficiency losses. One or more of the mass balance parameters may be fixed with a maximum and/or minimum value, while others may be desired estimates.

Example 1: A 50-ha irrigated crop is to have a gross daily water application depth of 5.0 mm. The water is to be applied to the crop in a 10.0 h time period. What is the required system flow rate?
Equation (1) is re-written to solve for Q_s using the known parameters of:

$T = 10.0$ h
$A = 50$ ha
$I_g = 5.0$ mm

$$Q_s = \frac{2.778 \times 50 \times 5}{10}$$

$$= 69.5$$

Therefore, the required system flow rate is 69.5 l/s.

Example 2: The peak available flow rate of a water supply is 20 l/s. How much area can be irrigated in a day if the maximum daily operating time for the irrigation system is 18.0 h and the gross peak daily irrigation depth is 5.0 mm?

Design and Components of Micro-Irrigation

Equation (1) is re-written to solve for A using the known parameters of:

$Q_s = 20.0$ l/s
$T = 18.0$ h
$I_g = 5.0$ mm

$$A = \frac{20 \times 18}{2.778 \times 5}$$

$$= 25.92$$

Therefore, the maximum area is $A = 25.92$ ha that can be irrigated daily.

Example 3: A crop has an effective root zone depth of 1,200 mm and monthly crop evapotranspiration of 260 mm. The effective rainfall during the 30 days' period is 20 mm. The field capacity and permissible soil moisture depletion on a volume basis are 16% and 8%, respectively. The irrigation interval in days for the crop will be?
Solution: Given

Effective root zone depth $(D_{root}) = 1,200$ mm
Evapotranspiration $(ET_{Crop}) = \dfrac{260}{30}$
$= 8.67$ mm/day
Effective rainfall per day $(ER_{per\ day}) = \dfrac{20}{30}$
$= 0.67$ mm/day
Peak moisture use rate of crop $= 8.67 - 0.67 = 8$ mm/day

Irrigation interval is $= \dfrac{(F.C. - \text{Permissible soil moisture depletion}) \times D_{root}}{\text{peak moisture use rate of crop} \times 100}$

$$= \frac{(16-8) \times 1,200}{100 \times 8}$$

$$= 12 \text{ days}$$

Example 4: The soil of a cropped field has a field capacity of 25% and wilting point of 13% on a weight basis. The effective root zone depth of the crop is 0.70 m and the consumptive use of water is 5 mm/day. The apparent specific gravity of the soil is 1.50. The allowable soil moisture depletion is 40%. What is the permissible moisture depletion between irrigations and frequency of irrigation?
Solution: Given

F.C. = 25%
W.P. = 13%
$D = 0.70$ m
$Z = 1.50$
$C_u = 5$ mm/day

$$D = \frac{(F.C. - W.P.) \times D \times Z}{100}$$

$$= \frac{(25-13) \times 0.7 \times 1.5}{100}$$

= 0.126 m = 126 cm

Permissible soil moisture depletion = 0.4 × 0.126
= 0.0504 m
= 5 cm

$$\text{Irrigation interval} = \frac{\text{Allowable soil moisture depletion}}{\text{daily water use}}$$

$$= \frac{5\,\text{cm}}{0.5\,\text{cm/day}}$$

$$= 10 \text{ days}$$

Example 5: Root zone depth of water crop grown in a sandy loam soil is 100 cm. The volumetric soil moisture contents at field capacity and permanent wilting point are 30% and 5%, respectively. Compute the soil moisture stress factor when the volumetric moisture content in the root zone reaches 15%. The critical value of available soil water is 50%.

Solution: Given

Root zone depth = 100 cm
Field capacity (FC) = 30%
Wilting point (WP) = 5%
Volumetric moisture content = 15%
Available soil water is = 50
Value of critical available soil moisture = Available soil moisture × critical value of available water
Available soil moisture = 30% − 5% = 25%
Therefore, value of critical available soil moisture = 25 × 0.5
= 0.125

$$\text{Soil moisture stress factor} = \frac{\text{critical available soil moisture}}{\text{volumetric moisture content}}$$

$$= \frac{0.125}{0.15}$$

$$= 0.833$$

Example 6: Find out the duty for the crop if the depth of water requirement is 50 cm. There is a 10 cm effective depth of rainfall in that region for 10 days.

Solution: Given

Base period = 10 days
Rainfall depth = 10 cm
Depth of water required = 50 cm

$\Delta = 8.64 \dfrac{B}{D}$

$\Delta = 50 - 10$ cm
= 40 cm

Therefore, $0.4 = 8.64 \dfrac{10}{D}$

$D = 8.64 \dfrac{10}{0.4}$

$D = 216$ ha/m³/s

Design and Components of Micro-Irrigation

Example 7: Calculate the amount of water required for irrigating a particular crop, if the consumptive use is 320 mm, application loss likely to occur in the field is 27.5 mm, and water needed for other purposes is about 3.57 mm.

Solution: Given

Consumptive use = 320 mm
Application loss = 2.75 mm
Water required for other needs = 3.57 mm
Therefore,
Water required for irrigation = Consumptive use + application loss
+ special needs
= 320 + 2.75 + 3.57
=326.32 mm

4.3 OBJECTIVES OF DESIGN

The following are the objectives of the perfect design of micro-irrigation:

- To maintain the optimum moisture level in the soil
- To keep both initial investment and annual cost at a minimum level
- To maintain the higher system and irrigation efficiency
- To design a system, which will last for a longer time and will perform well
- To design a manageable system
- To satisfy and fulfill the requirements of crop and farmer

4.4 DESIGN INPUTS PARAMETERS

Following are the design input parameters required before the installation of micro-irrigation in any field.

- Engineering survey
- Water source
- Agricultural details
- Climatological data
- Soil and water analysis

4.5 STEPS TO DESIGN MICRO-IRRIGATION SYSTEM

4.5.1 SYSTEM CAPACITY

System capacity is very important to apply the correct amount of water to all plants uniformly. It shall have the capacity adequate to fulfill 90% of plant water requirement (PWR) within the designed area and in an operation period of not more than 26 h. The system capacity should be matched with the available quantity of water in the source.

4.5.2 SELECTION OF EMITTING DEVICES

The selection of drippers is based on the peak water requirement of a crop, age, root zone, soil water holding capacity, and infiltration rate.

4.5.3 SELECTION AND DESIGN OF LATERALS

The size allowable length and frictional losses of laterals have been determined by charts and design guidelines provided by manufacturers for specific emitting devices.

4.5.4 Design of Submain

Calculate standard discharge rate in submain in l/h/m

SDR = (Total flow in submain (l/s) × 3,600)/submain length (m)

Then the size length and frictional head losses are to be determined by charts and design guidelines.

4.5.5 Design of Main Line

The size of the main line is to be determined by considering the quantity of water flowing through it, the length and path of the main line, and elevation of ground and charts provided by the manufacturer.

4.5.6 Selection and Design of Filtration Unit

The selection and design of the filtration unit are based on:

- Source of water
- Type, size, and concentration of physical impurities
- Type of irrigation system
- Filtration media

4.5.7 Selection and Design of Pump

The pump unit can be selected and designed by calculating the total head and discharge required for the efficient operation of the system. Then, we calculate the HP required by a pump for the efficient operation of a micro-irrigation system (MIS).

4.6 LAYOUT AND COMPONENTS

The field layout of a representative micro-irrigation system having three zones is given in Figure 4.1. The water supply and the control head are located in the center of the field (Figure 4.2). Several other arrangements and locations can be conceived. The pump stations and the associated controls, pressure gauges, system valves, and equipment for water treatment (filters and chemical injection systems) are contained in the control head. An initial screening or sand separator filter may be included in the system if the source of water has large particles or sand. For assessing line breaks and clogging, flow meters, and pressure gauges are essential. Several pressure gauges are used for monitoring pressure drops in the filtration units or chemical injection units. When chemicals are injected into the irrigation system, backflow prevention units are also required. The main line pipes stretch from the control head to the zone containing pressure regulators along with manual or automatic control valves.

4.7 DESIGN PROCESS

Since the design process can offer different system arrangements and outcomes, it is challenging to explain it with a single set of rules. Using the available data, the designer might need to generate one or more good starting scenarios. The primary steps of the design process are described below.

4.7.1 Preparatory steps of the micro-irrigation system design process:

1. Reach the site proprietor/operator to evaluate the desired outcomes and applications of the irrigation system

Design and Components of Micro-Irrigation

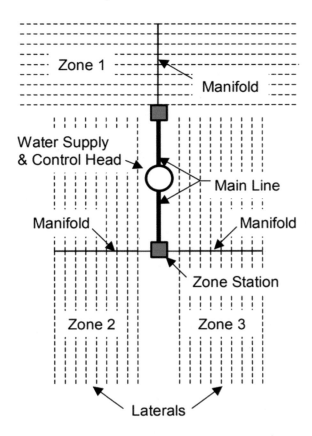

FIGURE 4.1 General field layout of a micro-irrigation system having three field zones

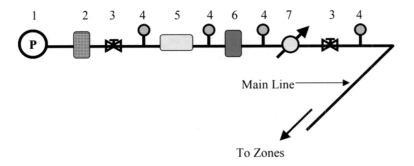

1. Pump or Pressurized Water Supply
2. Initial Filter for Large Particles & Sand (if needed)
3. Flow Control Valve
4. Pressure Gauge
5. Chemical Injection Station – Backflow Prevention
6. Main Filter Station
7. Flowmeter

FIGURE 4.2 Arrangement of the water supply and control head for a micro-irrigation system

2. Estimate the overall site characteristics, particularly the landscape, crop to be cultivated, and water supply
3. Prepare the "preliminary rough draft" of design scenarios on the basis of the results of Step 2 and mass balance analysis. Decide a preferred pump location if one does not presently exist. The field should be divided into a suitable number of zones if the whole area does not need to be irrigated at the same time. Place lateral lines along the contour or other wanted orientation, and lay manifold and main line pipes
4. Evaluate the economic, physical, and cultural constraints related to the water supply, the field, and the production system. Estimate the rough draft scenarios derived from that assessment
5. If additional information is required, go back to Step 1 and re-evaluate the site characteristics
6. If any of the rough draft scenarios are rational or acceptable, proceed to Step 8
7. If the existing system scenario can be altered to accommodate the added constraints, return to Step 1. If not, continue to Step 10
8. Present and systematically explain the rough draft design scenario(s) to the proprietor/operator
9. If any of the proposed design scenarios are acceptable to the owner/operator, proceed to Step 11
10. If other micro-irrigation system scenarios exist, move back to Step 1 to reconsider the site and system characteristics. If not, then stop
11. Continue with the formal design process

The following stage of the design process concentrates on sizing, selecting, and integrating the individual components into a working system that serves the constraints of the problem, approaches the established objectives, and reaches the coveted goal. This section of the process combines artful creativity with the designer's understanding of the system hydraulics, the features of the micro-irrigation emitters, and the biological and physical phases of the production system.

4.7.2 Steps for the General Micro-Irrigation System Design Process

1. Measure and enumerate the maximum flowrate capacity (Q_{max}) of the water supply system
2. Describe the total irrigated area, A, the anticipated number of zones, and the irrigated area per zone A_z
3. Evaluate and quantify the daily peak water requirements of the most water-demanding crop, ET_{cpeak}, and the net daily irrigation requirement. Systems used for multiple crops should be designed for the crop with the highest peak demand
4. Choose a micro-irrigation emitter configuration concerning the category of discharge device (dripper, spray jet, bubbler, etc.) and discharge device spacing that economically adapts to the cropping and production system constraints
5. Describe the irrigated volume of soil based on the characteristics of emitter water distribution. Then, ascertain the volume of water readily available at the field capacity of the soil. Define the adequate management allowed deficit (MAD) value for the soil and crop system to estimate the volume of readily available water to the crop
6. Based on the results of Steps 3, 4, and 5, describe and recognize the peak irrigation cycle scenarios
7. Determine the amount of water to apply per irrigation cycle for each zone given by Equation (2):

$$Q_p = \frac{2.778 \times A_z \times I_n}{E_a \times T} \qquad (2)$$

where the peak total flowrate for the system (Q_p, l/s) is established from the area irrigated per zone (A_z, ha) as defined in Step 2, the peak daily crop water use or net irrigation requirement (I_n, mm) from Step 3, the application efficiency of the irrigation system (E_a), and the irrigation operation time per cycle (T, h). The net irrigation requirement (I_n) must be modified for the number of days between cycles, the number of cycles per day, and/or for salinity leaching requirements.

8. If the requisite peak system flow rate surpasses the maximum available flow rate capacity ($Q_p > Q_{max}$), repeat Steps 2 through 7 by changing the variables (A_z and T) of Equation (2) as appropriate to meet the criterion of $Q_p \leq Q_{max}$
9. Layout and design a submain and lateral pipeline system that will maintain emitter discharge rate differences within acceptable limits
10. Develop an irrigation application and operating schedule
11. Assess whether the system is convenient to operate and conforms to the general crop production and field system's cultural constraints. If not, return to Step 2
12. Design an economical size for the main line pipe system, submain pipes, header pipes, manifolds, the pumping plant, and control head components (filters, injectors, valves, and controllers)
13. Generate an initial cost analysis of the existing design scenario
14. Estimate whether different, potentially acceptable system configurations and configurations exist. If yes, return to Step 2
15. Present the required system design and configuration to the system proprietor/operator and discuss options or final plans
16. Decide the system design and specifications, develop a bill of materials, and create a final product report. Give examples

4.8 SOURCES OF WATER

The design of a micro-irrigation system requires assessing the supply of water available from the source for use (total quantity and capacity or rate) and the amount necessary or used for the desired goal. The total amount of available water will depend on relevant laws or allocation schemes and the size of the source. Water limitations, use restraints, ownership, and uncertainty in water supply amounts will affect the total irrigated area, the irrigation scheduling decision process, and possibly the option of components and "permanence" of the system. A limited water supply may be used to irrigate a limited area adequately or to do "deficit" irrigation in a larger area. A few micro-irrigation systems are intended for long-term use (>10 years) on a single site. On the contrary, other methods might have some "portability" contained within the design to hold uncertainty in a local water source or land-lease agreements. These situations are not necessarily normal but can provide some exceptional and challenging design scenarios.

The water source quality must be evaluated as to how physical, biological, and chemical constituents in the water may affect or interact with the delivery system (pump, pipes, valves, and emitters), the soil, and the crop. Appropriate water treatment and amendment procedures must be used to avoid the clogging of micro-irrigation emitters. Water quality from both groundwater and surface water sources can vary from excellent to very poor, and characteristic quality concerns include suspended solids, dissolved solids, and biological organisms. Inadequate well screening and well development problems can result in undesirable quantities of suspended sand, silt, or clay particles. Surface water sources may contain suspended particles of silt and clay, aquatic plants, small fish, algae, larvae, or other organic debris. In most cases, these physical elements can be managed with proper filtration.

Dissolved solids, such as calcium, iron, or manganese, settle down under specific conditions and, consequently, clog emitters. Biological organisms include slimes (associated with iron and hydrogen sulfide), fungi, and algae. Chemical treatment of water is often necessary along with filtration to check the emitter clogging and restore clogged emitters. Critical occurrences of the preceding water quality problems may need costly remediation components and management practices to assure proper and continual operation of the micro-irrigation system. Because such conditions may result in a financially impractical design or poor system performance or failure, a thorough assessment of the water source quality must be performed before completing the final design.

4.9 TYPES OF MICRO-IRRIGATION SYSTEMS

Micro-irrigation systems are usually described with regard to the method of installation, the discharge rate of the emitter, wetted soil surface area, or operation mode. There are three basic types of micro-irrigation systems, namely bubbler, drip, subsurface drip, and sprinkler.

4.9.1 Bubbler Irrigation

In bubbler irrigation, a small stream or fountain from a hole or opening is used to apply water to the plants (Figure 4.3). The discharge rate of the point source is greater than 12 l/h but generally less than 250 l/h. The emitter discharge rate is higher than the infiltration capacity of the soil; as such, a basin is recommended to check the water distribution. Bubbler irrigation systems are used more extensively in landscape irrigation systems than in agriculture. Perennial crops, particularly orchards and vineyards having flat topography, are irrigated efficiently using bubbler irrigation systems.

Some of the significant advantages of this system of micro-irrigation include reduced repair and maintenance costs, lower filtration demands, and less energy requirement compared to other micro-irrigation systems. Nevertheless, these systems require larger size lateral lines for reducing pressure loss linked with larger discharge rates.

4.9.2 Application and General Suitability

Bubbler systems are utterly suitable for perennial crops, mainly orchards, and vines. This is because the irrigation system characteristically comprises buried pipes and small earthen basins around the

FIGURE 4.3 Bubbler irrigation

Design and Components of Micro-Irrigation 69

plants. Bubbler systems can also be applied to row crops that use furrows. After planting, laterals are positioned along furrows and are removed from the field after harvest. Sites with level topography or with a gentle, uniform slope are perfect for bubbler systems. A fine soil texture is also suitable. Bubbler systems can use low-head water supplies readily, similar to surface irrigation systems. The following sections outline the advantages and disadvantages of bubbler systems relative to other types of micro-irrigation systems.

4.10 ADVANTAGES AND DISADVANTAGES

Bubbler systems have certain potential advantages in comparison to other micro-irrigation systems:

1. *Low energy requirements:* The energy required to apply water by gravity flow bubbler systems is generally lower than for other systems. While other micro-irrigation emitters need a pressure between 100 kPa and 200 kPa to operate, pressures as low as 10 kPa are sufficient for bubbler systems
2. *Low maintenance:* Low-pressure bubbler systems utilize lesser equipment, such as filters and pumps, in the control head. Normally, chlorination and acidification are not required except when poor quality water leads to a build-up of aquatic plants in the system
3. *Lower chance of emitter clogging:* Compared to other micro-irrigation systems, both gravity and pressurized bubbler systems are not as vulnerable to emitter clogging. Clearances for bubbler emission tubes are comparatively large, and particulates can pass through these with much ease than for other micro-irrigation emitters
4. *Water containing higher suspended solids can be used:* The water for bubbler systems can have lower quality concerning solid content than that usually needed for most micro-irrigation emitters. Filtration is generally not required
5. *Low operating costs:* Though system hardware charges for bubbler systems are comparable to those of other micro-irrigation systems, the operating costs for bubbler systems are significantly lower because of the lesser energy and maintenance needs
6. *Irrigation intervals are long*: Bubbler systems use a basin flooding process, and hence the depth of water applied is generally more than 50 mm. The irrigation frequency is lesser than the other micro-irrigation systems since the depths of application are naturally smaller
7. *Duration of irrigation is short*: Since the rate of discharge for bubblers is greater than usual drip emitters (e.g., 250 l/h vs. 4 l/h), the application duration to attain a desired depth of water is substantially shorter
8. *Collected salts are uniformly leached*: The separate basins are entirely flooded with the bubbler system during individual irrigation. As this uniformly applied water infiltrates into and percolates through the soil, any accumulated salts are drawn downward with each irrigation
9. *Bubbler basins harvest rainfall*: In arid or desert regions, where precipitation is rare and occurs mainly through occasional large precipitation events, basins used for bubbler systems are perfect for harvesting intense rainfall and watering individual plants

Some potential disadvantages to bubbler systems are:

1. *A few agricultural bubbler systems have been installed or operational:* Though bubbler irrigation has been studied every so often over the past 30 years, farmers have not adopted the technology since other irrigation systems are generally more compatible with their geographical conditions or cultural practices
2. *Design criteria and recommended operating procedures are not well recorded:* Only a handful of farms use bubbler systems. Consequently, the chance to document practices

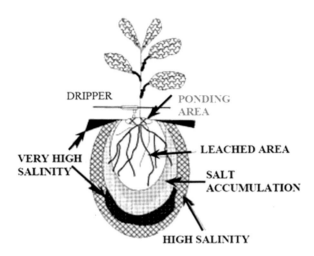

FIGURE 4.4 Movement of salts to the periphery in soils irrigated by drip irrigation

is inadequate. Because other micro-irrigation systems are well recognized and accepted by the agricultural community, there has been little motivation to chase technology with marginal benefits

3. *Trapping of air in the pipe network may cause blockages:* With comparatively low flow rates in gravity bubbler systems and with a normal rise in water temperature, dissolved air can leave the solution and collect in non-turbulent sections of the pipe network. Without proper hardware, the trapping of air in the pipe network may cause air blockages in these systems
4. *Farm topography should be nearly level:* The chief disadvantage for gravity bubbler systems is that the farm topography needs to be level or of a gentle, uniform slope. Slight elevation head changes cause relatively significant variations in total pressure head
5. *Bubblers are not adaptable for sandy soils:* Fine-textured soils with fairly low infiltration rates increase the quick spread of water over the whole basin. In contrast, sandy soils lead to small application efficiencies because of the irregular water distribution
6. *Small earthen basins are essential around plants:* Since bubbler emitters apply rather high rates of water, basins are required to hold the water till it has an opportunity to infiltrate. Earthen basins necessitate extra labor for preparation and maintenance
7. *Cultural practices are harder to perform around earthen basins:* Since the earthen basins positioned around the plants are fairly permanent, weed control and additional mechanical operations are limited to avoid damaging the basins

4.11 SYSTEM DESIGN

The bubbler system layout is similar to typical micro-irrigation systems with emitters, laterals, manifolds, and a control head. System components are usually built from polyvinylchloride (PVC), polyethylene (PE), or other plastic compounds.

4.12 DRIP IRRIGATION

Drip irrigation, also referred to as trickle irrigation, is an efficient irrigation method of the modern period. The technique is growing in popularity in areas where there is water scarcity and low-quality irrigation water. In drip irrigation, water is applied frequently at low rates through a low-pressure delivery system consisting of small diameter plastic pipes provided with outlets, termed emitters, or drippers, straight to the land surface close to the plant, where the roots grow.

Drip irrigation is extremely suitable for areas having water scarcity and regions where irrigation water quality is poor (saline water). Due to the frequent irrigation water application, salts are driven to the periphery of the moist zone, away from the root spread area. It is suitable for nearly all types of soils. In clay soils having low infiltration rates, water is applied gradually to check surface water ponding and runoff. In sandy soils with high infiltration rates, higher dripper discharge rates ensure adequate soil wetting. Drip irrigation is beneficial on lands with undulating topography without undertaking major operations and on slopes where the depth of soil is inadequate and the crop worth is high. The labor requirement is less in drip irrigation.

Drip irrigation is appropriate for nearly all plantation crops, orchard crops, and a majority of row crops. The method has been perceived to be extremely economical and useful in water scarcity areas to cultivate orchard and plantation crops. These include coconut, tea, coffee, cardamom, citrus, grapes, banana, papaya, mango, guava, pineapple, and pomegranate; row crops like sugarcane, cotton, groundnut, sapota, strawberry; and vegetable crops including tomato, potato, and other widely spaced vegetable crops, and flower plants (Figure 4.5). It is very suitable for nursery raising and the establishment of forestry plantations, especially under the wasteland development program. It is also suited to irrigate sugarcane, cotton, and groundnut. But, from economic factors, drip systems are not adapted to close growing field crops like cereals and pulses, as the number of laterals needed is high, which results in increased cost of the system.

4.12.1 Advantages

The significant advantages of drip irrigation, compared to other methods of irrigation, include the following:

1. *Higher crop yields:* Studies on crop production with drip irrigation have consistently registered yield increases from 20% to 50%. The rise in crop yield is primarily due to the maintenance of optimum soil moisture in the crop zone during the growing period of the crop with the greater control of irrigation water application in the drip system
2. *Improved harvest quality:* Drip irrigation results in enhanced quality of harvested crop owing to the preservation of optimum soil moisture conditions in the root zone during the growing season, allowing uniform crop growth. Besides, loss and injury due to water contact with foliage are eliminated, resulting in a better quality of produce
3. *Irrigation water saving:* With the exclusion of water application losses as, for instance, deep percolation, runoff, and evaporation, and decrease in the wetted volume of the soil (restricted to the crop root zone), there is a significant saving in the amount of water required for irrigation. Loss of water from the land surface, not occupied by the crop, is zero. Evaporation losses from the plant foliage are also reduced to a minimum. Moreover, the conveyance losses, like seepage, are eliminated with the appropriation of the drip method of water application. The savings of water resulting from drip irrigation usually range from 20% to 50% or more, when compared to surface methods of water application and to a lesser degree when compared to sprinkler irrigation
4. *Increased efficiency in fertilizer use:* A drip irrigation system is an efficient means of application of fertilizer and plant nutrients to the crop root zone where they are required. The performance in fertilizer application improves considerably when fertilizers and nutrients are administered with the drip system
5. *Reduced energy consumption:* Compared to sprinkler irrigation, drip irrigation results in decreased energy requirement in pumping, because of the low operating pressure required by the drip system. Moreover, the necessary quantity of water for irrigation is lesser under the drip system, which further decreases the energy requirement

FIGURE 4.5 Various crops under drip irrigation

6. *Tolerance to winds:* Wind has no unfavorable impact on drip irrigation. The laterals having emitters are placed at the land surface, and only the point of water application is wetted, which is not affected by the wind
7. *Decreased labor cost:* The labor requirements are less owing to the lessened cost of field preparation, exclusion of fertilizer application as a separate operation, the low need of weeding, and fewer harvesting rounds due to more uniform ripening of the crop
8. *Improved disease and pest control:* Plant diseases and pests are less in drip-irrigated crops, as compared to other methods of water application. Bacteria, fungi, and other organisms are lessened as the "above ground" plant parts remain dry. Also, the weed growth in the crop field is significantly reduced. With little or no weed growth, the severity of pest and disease occurrence is reduced
9. *Suitability for undulating terrain and sloping lands:* Hilly and undulating terrain cannot be irrigated using surface methods of irrigation without extensive land leveling operations.

Drip irrigation can be planned to operate effectively on nearly any topography without incurring the expense of land leveling. Drip irrigation can also be designed to suit varying field gradients

10. *Feasibility for problem soils:* Clay soils are heavy and have low infiltration rates. Reduced water application rates can be selected with the drip irrigation system to meet the soil characteristics. Conversely, for soils with high infiltration rates, for example, sandy soils, water can be applied in small quantities at frequent intervals, avoiding deep percolation losses
11. *Improved salinity tolerance:* Drip irrigation systems make it possible to use more saline water for irrigation. Since the root zone soil column is maintained at a higher moisture content, it results in decreased soil moisture stress. The contact of water with the plant is minimized, compared to other methods of irrigation. The frequent application of water also reduces salt concentration in the root zone by pushing salts away from the root zone to the periphery of the wetted zone
12. *Develops suitable soil physical conditions in the root zone:* The maintenance of soil moisture at approximately constant and optimum levels by restoring the water supply to the root zone at about the same rate as it is depleted by the plant results in low soil suction. This facilitates the uptake of water and nutrients by the plant. Moreover, in a well-designed and maintained drip irrigation system, the soil, on the other hand, is never saturated, and sufficient aeration is maintained during the growing period of the crop. Besides, no water is wasted due to the transpiration by weeds, as irrigation is confined to the crop. The drip system helps better control the irrigation water supply. There is no development of soil crust in drip irrigation
13. *Promotes more uniform irrigation water application:* A large number of emitters in a drip irrigation system provides more uniform water application. Low discharges and pressure heads in the water distribution network allow the use of smaller pipes of lower pressure ratings at a reduced cost. Fertilizers, pesticides, and other chemicals might be introduced into the system and applied in small quantities, as needed, with irrigation water
14. *Ease of operation:* Drip irrigation system is relatively easy to operate. Irrigation can be maintained throughout the day and the night, regardless of wind, daytime temperature, or cultural practices
15. *Promotes automation:* Automation can be easily included in the drip irrigation system
16. *Applied to irrigate crops in greenhouses:* The drip irrigation system is suitably adapted to irrigate crops planted in covered greenhouses with no wetting of the walls or the cover

4.12.2 Surface and Subsurface Drip Irrigation Systems

On the basis of installation and location of drip laterals concerning the installation depth, drip irrigation systems are classified as follows:

Surface drip irrigation systems: Surface systems are the standard drip irrigation systems. In these, the drippers and the lateral lines are placed on the land surface.
Subsurface drip irrigation systems: Subsurface drip irrigation is described as the water application beneath the soil surface using micro-irrigation emitters. The emitter discharge rate is usually less than 7.5 L/h (ASAE S526.2., 2001). Subsurface drip irrigation is different from subirrigation, where the root zone is irrigated by regulating the water table elevation. The proper depth for installing the subsurface driplines differs from crop to crop, type of soil, water source, pests, climate, and irrigator's decision. Some shallow subsurface systems (<20 cm depth) are replaced yearly and are quite similar to surface drip irrigation. Subsurface drip irrigation is suitable for a broad range of agronomic and horticultural crops and applies, in many aspects, to those crops presently under surface drip irrigation.

In subsurface irrigation systems, the lateral lines, including drippers, are placed below the soil surface within the crop root zone. Subsurface systems are seldom preferred in semi-permanent and permanent installations. The installation depth of subsurface laterals ranges typically from 20 cm to 60 cm. The mains and submains are set underground at depths below 60 cm to avoid any lateral pressure damage. The spacing of laterals ranges from 0.25 m to 5 m, depending on the crop spacing.

An advantage of the subsurface drip system is that the tubes remain buried for years and do not meddle with tillage or other cultural operations. Maybe the most significant advantage is that the system can be easily automated. Subsurface drip systems can adequately provide water and nutrients for several crops, even in high seasonal rainfall areas. Additional advantages claimed are labor-saving, avoiding evaporation losses from the soil surface, and better water distribution efficiency. The method stops the formation of soil surface crusts. Multiple cropping and rotation cropping will be easier because laterals remain in place. However, in India and various developing countries, surface drip irrigation systems are more common than subsurface systems. This is due to the ease of installation and also because de-clogging of drip systems with chemicals is not practiced widely. Farmers favor manual cleaning of emitters and laterals rather than inserting chemicals into the systems. Manual cleaning is not suited in subsurface drip systems.

4.12.3 Components of Drip Irrigation Systems

Water is pumped into drip irrigation systems and flows through valves, filters, main lines, submains, or manifold lines and laterals before being discharged into the field by point-source or line-source emitters. A check valve, usually installed downstream of the pump, is open during regular operation. During pump failure or breakdown, it gets closed to check the backflow of water toward the pump at times of pump failure or shutdown. The check valve blocks water containing suspended materials and dissolved substances (fertilizers and chemicals) from running back through the pump and contaminating the water source. Valves, flow meters, filters, pressure gauges, regulators, and flow control valves are the additional standard components used in drip irrigation systems. Figure 4.6 shows the layout of a typical drip irrigation system, including significant parts of the system.

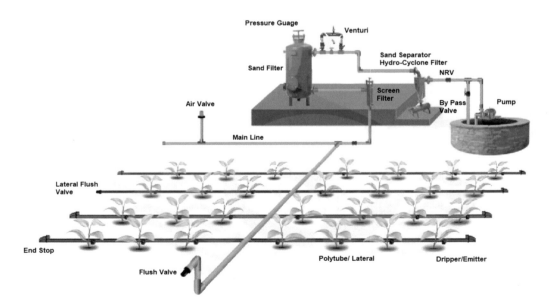

FIGURE 4.6 Layout of a drip irrigation system

Design and Components of Micro-Irrigation

4.12.3.1 Drippers

A dripper is an emitter fitted to drip irrigation lateral and is supposed to release water in the form of drops or continuous flow at emitter rates not more than 15 l/h, except during flushing (Figure 4.7). The emitter regulates the flow from the lateral. The emitter inlet is the point at which water enters the emitter. The emitter inlet end ("barb") is usually a sharp point reaching backward on an arrow inside the lateral pipe. The emitter outlet is the opening or group of slots in an emitter from which water is emitted to a specific location. There are several types and sizes of drippers, based on different operating principles.

Drippers function as energy dissipaters, decreasing the inlet pressure head in the lateral pipeline, which generally ranges from 0.3 to 1.5 atmospheres to 0 atmospheres at the outlet. The pressure loss by an emitter is obtained by small openings, long passageways, vortex chambers, manual adjustment, or other mechanical devices. The basis of the design of a dripper is to create an emitting device which will present a specified constant discharge, which does not vary much with pressure changes and does not get clogged easily. Drippers are usually manufactured from polypropylene or polyethylene material. Drippers may be classified on the basis of various criteria as given in Table 4.1.

4.13 DESIGN OF DRIP IRRIGATION SYSTEM

The following design criteria have been adopted for the design of the drip irrigation system:

- Collection of general information
- The layout of the field
- Crop water requirement
- Hydraulic design of the system
- Selection of components

4.13.1 DISCHARGE OF DRIPPERS

The discharge of any dripper may be expressed by the following relationship:

$$q = kh^x \tag{3}$$

where
q = discharge of the dripper (l/h)
k, x = constants for a specified emitter
h = operating pressure (m)

In Equation (3), x can be determined by measuring the slope of the log-log plot of pressure head versus discharge. With x known, h can be determined from Equation (3). The above relationship provides considerable insight into the performance of a particular emitter.

FIGURE 4.7 Different types of drippers

TABLE 4.1
Classification of Drippers

Criteria	Types
Operating pressure	i. Low pressure (below 0.8 m) ii. Medium pressure (2 to 5 m) iii. High pressure (8 to 15 m)
Path cross-section	i. Low (below 0.8 mm) ii. Medium (0.8 to 105 mm) iii. Wide (above 105 mm)
Discharge rates	i. Low (4 l/h), ii. Medium (4 to 10 l/h) and iii. High (15 l/h or more)
Flow regime	i. Laminar flow (emitters having long flow path and low discharges), ii. Partially turbulent long path and multi-exit emitters with relatively high discharges and in nozzle or vortex type emitters) iii. Turbulent flow (long path and multi-exit emitters with relatively high discharges and in nozzle or vortex type emitters)
Pressure dissipation	i. Long path emitters (pressure is dissipated during flow through a long and narrow path) ii. Nozzle or orifice type emitters and (pressure is dissipated as water discharges through a small opening) iii. Leaking lateral type (pressure is dissipated as the water is delivered through a large number of very small pores and perforations in the lateral pipe wall)
Lateral connection	i. On-line (mounted on the laterals) ii. Integrated (inserted in the laterals)
Water distribution	i. Single exit point, ii. Orifice and iii. Long path emitters
Cleaning Characteristics	i. Self-flushing types and ii. Emitters that need to be opened for cleaning
Pressure Compensation	i. Non-pressure compensating, NPC (the discharge of the emitters increases with the operating pressure) and ii. Pressure compensating PC (discharge is constant over a wide range of lateral operating pressure)
Material	iii. Polyvinyl chloride (PVC), iv. Low-density polyethylene (LDPE) and v. Linear low-density polyethylene (LLDPE)

Give an example.

4.13.2 Water Distribution Network

The water distribution network for a drip irrigation arrangement will generally consist of the following components: (i) main lines, which transport water to the fields; (ii) submains, which distribute water evenly to several lateral lines; and (iii) lateral lines, which distribute water uniformly along their length using drippers or emitters (Figure 4.7). The drip irrigation distribution system must be sturdy because of the harsh conditions under which they usually operate—the movement of heavy machinery, rats, squirrels, birds, ants, and chemicals.

4.13.3 Main and Submain Pipes

Rigid PVC and high-density polyethylene (HDPE) pipes are used as main pipes to minimize corrosion and clogging. It is a sound practice to set the main pipes underground. Pipes of 65 mm

Design and Components of Micro-Irrigation 77

diameter or higher with a pressure rating of 4 kg/cm² to 10 kg/cm² are advised for main pipes. Rigid PVC, HDPE, and low-density polyethylene (LDPE) are used as submain pipes. Pipes having an outer diameter of 32 mm to 75 mm with a pressure rating of 6 kg/cm² are usually recommended for use as submain pipes. The diameter of the main and submain pipes will be based on the quantity of water to be carried. Hose pipes (Figure 4.8) manufactured from LLDPE using extrusion technology and UV stabilization and carbon black content provide durable tubing with close dimensional tolerances. Larger diameter hoses are used in main lines and submains, and smaller diameter hoses are used as laterals in drip irrigation systems. Main lines are sized based on irrigation sectioning or sequencing.

A submain is normally connected to the main line through a control valve assembly. The purpose of the assembly is to control the water pressure in the submain, which is best attained by a throttling valve. The function of a submain is to distribute water uniformly to several lateral lines. The uniform distribution requires that the pressure along the submain be relatively constant, with a minimum pressure variation. Additional components, such as air and vacuum relief and pressure relief valves, pressure gauges, filtration, and automatic on-off control, are usually provided on a submain. At the end of each submain is fitted a valve for flushing the submain to remove contaminants and physical and chemical sediments

4.13.4 LATERALS

Laterals (Figure 4.9) are pipes on which drippers are mounted or inside which they are inserted. Laterals are made of polyethylene (LDPE and LLDPE), butylene, or PVC materials. They ordinarily range in diameter from 10 mm to 20 mm, with wall thickness varying from 1 to 3 mm with pressure ratings of 2.5 kg/cm². Lateral pipes should be flexible, non-corrosive, and resistant to ultra-violet (UV) rays. LLDPE gives more reliable against UV ray protection and longer service life than LDPE. Each lateral is connected to the submain pipe or the main pipe, usually through a manifold. Twin-line hoses are also used as drip irrigation laterals, using line-source emission devices. Dripline laterals are designed to sustain an acceptable variation of emission device discharge along their length.

FIGURE 4.8 Hose pipes used as mains and submains in drip irrigation systems

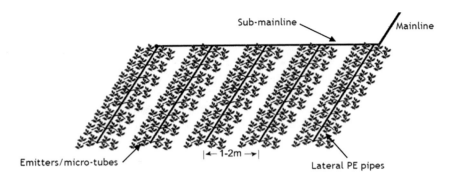

FIGURE 4.9 Layout of mains, submains, and laterals in a drip irrigation system

Pressure differences due to pipe friction, minor losses, and ground surface elevation changes are the significant reasons for along-the-lateral variation in emitter discharge.

4.13.5 Manifold

The manifold, or header, connects the main line to the laterals. Pressure loss in manifolds is dependent on the topography, pressure loss in laterals, and total variation permitted in the design of emitters. The submain's connection or main line to the manifold is in the manifold's center on flat terrain. In fields with an appreciable slope, the downhill elevation gains are balanced by reducing the pipe's size or increasing the connection point uphill to increase the number of laterals serving downhill. Typically, a combination of both means is used to balance the downhill elevation gain. Often, the manifold connection to the main line is regulated. On steep hills, one pressure-regulating valve can serve only one lateral. In such cases, several pressure and flow regulating valves might be required. Overall, one regulating point may serve 2 to 5 laterals, depending on the degree of slope.

The layout of lateral lines should be such that it provides the required emission points of the crop to be irrigated. Sometimes, two laterals per row of trees are needed. Other means of obtaining more emission points per tree are zigzag and "snake" layouts (Figure 4.10). Microtubes may also be used to obtain multioutlet emission points.

4.13.6 Pipeline Accessories and Fittings

PVC pipes can be joined by using PVC resin and HDPE pipes by using heat-butt welding. Polyethylene tubing, which is normally not cementable, is connected with barbed or compression fittings. The common fittings required in drip irrigation systems are tees, elbows, unions, crosses, and plugs or end caps (Figure 4.11). Likewise, for joining PVC/HDPE pipes, fittings of different sizes, such as couplers, tees, elbows, bends, threaded adapters, tailpieces, collars, and reducer rubber are required. Accessories include takeouts, grommet, end plug, joints, and manifolds. They are usually available in 4-, 10-mm, 12-mm, 16-mm, and 20-mm sizes. Takeouts are used for taking out laterals from the submain connector. Tees, bends, and end caps are used in drip laterals. They can be chosen based on the sizes of pipes used in the distribution network.

4.13.7 Valves

Valves are fundamental parts of pressurized irrigation systems, including drip irrigation. The valves used in drip irrigation systems comprise air and vacuum relief valves, pressure-regulating valves, flow regulation valves, nonreturn valves, and on-and-off valves.

Design and Components of Micro-Irrigation

4.13.7.1 Air Release and Vacuum Relief Valves

The presence of air in the pipeline restrains the water flow and enhances pumping costs. Air may enter the pipeline while filling, where water moves into the pipeline at the pump or gravity inlet point and by broken or cracked joints and fittings. To remove any air that may enter the pipeline, adequately sized air and vacuum relief valves are installed at all high points of main lines, submains, control risers, and other points in the pipeline system (Figure 4.12). Air and vacuum release valves (Figure 4.13) are intended to exhaust air at various pressure conditions during regular pipeline operation while restricting water's outflow. Exhaust ports for such valves are usually of diameter 1.6 mm to 6.4 mm. Air release valves allow air to escape when filling pipelines with water and

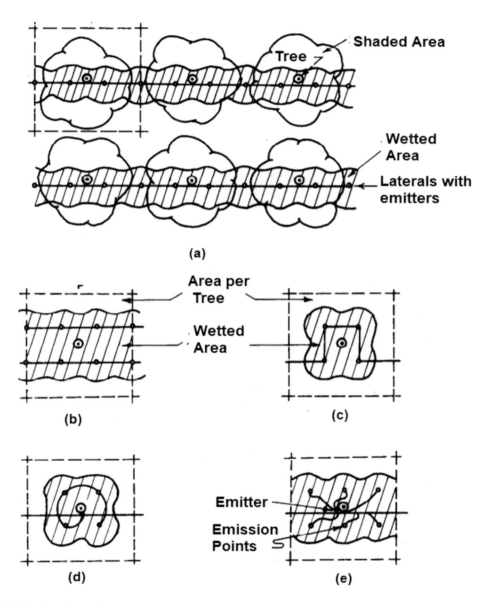

FIGURE 4.10 Views of layout of laterals for widely spaced tree crops: (a) Single lateral for each row of trees, (b) Two laterals for each row of trees, (c) Zigzag layout of a single lateral for each tree in the row, (d) "Pigtail" layout of emitters at each tree. (e) Multioutlet emitter with distribution tubing (USDA, 1984)

FIGURE 4.11 Different types of fittings used in drip irrigation

FIGURE 4.12 Air release valve

FIGURE 4.13 Vacuum relief valve

Design and Components of Micro-Irrigation 81

remove air pockets at high points in the system. Vacuum relief valves, which have orifices of size 25 mm to 200 mm in diameter, are designed to exhaust large volumes of air during pipe filling and close when filling stops.

4.13.7.2 Pressure Relief Valves

Pressure relief valves (Figure 4.14) are installed at any point where the chance exists for extremely high pressures, both static and surge pressures, to occur. Abnormally high-pressure situations may occur from abrupt opening or closing of a valve, starting or stopping a pump, breakdown of a pressure regulator valve, and sudden closure of an air vent under high pressure.

4.13.7.3 Pressure-Regulating Valves

Pressure-regulating valves are automated valves that maintain constant downstream pressure despite changing flow and pressure conditions. They are used where pressure variations in the field make it hard to apply water uniformly. Usually, pressure-regulating valves give surge control by protecting downstream pipes from surges originating upstream.

4.13.7.4 Nonreturn Valves

Flow in the backward direction through an irrigation system is termed reverse flow. Nonreturn valves, also referred to as check valves, prevent reverse flow or backflow. Reverse flow is caused by a pump, pipeline, or valve malfunction resulting in irrigation system damage and water source contamination.

4.13.7.5 Flow Control Valves

Flow control valves, also termed off valves, can be manually operated (Figure 4.15) or automatic (Figure 4.16). They are employed for a variety of purposes. When these are placed at the upstream end of a pipe, they present an on-off service to the downstream pipes. This enables water to be cycled to various parts of the system to meet all portions of the farm's irrigation requirements. It also enables parts of the system to be detached for maintenance and repair works. When located at the downstream end of permanently laid pipelines, control valves allow flushing of pipes of sediment and debris.

4.13.8 FILTRATION SYSTEMS

Filtration is essential to the success or failure of a drip irrigation system. The clogging of the drippers and different components is mostly due to the tiny diameter water passage in the drippers, dirt

FIGURE 4.14 Pressure relief valves

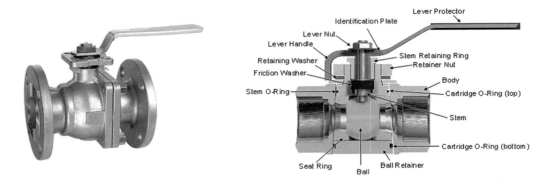

FIGURE 4.15 Manual flow control valve and its cut section showing different parts

FIGURE 4.16 Automatic flow control valve

in the irrigation water, and the system's relatively low operating pressure. When designing a drip irrigation system, it should be kept in mind that the water quality does not remain constant and may change over a period or different seasons in a year. The removal of the particles resulting in clogging is necessary for the satisfactory performance of a drip irrigation system. Settling basins, sand or media filters, screen filters, centrifugal sand separators, and disc filters are the primary devices utilized to separate suspended and other materials jamming drip irrigation systems.

4.13.8.1 Settling Basins

Settling basins or ponds can remove considerable volumes of sand and silt transported by streams, used as water sources for drip irrigation, commonly through diversion works. The minimum particle size that can be removed depends on how the sediment-laden water is detained in the basin. The pond should be planned to give the required settling time. Long and narrow ponds are favored. If the construction area is insufficient, baffles may be constructed (Figure 4.17). At least 15 minutes will be necessary for the settling of mineral particles bigger than 80 microns. More prolonged detention time is required to remove smaller particles. Clay-sized particles will need many days to

Design and Components of Micro-Irrigation

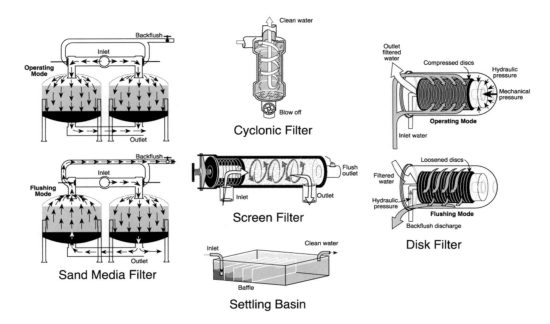

FIGURE 4.17 Composite drawing of several filter types used in drip irrigation system. (Kansas State University)

settle except when flocculating agents like alum or polyelectrolytes are used. For settling tanks to be effective, the intake pipe to the drip irrigation system should be located in such a way that water from the pond's upper level enters the irrigation system.

Nevertheless, sedimentation solely may not provide the wanted water quality. Algal growth and additional wind-blown contaminants in the settling pond may induce serious problems. Chemical treatments with chlorine or copper sulfate are required to control algae. Because of these predicaments, settling basins are not employed in drip irrigation using groundwater. In canal irrigation, settling basins are used only under conditions of heavy sediment loads in the water source.

4.13.8.2 Gravel/Sand Media Filters

Sand media filters, generally termed media filters, consist of layered beds of graded sand and gravel placed in one or several cylindrical tanks. Media filters are necessary for the principal filtration of irrigation water. The diameter of filter tanks ranges from 20 cm to 200 cm, depending on the system capacity (Figures. 4.18–4.20). They are of vertical or horizontal types and are typically made of steel and coated with 70 to 80-micron-thick blue polyester powder from both inside and outside surface for protection facing corrosion and weather effects. Media filters are relatively low cost and easy to operate. Every media filter is housed inside a pressure vessel or tank with service openings at the top and side, inlet, outlet, and back-flush connections, generally with a gravel bed to aid back-flushing. The filter tank houses all filter components, except for the control valves and the connecting pipe.

Usually, the filter media consists of crushed granite, silica, or quartz sieved to particular particle sizes (generally of 0.7 mm to 1.2 mm size). Numbers designate the sand media used in most drip irrigation system filters. Numbers 8 and 11 are crushed granite, and numbers 16, 20, and 30 are silica sands. The mean granule size is about 1,900, 1,000, 825, 550, and 340 microns, respectively, for numbers 8, 11, 16, 20, and 30 (USDA, 1984). It is a standard practice to choose the smallest medium possible. Still, a larger medium may seldom be acceptable as it creates less pressure drop and has a more gradual build-up of particles. The degree of filtering needed in a given drip irrigation system will depend on the system's arrangement and the type and size of the discharge devices used.

FIGURE 4.18 Sand media filter and screen filter installed in a drip irrigation system

FIGURE 4.19 filter installed in a drip irrigation system

Media filters efficiently separate suspended minerals, organic materials, and maximum suspended materials originating from surface or groundwater sources. Yet, they do not exclude very fine materials like silt and clay, as well as bacteria. Filter cleaning involves removing the filtered substances that have accumulated in the pores of the filter material.

4.13.8.3 Screen Filters

The screen filter (Figure 4.20) is equipped in series with the media filter to separate the solid impurities like fine sand and dust from the irrigation water after passing through the media filter. The screen filter parts consist of the body, one or two filtration elements, gaskets, cover, inlet, outlet, and drainage valves. The necessary parts of a screen filter are the filter screen or strainer (element)

Design and Components of Micro-Irrigation

FIGURE 4.20 Different types of Filters used in drip system

and the filter housing. The strainer in a screen-type filter includes filtering elements used to segregate suspended solids from the water flowing through it and collect them on the filter element's face. The screen is made of stainless steel, nylon, or polyester mesh. The filter element consists of a perforated plate and fine mesh screen to hold solid contaminants larger than a given size from the water flowing through the component. The approved mesh size of screen filters is 100 microns. Other sizes are also available. The "turbo-clean" mechanism and endless vortex inside the elements block contaminants from settling on the filtration surface. The flow direction is from the inside of the element to the outside. The strainer housing is the strainer's portion that holds all the strainer components, besides the control valves. The construction material of the filter housing and the filter element usually may be mild steel or plastic. When used as a filter housing, mild steel is coated with 70–80-micron-thick deep blue colored polyester, both inside and outside, as protection against corrosion and weather effects.

Screen filters need periodic cleansing of the screen element to separate the filtered materials from the screen's surface and inside the housing. Arrangements for drain valves and ports for measuring pressure on both inlet and outlet of screen filters are necessary. Careful thought to maintenance and cleaning is essential with screen filters. Manually cleaned screens must be frequently disassembled to clean the elements. Failure to do so will lead to a build-up of stuff on the screen and an extreme pressure drop across the filter. An extreme pressure loss across the screen may push out particles, especially organic material, through the screen, where they can pass into the irrigation system and create clogging problems.

4.13.8.4 Hydrocyclones or Centrifugal Sand Separator

Centrifugal filters, also named hydrocyclones or cyclonic separators, apply vortex action, and centrifugal force to separate suspended materials with specific gravities greater than 1.2 (Figure 4.21). The filtration capacity of hydrocyclones is more for high-density particles than for lower-density ones. Since these filters efficiently discard large quantities of sand particles, they are installed upstream of other filter types. The water is allowed tangentially at the top of the cone. It produces a circular motion appearing in a centrifugal force, which throws the massive suspended particles toward the filter body's wall. The isolated particles travel toward the bottom of the collecting tank (Figure 4.22). The accumulated filtrate (sand) is cleared manually or by a flushing valve. The clean water rises in a spiral movement to the outlet. The diameters of the top and bottom of the cone are designed according to the water flow rate, which varies from 3 m^3/h to 300 m^3/h. The pressure loss inside the filter ranges from 4 m to 8 m. If the sediment concentration is high in the water source, then more filters should be installed in series.

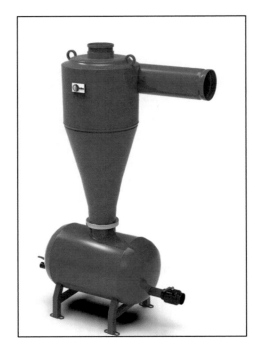 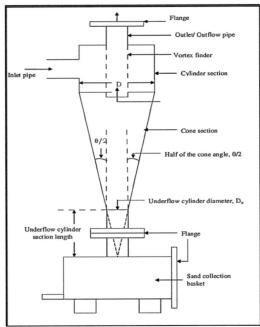

FIGURE 4.21 Hydrocyclone filter and its sectional view

FIGURE 4.22 Disc filter assembly

4.13.8.4.1 Disc Filters

Disc filters can remove organic material and algae, and other extremely fine particles that other filters cannot remove. Disc filters are usually the final filtering element before the water enters the drip line. Disc filters (Figure 4.22) comprise stacks of grooved, ring-shaped discs that catch the contaminating particles. During filtration, the discs are clasped together. In the back-flushing mode, the direction of flow is inverted. The discs are relaxed and cause a spinning motion, assuring absolute removal of retained particles.

4.14 APPLICATION OF FERTILIZERS AND CHEMICALS USING DRIP IRRIGATION SYSTEMS

Water-soluble fertilizers can be efficiently and effectively administered through drip irrigation systems. This is termed fertigation. Decreased labor, equipment, energy costs, and higher fertilizer use efficiency are the significant advantages of fertigation compared to traditional water application methods. The success of drip irrigation, to a reasonable degree, is due to the enhanced supply of nutrients to plants, which is a novel capability of fertigation. Precise control of the time of application, the concentration of fertilizers/chemicals, the balance of the nutrients, and the location of the point of application are possible with the fertilization systems used in pressurized irrigation systems. Plant protection chemicals can also be applied efficiently using the same equipment. However, materials such as inorganic phosphorus forms that form chemical precipitates and produce clogging should not be injected into drip systems.

Pressurized irrigation systems (both sprinkler and drip) present a suitable medium for providing nutrient materials to the crop. The term fertigation is frequently used for fertilizer application through pressurized irrigation systems. Besides fertilizers, other chemical products can also be supplied by pressurized irrigation systems for the following purposes:

1. For enhancing the chemical properties of water, (e.g., reducing the pH) or of the soil (use of chelates to control certain chloroses of fruit trees)
2. For preventing certain crop diseases (insecticides, nematicide, systemic fungicides)
3. Flushing the irrigation equipment network and removing calcareous deposits from it or cleaning it to inhibit clogging resulting from the proliferation of certain microorganisms

In almost all pressurized irrigation systems, the fertigation component is an essential component. Fertigation allows more effective use of fertilizers (water application limited to the crop root zone), resulting in higher crop yields. Nearly all water-soluble fertilizers and liquid fertilizers can be applied efficiently to most crops through irrigation water during the growing season. The reason why fertigation has grown as the "state of the art" in vegetable farming is that nutrients can be applied in the exact dosage at the time suitable for the particular stage of plant growth. The most frequently used fertilizers are nitrogen, phosphorus, and potassium (N, P, and K). Micronutrients are applied when they are insufficient in the crop root zone.

4.14.1 EQUIPMENT AND METHODS FOR FERTILIZER INJECTION

Fertilizers are injected into drip irrigation systems by choosing a wide variety of pumps, valves, fertilizer tanks, ventures, and aspirators. Fertilizers may be introduced into the irrigation system by a differential pressure system, a venturi injector, or by pumping under pressure into the irrigation pipeline. Chemicals which demand a uniform concentration rate of the fertilizer solution entering the irrigation water are injected by constant rate injection devices only. The generally utilized elements of the equipment to apply fertilizers in pressurized irrigation systems, besides the irrigation system, constitute a fertilizer tank, fertilizer dissolver, fertilizer injection device, a filter, a pressure gauge, check valves, and a pressure regulator (Figure 4.23).

4.14.2 FERTILIZER TANK

The fertilizer tank is employed for mixing the fertilizer (Figure 4.23). The fertilizer tank should have an adequately substantial capacity to hold the entire fertilizer solution needed for the cropped

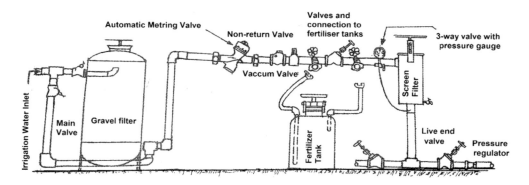

FIGURE 4.23 Diagram showing control head assembly of drip irrigation system with fertigation equipment

area for any application. It should be made from materials that endure the corrosivity of the fertilizers. Large, low-cost tanks built from epoxy-coated metal, plastic, or fiberglass are helpful when injection pumps are used.

4.14.3 Fertilizer Dissolver

A fertilizer dissolver is an appliance used for preparing fertilizer solution. After filling the tank with water, the pump injects the water into an internal strainer that holds the dry fertilizers. The water slowly dissolves the fertilizer into a solution by distributing the solution within the dissolver.

4.14.4 Fertilizer Injection Devices

Pressurized irrigation methods need fertilizer injectors for the injection of fertilizers into the irrigation water. The two basic concepts based on which fertilizer injectors are designed have been given below:

a) *Proportional concept* is characterized by the constant concentration of the fertilizer solution in irrigation water throughout the irrigation duration. Fertilizer injection devices such as venturi pumps and fertilizer injection pumps operate on this principle which enables the delivery of constant concentration during the entire irrigation duration
b) *Quantitative or non-proportional* concept is characterized by a change in the concentration of fertilizers in irrigation water. The concentration of fertilizer decreases gradually with the irrigation duration. Fertilizer tanks work on this principle. The total amount of fertilizers applied in both the cases should be equal since the requirement of nutrients of the plants is independent of the injection device and the method of fertilization

   i. **Fertilizer injection by a volute centrifugal pump**

Fertilizer is injected into the system from the suction side of the pump by a pipe and controlled by a valve (Figure 24). For maintaining the water supply in the tank, a pipe is joined from the discharge side of the pump to the fertilizer container. This system is comparatively simple, but pump impellers are prone to get corroded due to the fertilizer solution except when the impellers are made of corrosion-resistant materials.

   ii. **Fertilizer injection pump system**

FIGURE 4.24 A sprinkler irrigation system irrigating a row crop

In this method, a pump is used to extract the fertilizer stock solution from the storage tank and introduce it under pressure into the irrigation stream. The injection rate can be set to produce a desired mixing ratio. The fertilizer solution is generally pumped from an unpressurized storage tank.

    iii. **Pressure differential injection system**

Pressure differential (PD) units are another means of adding fertilizer into drip irrigation systems. The PD unit takes benefit of the system's pressure-head differences. Valves, venturi, elbows, or pipe friction can generate pressure differences.

    iv. **Venturi injection system**

The venturi injection system consists of a converging section, a throat, and a diverging section. In a pressurized irrigation system using filters, the venturi unit is located between the sand filter and the screen filter. The venturi generates the differential pressure (about 20% from one side of the device to the other) and lets the fertilizer solution flow in the main water line.

4.14.4.1 Location of Fertilizer Injection Systems

In a drip or sprinkler irrigation system, the fertilizer injection unit is positioned between the sand filter (when used) and the screen filter. The usual recommendation is that the fertilizer solution should pass through at least two 90-degree turns to assure sufficient time for thorough mixing and any residue to come out in front of the screen filter. The fertilizer applicator must be installed at the upstream end of the screen filter.

4.14.5 COMPUTATION OF THE QUANTITY OF FERTILIZER TO BE APPLIED

To estimate the amount of fertilizer to be introduced into the system per setting, the area irrigated in each setting of the lateral line is obtained by multiplying the length of the lateral coverage and the move of the lateral. The amount of fertilizer to be inserted is determined for this area according to the prescribed rate of fertilizer application in kilograms per hectare, using the following formula given in Equation (4).

$$W_F = \frac{D_s \times D_e \times N_s \times W_f}{10,000} \qquad (4)$$

where
W_F = amount of fertilizer per setting, kg
D_s = distance between sprinklers, m
D_e = distance between laterals, m
N_s = number of sprinklers
W_f = recommended fertilizer dose, kg/ha

Example 8: A lateral has 15 sprinklers spaced 12 m apart. The laterals are spaced 16 m on the main line. Determine the amount of fertilizer to be applied at each setting when the recommended fertilizer dose is 50 kg/ha.
Solution: Using Equation (4),

D_s = 12 m
D_e = 16 m
N_s = 15
W_f = 50 kg/ha

$$W_F = \frac{12 \times 16 \times 15 \times 50}{10,000} = 14.4 \text{ kg/ha}$$

Example 9: Fifteen sprinklers with twin nozzles of 5 mm diameter and 4 mm diameter each with a coefficient of discharge 0.98 are operating at a pressure of 2.5 kg/cm². The sprinkler spacing is 12 m × 16 m. Find out the

a) Total sprinkler capacity of the system
b) Discharge of sprinkler

Solution: System capacity (q) = $A \times V \times C_d \times N$
Here, $A = a_1 + a_2$

$$= \frac{\pi}{4}\{(0.005)^2 + (0.004)^2\}$$

$$= \frac{\pi}{4}(0.000025 + 0.000016)$$

= 0.000032 m²
A = 0.000032 m²

$$V = \sqrt{2gh}$$

V = (2 × 9.8 × 25)$^{0.5}$
 = 22.135 m/s
V = 22.135 m/s
Q = 0.000032 × 22.135 × 0.98 ×15
 = 0.01041 m³/s
Q = 10.41 l/s

Design and Components of Micro-Irrigation

Example 10: Keeping 100 cans uniformly spaced in an area covered by 4 sprinklers, the average depth caught in a given time was 27 mm with an average deviation of 2.4 mm from the mean. What is the uniformity coefficient? Assuming that the infiltration rate was not exceeded and the percolation loss was 5%. Find the application efficiency.

Solution: Given
$\bar{X} = 2.4$ mm
$\bar{d} = 27$ mm

Uniformity coefficient $= 100(1 - \frac{\bar{X}}{\bar{d}})$
$= 100(1 - \frac{2.4}{27})$
$= 91\%$

Percolation loss = 5%. Therefore,
Application efficiency $= (1 - \frac{0.05 \times 27}{27}) \times 100$
$= (1 - 0.05)100$
$= (0.95)100$
$= \mathbf{95\%}$

Example 11: Determine the capacity of the sprinkler system which has 12 sprinklers spaced at 10 m intervals on each of the 2 laterals spaced at 15 m apart. The application rate of water is 1.5 cm/h.

Solution:
Application rate = 1.5 cm/h
Distance between sprinklers = 10 m
No. of sprinklers = 12
No of laterals = 2
Distance between laterals = 15 m

$$\text{Capacity of the system} = \frac{12 \times 2 \times 10 \times 15 \times 1.5 \times 10^{-2}}{3,600}$$

$$= 0.015 \text{ m}^3/\text{s or} = 15 \text{ l/s}$$

Example 11: In a citrus orchard, planting is done at a spacing of 5 m × 5 m. The daily pan evaporation of the orchard is 6 mm. The pan coefficient, wetting factor, and crop coefficient are 0.8, 0.6, and 0.6 respectively. Four drippers each of 4 l/h discharge are used to irrigate each plant. The time of operation of drip irrigation system in hours will be?

Solution: The daily irrigation requirement is
$V_r = K_C \times K_P \times E_P \times C_C \times A$
$= 0.6 \times 0.8 \times 0.6 \times 6 \times 25$
$= 43.2$ l
We have 4 drippers having 4 l/h discharge
Therefore, $4 \times 4 = 16$ l/h

$$\text{Time} = \frac{\text{Daily irrigation requirement}}{16}$$

$$= \frac{43.2}{16}$$

Time = 2.7 h

Example 12: A permanent matured orchard has a tree spacing of 4 m × 5 m. Each tree has a shading area of 40% to be irrigated with multi-exit drip emitters. The effective wetting geometry of each emitter is 2 m × 2 m. The emitters have a discharge constant and exponent of 0.3 and 0.6, respectively. The coefficient of variation of emitter discharge is 0.06. The average and minimum operating pressures are 120 kPa and 100 kPa, respectively. The emission coefficient of the emitters is?

Solution: Given
Plant area = 4 m × 5 m; $A = 20$ m²
Area to be irrigated = 40%
i.e., $0.40 \times 20 = 8$ m²
Wetting area of each emitter = 4 m²

$$\text{No. of emitters required} = \frac{\text{area to be irrigated}}{\text{wetting area of each emitter}}$$

$= \frac{8}{4} = 2$

Given, $P_{average} = 120$ kPa; $P_{minimum} = 100$ kPa; coefficient of variation $(c_v) = 0.06$
Discharge const $(k_d) = 0.3$
Exponent factor $(m) = 0.6$
$Q_{average} = k_d \times (P_{average})^m$
$= 0.3 \times (120)^{0.6}$
$= 5.30$ l/h
$Q_{minimum} = k_d \times (P_{minimum})$
$= 0.3 \times (100)^{0.6}$
$= 4.76$ l/h

$$E_u = 100 \left[1 - \frac{1.27 \, (\text{Coff. of variation})}{\sqrt{N}} \right] \frac{Q_{minimum}}{Q_{average}}$$

$$= 100 \left[1 - \frac{1.27 \, (0.06)}{\sqrt{2}} \right] \frac{4.76}{5.30}$$

$E_u = 84.82\%$

Example 13: One-line drip emitters are placed at 1 m spacing on a 50-m-long 16-mm-diameter lateral. The design discharge of each emitter is 1 l/h. Using density of water as 0.998 g/cm² dynamic viscosity of water 1.002 ×10⁻³ Ns/m² at 20°C temperature, the reduction factor due to 50 outlets as 0.343 and equivalent length due to loss from the emitter bard as 1 m, determine the (a) Reynolds number (b) friction factor (e) head loss due to friction in the areal pipe?
Solution:

i) Reynolds no $(R_e) = \frac{\rho \, v \, dl}{\mu}$ $\rho = 0.998$ g/cm³

$= \frac{0.998}{1,000}$ kg/cm³

$= 0.998 \times 10^3$ kg/m³
$Q = A \times V$

$V = \frac{1 \text{lt/h}}{\frac{\pi}{4}(16 \times 16)}$

$= \frac{4 \times 106}{1,000 \times 3,600 \times 3.14 \times 16 \times 16}$

Design and Components of Micro-Irrigation

$$V = 1.382 \times 10^{-3} \text{ m/s}$$

$$R_e = \frac{998 \times 50 \times 1 \times 10 - 3 \times 0.016 \times 1{,}000}{3{,}600 \times \frac{\pi}{4} \times 0.016 \times 0.016 \times 1.002}$$

$R_e = 1{,}098.\,304$ or $R_e = 1{,}100$

ii) Friction factor $(4f) = \dfrac{64}{Re}$

$$= \frac{64}{1{,}100}$$

$$= 0.058$$

iii) Head loss due to friction in pipe $(H_f) = \dfrac{(4f)lv^2}{2gd}$

$$= \frac{0.058 \times 50 \times 1.38 \times 1.38}{2 \times 9.8 \times .016 \times 1{,}000{,}000} + 0.345 + 1$$

$$= 1.015 \text{ m}$$

Example 14: The drip irrigation system is installed for 1 ha area of guava plantation with a spacing of 4 m × 4 m. The percentage wetted area crop coefficient are 40% and 0.7, respectively. The maximum pan evaporation during summer months and pan coefficient are 8 mm/d and 0.7, respectively. The pump, used to irrigate the plantation, has a discharge of 2.18 × 10⁻³ m³/s). The total dynamic head required to operate the pump is 20 m. The pump efficiency is 60%. Find the total daily water requirement of guava plantation and HP for the pump.

Solution: Given
Total area = 1 ha = 10,000 m²
Spacing = 4 m × 4 m
Crop coefficient $(K_C) = 0.7$
Wetted area = 40%
Pan coefficient = 0.7
Pan evaporation = 8 mm/day
Daily water requirement = pan evaporation × pan coefficient × crop factor × area × wetted area
= 8 ×10⁻³ × 0.7 × 0.7 × 10⁴ × 0.4
Daily water requirement = 15.68 m³
Power of the pump = discharge × ΔP
ΔP = 20 m = 2 kg/cm²
ΔP = 19.62 × 10⁴ pascal or N/m²
Therefore, power = 2.18 × 10⁻³ × 19.62 × 10⁴
Power = 427.28 w
Power = $\dfrac{427.28}{746}$ = 0.573 HP
Given efficiency = 60%
Total HP = $\dfrac{0.573}{0.60}$
Total HP = 0.95

4.15 IRRIGATION WATER REQUIREMENT

The irrigation water requirement of a field is dependent on the crop type, weather conditions, soil type, and area under cultivation. The monthly irrigation water requirement is estimated based on monthly pan evaporation data and crop coefficient. The monthly irrigation water requirement can be calculated based on monthly pan evaporation data and crop coefficient by using Equation (5):

$$V_m = K_c \times K_p \times C_c \times E_p \times A \qquad (5)$$

where
  V_m = Monthly irrigation water requirement, l
  K_c = Crop coefficient
  K_p = Pan evaporation factor (generally 0.8)
  C_c = Canopy factor
  (C_c = 1.0, for closely spaced field crop, C_c = wetted area/plant area for orchards and vegetable crops)
  E_p = Normal monthly pan evaporation, mm
  A = Area to be irrigated, m²

4.16 CAPACITY OF DRIP IRRIGATION SYSTEM

The capacity of a drip irrigation system is based on the irrigation water requirement of the irrigated area, the daily period of system operation, irrigation interval, and irrigation efficiency. The operation time of each section (block) of a drip irrigation system is mostly confined to 1.5 to 2 h, to avoid deep percolation losses. In most cases, the irrigation interval may be based on irrigating daily, every alternate day, or once in 3 days. Irrigating intervals greater than three days may cause soil moisture stress in some crops irrigated by the drip system. The capacity of a drip irrigation system may be computed using the relationship in Equation (6):

$$Q = V_d \times \frac{T}{(\eta_a \times t)} \qquad (6)$$

where
  Q = Capacity of drip system, l/h
  V_d = Daily water requirement, l
  T = Irrigation interval days
  η_a = Water application efficiency (in fraction)
  t = Duration of each irrigation, h

The discharge required per plant (Q_p) can be estimated by dividing the capacity of the drip system (Q) by the number of plants (n) in the area to be irrigated (i.e., $Q_p = Q/n$).

Example 15: Calculate the capacity of a drip irrigation system designed for a 5 ha farm. The maximum pan evaporation is recorded in the month of May as 225 mm. The crop cultivated has a crop coefficient value of 0.6 and the canopy factor is 0.6. The soil is sandy loam with an application efficiency of 88%.

Solution: Using Equation (5) the monthly water requirement for the given field can be calculated as:

  K_c = 0.6
  K_p = 0.8
  C_c = 0.6
  E_p = 225 mm
  A = 5×10⁴ m²

$$V_m = 0.6 \times 0.8 \times 0.6 \times 225 \times 50,000$$

Design and Components of Micro-Irrigation

$$= 3,240,000 \, l$$

Therefore, daily water requirement is

$$V_m = \frac{3,240,000}{31}$$

$$= 104516.2 \, l$$

The capacity of the drip irrigation system is estimated by using Equation (6).

$V_d = 104,516.2 \, l$
$T = 1$ day
$\eta_a = 0.88$
$t =$ Assuming that the system is operated for a daily duration of 2 h.

$$Q = 104,516.2 \times \frac{1}{0.88 \times 2}$$

$$= 59,384.2 \, l/h$$

$$= \frac{59,384.2}{60 \times 60}$$

$$= 16.5 \, l/s$$

4.17 SPRINKLER IRRIGATION

The sprinkler irrigation system is used extensively for grain crops, such as wheat. The system designs are comparable to other micro-irrigation systems, but sprinkler systems usually need a higher flow rate per unit area. Typically, sprinkler systems have a single 40–75 l/h flow rate emitter per tree. Common (spray) emitters have slotted caps or deflector plates which circulate water in different streams. Other designs (spinners) have a moving component that turns to scatter the water stream more evenly over the wetted diameter.

A sprinkler irrigation system constitutes a network of pipelines and sprinklers. The pipelines carry water and supply it to all the working sprinklers at the proper pressure (Figure 4.24). The pressure head gets converted to velocity head at the nozzles of sprinklers. Water rushes out of the nozzle in the form of a stream. The stream or jet then breaks down into drops of water falling on the land surface and the foliage of plants like raindrops. The area of land wetted by a sprinkler depends on the velocity head given to the water jet, the angle of flow of the jet, the type of sprinkler and its design and the wind conditions during irrigation. The amount of water used is set to meet the water requirement of the crop precisely. When accurately planned, sprinkler irrigation can obtain high values of water distribution and water application efficiencies. It is possible to apply fine and frequent irrigations to keep a favorable soil moisture regime in the crop root zone, resulting in higher crop yields.

4.17.1 ADAPTABILITY OF SPRINKLER IRRIGATION

Sprinkler irrigation systems are suitable for nearly all soil types, topographic conditions, and all varieties of crops, particularly close growing crops, except rice and jute. These are, however, not ordinarily suitable for heavy clay soils with meager infiltration rates. Widespread field surface channels, which occupy a substantial land area (often 2% to 3%) for water distribution, are not needed. There is a significant saving of labor for irrigation. The system is especially suitable for sandy soils

FIGURE 4.25 A sprinkler irrigation system working in a hilly field

that have a high infiltration rate. Shallow soils that need to be appropriately leveled for surface irrigation methods can be irrigated harmlessly by sprinklers. The versatility of the sprinkler equipment and its effective control of water application makes this method applicable to most topographic conditions without comprehensive land preparation. It particularly fits for steep slopes or uneven topography (Figure 4.25). If soil erosion is a possibility, sprinkler irrigation can be used in combination with contour bunding, terracing, mulching and strip cropping.

Land leveling is not required for sprinkler irrigation systems. But some smoothing or grading is desirable if surface drainage is an obstacle or to produce a uniform surface for sowing, tillage, and harvesting. Small irrigation water streams may be used efficiently, and well-designed sprinklers dispense water properly. Surface runoff of irrigation water can be reduced. The quantity of water can be regulated to meet crop water requirements, and light applications can be made effectively on seedlings and young plants.

Water-soluble fertilizers and other chemicals like herbicides and fungicides can be economically introduced in the irrigation water and with limited additional equipment. Penetration of fertilizers into the soil can be regulated by administering the fertilizer at chosen times during the application of water. Sprinkler irrigation can be used to shield crops against frost and high temperatures that decrease the quantity and quality of the harvest. Labor expenses usually are less than for surface methods on soils with high infiltration rates and steep and rolling land. The irrigation method does not interfere with the movement of farm machinery.

4.17.2 Advantages

Sprinkler systems have certain benefits over other irrigation methods. Some of these have been listed below:

1. *Water savings:* Sprinkler systems utilize limited water than surface irrigation methods owing to their greater application efficiencies

Design and Components of Micro-Irrigation

2. *Freeze protection:* Sprinklers are frequently the favored micro-irrigation systems for tree crops because these give a higher degree of freeze protection than do drip systems
3. *Frequent applications:* The capacity to keep sufficient soil water levels by applying small quantities of water at frequent intervals during crucial growth stages is advantageous for various tree crops. Sprinkler systems can frequently be economically planned for automatic operation to incorporate features, such as real-time ET-based scheduling or include sensors to start or stop irrigations
4. *Wetted area:* Sprinklers are usually favored over drip systems in areas with coarse-textured soils where the lateral movement of soil water is restricted. The larger wetting patterns of sprinklers cover a larger root area, and these systems need fewer management efforts than drip systems
5. *Fertigation:* Sprinklers present an economical means of administering fertilizer, pesticides, and other agricultural chemicals on a well-timed basis
6. *Reduced evaporation:* Sprinkler systems on tree crops significantly reduce yearly water applications in contrast with flood irrigation, mainly due to higher application efficiencies without application to non-productive areas. These water savings are particularly apparent in a young orchard when trees are small and have restricted root zones
7. *Weed control:* The diminished wetted area of sprinklers in comparison to flood irrigation methods usually results in the added benefit of lower weed growth
8. *Flexibility:* Although sprinkler systems need higher flowrates/areas than do drip systems, the run time needed to apply an equal amount of water is less

4.17.3 Limitations

Sprinklers have some definite limitations compared with other irrigation methods, including:

1. *Management level:* Sprinkler irrigation systems demand a greater level of management skill than conventional irrigation methods owing to repeated operation, lower application rates, minute orifices, and wetting patterns that reach only a fraction of the root zone
2. *Maintenance:* Sprinkler systems need higher maintenance than surface irrigation systems. Besides, sprinkler stake arrangements are likely to be damaged or be relocated during regular orchard cultural operations, such as mowing, pesticide application, and harvesting
3. *Frequent operation:* Irrigation for sprinkler systems need to be scheduled more regularly than surface irrigation systems because the sprinklers dampen just a fraction of the root zone. Automation might be necessitated to use labor efficiently and meet crop water demands, where irrigation durations are comparatively small, and many irrigations per day are required
4. *Discharge into the air:* Since sprinklers release water jets into the air, the application losses usually are more significant than drip irrigation systems. In regions utilizing high-salinity irrigation water, trees may be salt-damaged from wind-blown water that reach the trunks, leaves, and flowers
5. *Cost:* Sprinkler systems are more expensive in terms of initial costs than similar drip systems because the discharge rate per unit area of the emitters is higher. Consequently, pumps, filters, and the piping network must be of greater capacities

4.17.4 Components of Sprinkler Irrigation

The various components of a sprinkler irrigation system are the pump to provide the needed pressure, the main pipelines and laterals, risers, and sprinkler heads (Figure 4.26). The construction materials of a sprinkler irrigation system vary. Among the principal parts, mains and laterals are made from aluminum, polyethylene and steel pipes. Aluminum has the advantage of being

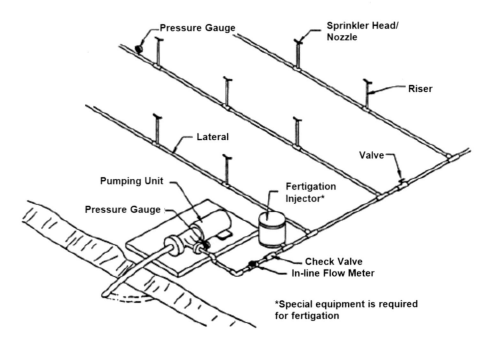

FIGURE 4.26 Components of a sprinkler irrigation system

lightweight and portable. Polyethylene pipes are more durable and flexible. For permanent mains, steel pipes are preferred. Sprinklers are mounted on riser pipes which are usually made of mild steel. Brass or similar alloys are used for the construction of sprinkler heads.

i. **Pumping Unit**

The pump lifts water from the source and forces it into the distribution system and the sprinklers. It might also be used to boost the flow of water in an existing water distribution pipeline (Figures 4.27–4.28) to push it into the sprinkler system at the necessary pressure. In all situations, the pump must be designed to lift the expected quantity of water from the supply source to the highest position in the field and maintain sufficient operating pressure. Volute centrifugal pumps, deep well turbine pumps, and submersible pumps can operate sprinkler systems. Electric motors and internal combustion engines drive the pumps operating the sprinklers. Electric motors are more suitable for fixed installations. Engines are used for portable pumping units and at sites where electricity is not available.

ii. **Main Lines**

Mains can be permanent or movable. Permanent main lines are best suited for fields having fixed boundaries and where crops need irrigation throughout the growing season. When a sprinkler system is used in several fields, portable systems are more economical. Permanent lines are buried in such a way that these do not interfere with farm operations. Steel, asbestos, and PVC pipes are used as permanent mains, while aluminum pipes are suitable for portable systems. The water from the mains enters the laterals at a junction through a valve or an L- or T-section placed on the mains.

iii. **Laterals**

The lateral lines usually are portable. In many orchards or tree nurseries, buried permanent laterals are used. Most portable laterals are made from quick-coupled aluminum pipes. Figure 4.28 shows a sprinkler mounted on the riser pipe fixed on a lateral pipe.

Design and Components of Micro-Irrigation

FIGURE 4.27 A fixed pumping unit attached to the sprinkler system

FIGURE 4.28 A lateral having risers and sprinklers irrigating a row crop

iv. **Sprinkler Heads**

The sprinkler head is the most crucial part of the sprinkler irrigation system (Figure 4.29). The sprinkler head's operational characteristics under optimum water pressure and climatic conditions define the efficiency of the system. The two primary techniques used to produce the spray expected for sprinkling are (i) revolving head sprinklers having one or more nozzles, depending essentially on the diameter of the wetted circle and (ii) pipe comprising lines of minute holes along its top. Agricultural sprinklers are mostly of the slow rotation type. They range from small single-nozzle sprinklers to giant multiple nozzle sprinklers that operate at high pressures. The main parts of a twin nozzle rotating sprinkler head are shown in (Figure 4.30).

Single-nozzle sprinklers (Figure 4.31) are used for low water application rates. The most commonly used sprinkler heads have two nozzles, one to apply water at a significant distance from

FIGURE 4.29 A sprinkler head in action

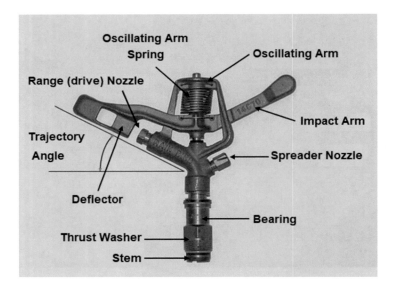

FIGURE 4.30 Details of a twin nozzle rotating sprinkler head

the sprinkler and another to cover the area near the sprinkler center. Two-nozzle sprinkler heads (Figure 4.31) apply water at higher rates than single-nozzle sprinkler heads; however, they offer good application uniformity at low pressures. "Pop-up" sprinkler heads (Figure 4.32) are especially acclimated for lawn irrigation. The sprinkler head is housed in a casing which is provided with a cover at the top. The top of the cover is nearly at the land surface. The cover opens using a retract spring when the sprinkler is in operation. The cover is closed when the sprinkler is shut off. Giant sprinklers (Figure 4.33) usually have one large nozzle with smaller extra nozzles to fill in pattern gaps. They are adaptable to close growing high plantings, such as sugarcane where the closer spacing of sprinklers is hard. They are confined to soils with high infiltration rates and are not fit in regions of considerable wind. Giant sprinklers have a low water distribution efficiency.

Design and Components of Micro-Irrigation

FIGURE 4.31 Single-nozzle sprinkler head

FIGURE 4.32 Pop-up sprinkler head

Revolving head sprinklers are classified according to the pressure range and their location with respect to irrigated crops. The classification of common types of sprinkler heads is given in Table 4.2. The optimum pressure must be provided for the sprinkler heads to be used. Very low or very high pressures for a particular sprinkler head may lead to poor water distribution patterns. Commonly, a device is used to tap the sprinkler head with a small hammer to rotate it. The hammer is activated by the force of water striking against a little vane connected to it. Certain sprinklers, particularly the high-pressure giant sprinklers, are rotated by water-activated gear drives.

FIGURE 4.33 Giant nozzle high-pressure sprinkler head

4.18 DESIGN OF SPRINKLER SYSTEM

The points considered during the designing of the sprinkler irrigation system are given as

- Inventory of area
- Depth of irrigation
- Irrigation interval
- Effect of wind
- Application rate for different soils and slopes
- Selection of sprinkler nozzles
- Spacing of sprinkler nozzles
- Number of sprinkler nozzles
- Total discharge of sprinkler system
- Layout of sprinkler system

4.18.1 Uniformity Coefficient of Sprinklers

A measured index of the degree of uniformity achievable for any sprinkler size working under given conditions is termed the uniformity coefficient (C_u). This uniformity coefficient is influenced by the pressure-nozzle size relations, the spacing of sprinklers, and wind conditions. The uniformity coefficient is calculated from field measurements of the depths of water collected in open cans placed at fixed intervals within a sprinkled area. It is represented by Equation (7) (Christiansen, 1942).

$$C_u = 100\left(1.0 - \frac{\sum X}{mn}\right) \qquad (7)$$

where
  X = numerical deviation of individual observations from the average application rate, mm
  m = average value of all observations (average application rate), mm
  n = total number of observation points

TABLE 4.2
Classification of Rotating Head Sprinklers

Type of Sprinkler	Pressure Range (kg/cm^2)	Sprinkler Discharge (l/s)	Diameter of Nozzles (mm)	Diameter of Coverage (m)	Range of Sprinkle Spacings (square) (m)	Recommended Speed of Sprinkler Rotations	Adaptability
Gravity-fed under-tree sprinkler systems	0.7 to 1.0	0.06 to 0.25	1 to 6	10 to 14	—	—	Usually uses single-nozzle sprinkler heads, used as under-tree systems in uplands, has low uniformity of coverage
Normal under-tree sprinkler systems	1 to 2.5	0.06 to 0.25	1.5 to 6	6 to 23	—	0.5 to 1 rpm	Usually uses single-nozzle sprinkler heads, used as under-tree systems in uplands, has low uniformity of coverage
Permanent overhead systems	3.5 to 4.5	0.2 to 0.6	3 to 6	30 to 45	18 to 30	1 rpm	Used for orchards. Triangular spacing necessary for low application rates (1.5mm/h to 3 mm/h)
Small overhead systems	2.5 to 4	0.6 to 2.0	6 to 10	25 to 35	9 to 24	0.67 rpm to 1 rpm	Commonly used for a low rate of application (3.5mm/h to 6 mm/h) and to help reduce the effects of wind. High risers are used for orchards and lower risers for field crops
Low-pressure systems	1.5 to 2.5	0.3 to 1	3 to 6	20 to 35	9 to 18	0.5 rpm to 1 rpm	Two-nozzle sprinklers can be used with lower pressures than single-nozzle sprinklers. More overlap is required. Rate of application tends to be high
Intermediate pressure systems	2.5 to 5	2 to 10	10 to 20	40 to 80	24 to 54	0.7 rpm	Usually, single-nozzle sprinklers, rates of application range from 6 to 12 mm/h, suitable for supplemental irrigation, unsuitable under windy conditions
High-pressure systems	5 to 10	10 to 50	20 to 40	80 to 140	54 to 100	0.5 rpm	Used in dense high plantings. Unsuitable where there is appreciable wind

A uniformity coefficient of 100% (obtained with overlapping sprinklers) is suggestive of completely uniform application, whereas the water application is less uniform with a lower percentage. A uniformity coefficient of 85% or more is supposed to be satisfactory.

The data on the uniformity coefficient is helpful as a base for deciding the combination of spacings, discharge, nozzle size, and operating pressure to achieve high values of irrigation efficiency at particular operating conditions. The preeminent producers of sprinkler irrigation equipment typically provide data on the diameter of coverage of each nozzle size and the pressure needed at the sprinklers.

Example 16: Determine the uniformity coefficient from the following data obtained from a field test on a square plot bounded by four sprinklers:

Sprinkler: 3.135×2.261 mm nozzles at 2.6 kg/cm
Spacing: 25 m × 25 m
Wind: 3.8 km/h from southwest
Humidity: 42%
Testing time: 1.0 h

S	7.3	8.5	9.6	S
7.6	9.5	9.0	8.6	8.0
7.4	8.9	9.3	10.8	10.2
8.0	7.2	8.7	8.7	9.7
S	10.4	8.1	10.9	S

Solution: The calculations are given in the following table.

Observation	Numerical Deviation
7.6	1.3
7.4	1.5
8	0.9
7.3	1.6
9.5	0.6
8.9	0
7.2	1.7
10.4	1.5
8.5	0.4
9	0.1
9.3	0.4
8.7	0.2
8.1	0.8
9.6	0.7
8.6	0.3
10.8	1.9
8.7	0.2
10.9	2
8	0.9
10.2	1.3
9.7	0.8
mn=**186.4**	**ΣX=19.1**

$$\text{Mean} = \frac{186.4}{21} = 8.9$$

Using Equation (7)

$$C_u = 100\left(1.0 - \frac{19.1}{186.4}\right)$$

$$= 89.75\%$$

Design and Components of Micro-Irrigation

v. Debris Screens

Debris screens are typically required when surface water is utilized as the source of irrigation (Figs. 4.34–4.35). The purpose of screens is to keep the arrangement clear of trash that might plug the sprinkler nozzles. Screens should be sufficiently fine to trap weed seeds and other small particles. They may also be needed to check sticks, plant stalks, and other large debris from entering

FIGURE 4.34 Debris screen for irrigation water

FIGURE 4.35 Settling pond or desilting basin

the system. Two or more screens of progressively finer mesh can be employed when heavy loads of trash are expected. The collected waste needs to be removed from the screens before water flow to the pump is blocked.

vi. Desilting Basins

Desisting basins are needed to catch sand or suspended silt when the water sources are streams, open ditches, or well water containing silt (Figure 4.35). Often desilting basins and debris screens are made as a combined structure. Desilting basins should be sufficiently large to protect for at least one full day. Higher capacities are desirable.

vii. Booster Pumps

Booster pumps are used when a sprinkler irrigation system is used with an existing pumping system established in a well and the pump capacity is inadequate to force the water through sprinklers. These provide sufficient pressure for small fields that are located at elevations above the main area to be irrigated. The use of the booster pump in these conditions eliminates the necessity to provide high pressures from the main pumping plant for the comparatively smaller part of the total discharge that is required on the elevated area.

viii. Take-off-Valves

Take-off valves usually are required to regulate pressures in the laterals. The valves should perpetually be used in systems where there are notable differences in main line pressures at the different lateral take-off points. In multiple-lateral systems, they also allow moving a single lateral without shutting down the complete system.

ix. Flow Control Valves

Flow control valves to control the pressure and flow of individual sprinklers may be necessary when the unevenness of the terrain causes an irregular distribution of pressure along the lateral. They are rarely needed on level areas or flat slopes.

4.18.2 CAPACITY OF SPRINKLER IRRIGATION SYSTEMS

The required capacity of a sprinkler system depends on the size of the area to be irrigated (design area), the gross depth of water applied in each irrigation, and the net operating time allowed to apply water to this depth. The capacity of the system may be calculated by Equation (8).

$$Q = 2,780 \frac{A \times d}{F \times H \times E} \tag{8}$$

where
  Q = discharge capacity of the pump, l/s
  A = area to be irrigated, ha
  d = net depth of water application, cm
  F = number of days allowed for the completion of one irrigation
  H = number of actual operating hours per day
  E = water application efficiency, %

From Equation (8), it is clear that the greater the product of these two factors (operating time) the smaller is the system capacity (and thereby the cost) for a given area.

Design and Components of Micro-Irrigation

Example 17: Determine the system capacity for a sprinkler irrigation system to irrigate 15 ha of a certain crop. The design moisture use rate is 4 mm per day. The moisture replaced in soil at each irrigation is 6 cm. The irrigation efficiency is 78%. The irrigation period is 8 days in a 10-day interval. The system is to be operated for 15 h per day.

Solution: Using Equation (8) the capacity of the sprinkler system be calculated as:

A = 15 ha
d = 6 cm
F = 8 days
H = 18 h per day
E = 78%

$$Q = 2{,}780 \frac{15 \times 6}{8 \times 18 \times 78}$$

$$= 22.3 \text{ l/s}$$

QUESTIONS

1. Define micro-irrigation.
2. Discuss the potential of micro-irrigation in India.
3. Enlist different methods of micro-irrigation.
4. Enlist the components of drip and sprinkler irrigation methods.
5. Write a short note on:
    i) Uniformity coefficient
    ii) Design criteria of micro-irrigation
6. Describe the advantages of drip irrigation. Also, explain the design steps of drip irrigation.
7. What is a sprinkler irrigation system? Describe design steps of sprinkler irrigation system.
8. Write a short note on:
    i) Rain gun
    ii) Fogger
    iii) Emitter
    iv) Lateral
    v) Filter
    vi) Venturi meter
9. Determine the system capacity for a sprinkler irrigation system to irrigate 10 ha of maize crop. Design moisture use rate is 5 mm per day. The moisture holding capacity of soil is 18 cm/m depth and the depth of irrigation is 12 cm. Irrigation efficiency is 70% and the irrigation period is 10 days in 12 days interval. The system is to be operated for 20 h per day. **(0.01190 m³/s)**
10. A sprinkler irrigation system is designed to deliver a design daily irrigation requirement of 7 mm and a desired depth of 15 mm. Ten 300-m-long intervals with sprinklers in a 15 m square spacing pattern are operated simultaneously to irrigate 25 ha field. Determine the maximum time between successive irrigations and the sprinkler system capacity required for a set length of 8 h. Assume that 1 h in each set is required to move each lateral and an application efficiency is 80%. **(54 h)**
11. Find application uniformity and system efficiency. The given data are coefficient of variation is 0.12, no of emitters is 1, and application efficiency is 88.5%. **(75.04%)**
12. At a lateral length of 50 m, a total of 5 sprinkler heads have been installed. Determine the discharge at the last sprinkler head, if
    a) Discharge rate of the first sprinkler is 0.015 m³/s

b) Pressure at the first sprinkler is 3.0 kg/cm^2
c) Pressure at the last sprinkler is 2.25 kg/cm^2
13. Determine the required capacity of a sprinkler system to apply water at the rate of 1.25 cm/h. Two 186-m-long sprinkler lines are required. Sixteen sprinklers are spaced at 12 m intervals on each line. The spacing between lines is 18 m.
14. Discuss micro-irrigation in detail. What are the common causes for cogging of drippers?

MULTIPLE-CHOICE QUESTIONS

1 The rainfall intensity is the ratio of
   a) Time to rainfall amount
   b) Rainfall amount to time
   c) Rainfall amount to falling velocity
   d) Both (b) and (c)
2 The suction lift of a pump depends upon
   a) Atmospheric pressure
   b) Water temperature
   c) Velocity of water in the suction pipe
   d) All the above
3 The pH value of water fit for drinking is
   a) 13
   b) 11
   c) 9
   d) 7
4 Air inlet valve in water mains is generally provided at
   a) Summit of the pipe
   b) Upstream of sluice valve
   c) Downstream of sluice valve
   d) Both (a) and (c) of above
5 If the speed of a centrifugal pump is doubled, the power required will be increased by
   a) 2 times
   b) 4 times
   c) 6 times
   d) 8 times
6 What is the net depth of irrigation in a sprinkler system with a root zone depth of 60 cm and available water holding capacity of 9 cm/m depth? If irrigation will be done when the moisture depletion 50% of available moisture
   a) 2.7 cm
   b) 2.5 cm
   c) 2.0 cm
   d) 2.3 cm
7 The overlap of sprinklers increases with _____ in wind velocity
   a) Decreases
   b) Increases
   c) Lowering
   d) None of the above
8 The water saving to the tune _____ of percent is possible in sprinkler irrigation system
   a) 30 to 50
   b) 60 to 65
   c) 55 to 85
   d) 70 to 80

9 The first international drip irrigation meeting was held at _____ in 1971
   a) Israel
   b) France
   c) India
   d) USA
10 In water saving to the tune of _____ % is possible in drip irrigation system
   a) 30–70
   b) 80–90
   c) 20–40
   d) 10–30

ANSWERS

1	2	3	4	5	6	7	8	9	10
b	d	d	d	d	a	c	a	a	a

REFERENCES

ASAE. 2001. *ASAE Standard S526.2 JAN01, Soil and Water Terminology.*
Christiansen, J. E. 1942. *Irrigation by Sprinkling.* University of California Agricultural Experiment Station Bulletin.
Clark, N. 1874. *Improvements in Irrigation Pipe.* U.S. Patent Office. Patent No. 146,572, issued January 20. Washington, DC.
USDA. 1984. *Trickle Irrigation.* In *National Engineering Handbook.* US Department of Agriculture, Washington, DC, p. 129.

5 Application of Plastic in Water Management

5.1 INTRODUCTION

Water is the most precious gift of nature, is essential for human and animal life (Clothier, 1990), and plays an important role in plant growth (Kumar *et al.*, 2012). Unlike most resources, there is no substitute for water, it plays a crucial role in human life and for all kinds of development. Water resources play an important role as a catalyst for the economic development of a nation. Therefore, it is necessary to develop, conserve, utilize, and economically manage this critically important resource on an integrated basis so as to meet the ever-growing demand for agriculture, industry, domestic use, and power generation. To meet the food security and income and nutritional needs of the projected population in 2020 the food production in India will have to be almost doubled (Meena *et al.*, 2019). Out of 400 million ha-m precipitation in India, 70 million ha-m is consumed as evapotranspiration and 215 million ha-m infiltrates into the soil, whereas 115 million ha-m is lost as runoff, resulting in drought in the catchment, on the one hand, and flooding downstream, on the other hand. Harvesting of one-fourth of this 115 million ha-m runoff water in farm ponds can provide three irrigations to the entire rain-fed area, which constitutes two-thirds of the cultivated land of the country.

The need of the hour is, therefore, maximizing the production per drop of water. Hence in the present context, a lot of emphases are given to water management by improving the irrigation practices to increase crop production and sustain productivity levels. For increasing the water use efficiency by crops, plastic plays a very important role in micro-irrigation, moisture conservation, canal lining, water harvesting, and underground pipeline system. In agriculture, the applications of plastics are endless.

5.2 MICRO-IRRIGATION

It is introduced primarily to save water and increase the water use efficiency in agriculture. Reduction in water consumption due to micro-irrigation over the surface method of irrigation varies from 30% to 70% and gain in productivity in the range of 20% to 90% for different crops. By introducing micro-irrigation, it is possible to increase the yield potential of crops threefold with the same quantity of water (Goel *et al.*, 2002; Meena *et al.*, 2019; Meissner *et al.*, 2020). The plastic material is widely used in making micro-irrigation setups because it has a lot of advantages over other materials like being light in weight, easily transportable, resistant to corrosion, readily available, less friction at the pipe walls for water flow, etc. Micro-irrigation systems ensure uniform water application. Therefore, all plants in a field receive an equal amount of water. Higher uniformity results in efficient irrigation, thereby, causing less wastage of water, power, and fertilizers. Consistent water application results in better and uniform crop yields as each plant is given the required amount of water and nutrients for optimum growth.

Micro-irrigation systems have many potential advantages when compared with other irrigation methods. Micro-irrigation techniques have materialized the concept of "more crop per drop", by ensuring the availability of adequate quantity and quality of water especially in dryland agriculture where water is the most limiting factor in crop production. The advantages and the design of micro-irrigation in detail have been given in Chapter 4.

5.3 MOISTURE CONSERVATION

Biodegradable plastics, as defined by the American Society for Testing and Materials, are "degradable plastic in which the degradation results from the action of naturally occurring microorganisms such as bacteria, fungi and algae" (Mooney, 2009). Plastic film mulch is called "mulch" even though its appearance as a thin sheet of plastic film does not resemble traditional types of organic mulch, such as straw or leaves. It is purchased in rolls and applied to the length of field rows to seal the upper layers of the soil. Crops are allowed to grow up through the holes that are cut into the plastic. The methods used range from fully mechanized installation to hand tools. Some of the desired outcomes from plastic mulch include increased yield, conserving water, weed suppression, rain shedding, promoting earliness in ripening or compressed cropping cycles, enhanced germination, agrochemical fumigation, greater variety of crop types, and reduced soil erosion.

Plastic mulch is practically impervious to carbon dioxide, a gas that is of prime importance in photosynthesis. Very high levels of carbon dioxide buildup under the plastic, because the film does not allow it to escape. It has to come through the holes made in plastic for the plants and a "chimney effect" is created, resulting in localized concentrations of abundant carbon dioxide for actively growing leaves. The evaporation loss from plastic-mulched soil can be half that of non-mulched soil. Therefore, the total remaining soil moisture loss under plastic-mulched conditions is primarily through transpiration and percolation. However, the growth rate of plants in mulched-bed systems can be faster than plants growing in bare soil, and the larger plants will use more water per unit area. Even so, the overall amount of water used per unit of production will be less in a mulched system. Opaque plastic mulches also suppress weed growth and reduce the use of water by weed plants, further improving the water use efficiency. The use of plastic mulch conserves soil moisture with the conjunction of the drip irrigation system. To observe crop yield and water conservation concurrently in a field trial, a hole must be made in the plastic either through mechanically injecting a seed through the plastic and into the soil, through manual cutting of the plastic to allow for the seeds that were previously planted to emerge, or by cutting transplant holes into the plastic.

The physical features of the mulching film are obviously supposed to keep the crop root zone moist and protect the soil from wind and water erosion (Cooper *et al.*, 1987). The most important aspect of promoting a crop's water use efficiency derives from changing the balance between evaporation and transpiration under conditions of limited water (Rockstrojm, 2003). Peters and Johnson (1962) showed that mulching film reduced evaporation water losses in groundnut. Abu-Awwad (1998) has stated that mulching the soil surface lessened the irrigation requirement in pepper by about 14–29% with the purpose of eliminating soil evaporation. Mulching the soil surface beneficially influences the soil-water content by controlling the water evaporation by management of soil surface, which improves soil-water retention (Anikwe *et al.*, 2007). Different researchers have indicated that plastic mulching can favorably influence soil evaporation among plants during their seeding stage. As the coverage increases from 35% to 90%, the soil evaporation reduces by 51.8% when compared with that of naked soil (Men *et al.*, 2003).

5.4 CANAL LINING

Placing a plastic liner on the canal bed can prevent seepage of irrigation water in the earthen canals prior to irrigation season (Salaev *et al.*, 2014). The lining of canals is a well-tested remediation method for minimizing seepage loss from canals. Canal lining is an impermeable layer provided for the bed and sides of the canal to improve the life and discharge capacity of the canal. Approximately 60% to 80% of water lost through seepage in an unlined canal can be saved by construction canal lining (Figure 5.1). The plastic lining of the canal is a newly developed technique and holds good promise. There are three types of plastic membranes which are used for canal lining, namely:

Application of Plastic in Water Management

FIGURE 5.1 Canal lining using plastic (Salaev *et al.* 2014)

- Low-density polyethylene
- High molecular high-density polythene
- Polyvinyl chloride

The advantages of providing plastic lining to the canal are many such as plastic is negligible in weight; is easy to handle, spread, and transport; is immune to chemical reaction; and helps speedy construction. Plastic canal lining has the following advantages:

- Seepage reduction
- Prevention of waterlogging
- Increase in commanded area
- Increase in channel capacity
- Less maintenance
- Safety against floods

5.4.1 Seepage Reduction

The main purpose behind the lining of the canal is to reduce seepage losses. In some soils, the seepage loss of water in unlined canals is about 25% to 50% of the total water supplied. The cost of canal lining is high but it is justifiable for its efforts in saving most of the water from seepage losses.

5.4.2 Prevention of Waterlogging

Waterlogging is caused due to phenomenal rise in the water table due to uncontrolled seepage in an unlined canal. This seepage affects the surrounding groundwater table and makes the land unsuitable for irrigation. So, this problem of waterlogging can be surely prevented by providing proper lining to the canal sides.

5.4.3 Increase in Commanded Area

The commanded area is the area which is suitable for irrigation purposes. The water-carrying capacity of the lined canal is much higher than the unlined canal and hence more area can be irrigated using lined canals.

5.4.4 Increase in Channel Capacity

Canal lining can also increase the channel capacity. The lined canal surface is generally smooth and allows water to flow with high velocity compared to the unlined channel. The higher the velocity of flow, the greater is the capacity of the channel, and hence channel capacity will increase by providing lining. On the other side, with this increase in capacity, channel dimensions can also be reduced to maintain the previous capacity of the unlined canal, which saves the cost of the project.

5.4.5 Less Maintenance

Maintenance of lined canals is easier than unlined canals. Generally, there is a problem of silting in unlined canals, which requires huge expenditure; but in the case of lined canals, because of high velocity of flow, the silt is easily carried away by the water. In the case of unlined canals, there is a chance of growth of vegetation on the canal surface but not in the case of lined canals. The vegetation affects the velocity of flow and the water-carrying capacity of the channel. The lined canal also prevents damage to the canal surface due to rats or insects.

5.4.6 Safety Against Floods

A lined canal always withstands floods while an unlined canal may not resist and also there is a chance of a breach, which damages the whole canal as well as surrounding areas or fields.

5.4.7 Advantages of Canal Lining

By lining the canal, the velocity of the flow can increase because of the smooth canal surface. For example, with the same canal bed slope and with the same canal size, the flow velocity in a lined canal can be 1.5 to 2 times that in an unlined canal, which means that the canal cross section in the lined canal can be smaller to deliver the same discharge. The possible benefits of lining a canal include:

- Water conservation
- No seepage of water into the adjacent area
- Reduced canal dimensions and
- Reduced maintenance

An important reason for lining a canal can be the reduction in water losses, as water losses in unlined irrigation canals can be high. Canals that carry from 30 l/s to 150 l/s can lose 10% to 15% of this flow by seepage and water consumption by weeds. Lining a canal will not completely eliminate these losses, but roughly 60% to 80% of the water that is lost in unlined irrigation canals can be saved by a hard-surface lining. If canal banks are highly permeable, the seepage of water will cause very wet or waterlogged conditions or even water stagnation on adjacent fields or roads. The lining of such a canal can solve this problem, since the permeability of a lined canal bank is far less than that of an unlined bank, or may even be zero, depending on the lining material. The plastic lining may prove most effective in the soils where concrete lining would result in cracks.

Using canal lining technology, the average performance ratio can be increased from 40% to 90% during the growing season. The results showed that plastic lining is effective in reducing water loss due to seepage, and groundwater levels decreased in the plastic-lined canal. However, though the plastic lining remained effective in reducing water seepage, as seen in the consistently lower groundwater level, it could not counteract the movement of groundwater from nearby areas. Reduced seepage or infiltration of irrigation water allowed saving an enormous amount of water.

5.5 WATER HARVESTING

Drought-prone areas have to be provided with water not only for human and cattle consumption but also for irrigation. However, even after good monsoons, water is not available only because of the lack of proper water management and storage. The cheapest and purest source of water is rainwater. Harvesting of the water in ponds, lakes, wells, tanks, and reservoirs helps to preserve this water so that it can be used during the lean period. Vast amounts of water are lost through seepage, especially where the soil is gravelly and porous. It is estimated that 70% of water is lost between the storage and usage point. Many states have been experiencing drought resulting in a shortage of water, particularly during the summer months.

The introduction of plastic sheets in the form of low-density polyethylene (LDPE) film below the hardcover lining is effective in improving the water tightness of the lining. The need for water during the rainy season is minimal, and therefore, the water available from precipitation from rivers/streams is stored in ponds and reservoirs for various domestic, agricultural, and industrial purposes over a period of time. In the areas of acute shortage of water, ponds and reservoirs sometimes form the only source of water. However, a good portion of water stored in these reservoirs or ponds is lost by evaporation, seepage, and temperature.

To conserve water stored in ponds and reservoirs, means must be found to minimize seepage and evaporation losses. Even in the areas receiving very low rainfall, about 200 m^3 to 300 m^3 of water is reported to be possible for collection from each hectare of land, which otherwise goes waste by runoff. Harvesting and storing of this water if done in dugout farm ponds will not only help in reducing the runoff but will also help in providing protective irrigation to a portion of the catchment area. If protected against seepage, it would have the potential to support a short-duration crop. The seepage and other losses in normal unlined ponds in sandy loamy soils are reported to be to an extent of 5 cm/day to 15 cm/day. If this can be minimized by way of lining the walls and floor of the ponds with LDPE films or any other such material, the water could be well utilized. Conventional lining methods like brick and tile lining are either too expensive or not sufficiently effective. Since the water in the reservoir/pond is stagnant, LDPE film lining with 30–60 cm soil cover is adequate for bed lining. Depending on the circumstances and soil conditions, single-tile lining with the LDPE film or double-tile lining or LDPE film lining with soil cover can be used on the sides. Polyethylene film lining alone or in combination with conventional lining has proved to be a seepage-proof barrier between the soil and the water.

Owing to distinct topology, soil characteristics, and rainfall patterns, low-density polyethylene (LDPE) lined tanks are more suitable for hilly regions. Majority of hill farmers are small and marginal with scattered land holding who cannot afford to construct cement tanks. Poly tanks are low-cost structures as compared to cement tanks. The cost of construction of cement tanks is typically 5–6 times more as compared to a similar capacity poly tank. Landslides and earthquakes are common features of the hilly region and the risk of cracking in cemented tanks is very high. Poly tanks adjust their shape according to minor disturbances in land due to their flexible nature and are ideal under this kind of setup. Huge loss of water is involved in the long-term storage of water in the "kachha" tank. The lining of the tank with LDPE film provides a better option to reduce the seepage losses from the storage.

5.6 FARM PONDS

Farm ponds are small tanks or reservoirs constructed for the purpose of storing water essentially from surface runoff. Farm ponds are useful in providing supplemental irrigation to the crops and supply of water for domestic use and for livestock. A large number of ponds constructed in the catchment area will certainly reduce the flood flows downstream. The design and construction of farm ponds require a thorough knowledge of the site conditions and requirements. Some sites are ideally suited for locating the ponds and the advantage of natural conditions should always be taken.

5.6.1 Types of Ponds

Depending on the source of water and their location with respect to the land surface, farm ponds are grouped into four types as given below.

1. Dugout ponds or excavated ponds
2. Embankment-type farm ponds
3. Spring or creek-fed ponds
4. Off-stream storage ponds

5.6.1.1 Dugout Farm Ponds

Dugout ponds are excavated at the site and the soil obtained by excavation is formed as an embankment around the pond. These ponds are generally constructed in the area having flat topography. It is advantageous in the area where evaporation losses are high because it can be constructed to expose minimum water surface area in proportion to volume. These farm ponds are generally used where only a small supply of water is required. The size of the pond depends on the catchment area, soil type, and expected runoff. The pond could be fed by either surface runoff or groundwater wherever aquifers are available. In the case of dugout ponds, if the stored water is to be used for irrigation, the water has to be pumped out. The side slopes of the dugout pond should not be steeper than the natural angle of repose of material being excavated. Generally, the side slopes should be flatter than 1:1. The dugout ponds are usually constructed in a rectangular shape.

5.6.1.1.1 Estimation of Volume of a Pond

The volume of excavation required can be estimated accurately by

$$V = \frac{A + 4B + C}{6} \times D$$

where
 V = Volume of excavation (m³)
 A = Area of excavation at the ground surface (m²)
 B = Area of excavation at the mid depth point (D/2) in m²
 C = Area of excavation at the bottom of the pond (m²)
 D = Average depth of the pond (m)

Example 1: Estimate the volume of excavation required to construct a dugout pond with an average depth of 2.5 m, a bottom width of 9.0 m, and a bottom length of 20 m. The side slope adopted is 2:1.
 Solution
 The volume of excavation required

$$V = \frac{A + 4B + C}{6} \times D$$

Application of Plastic in Water Management

Top length = 20 + (2.5 × 2) × 2 = 30 m
Top width = 9 + (2.5 × 2) × 2 = 19 m
A = 30 × 19= 570 m²
Mid length = 20 + (1.25 × 2) × 2 = 25 m
Mid width = 9 + (1.25 × 2) × 2 = 14 m
4B = 25 × 14 × 4= 1,400 m²
C = 20 × 9 =180 m²
A + 4B + C = 570 + 1,400 + 180 = 2,150 m²
V = 2,150/6 × 2.5 = 909.16 m³

5.6.1.2 Embankment-Type Farm Ponds

This type of pond consists of an earthen embankment or earthen dam which is partly excavated and an embankment is constructed to retain the water. It is the most common type of farm pond and generally constructed across the stream or nalla. Generally, a site which has a depression already is chosen for pond construction. The pond is fed by surface runoff from its catchment area. The earthen embankment, mechanical spillway, and emergency spillway are parts of the pond. It is desirable to take the water out of the pond through a gravity outlet for irrigation. Selection of a suitable site for this purpose is, therefore, important. These farm ponds are usually constructed in areas where land slope ranges from gentle to moderately steep and also where stream valleys are sufficiently depressed to permit a maximum storage volume with the least amount of earthwork. It is the most common type of farm pond. All the embankment-type ponds are constructed with suitable spillway arrangements. These farm ponds are further divided depending upon their adaptability, topographical features, and usefulness.

5.6.1.3 Spring- or Creek-Fed Ponds

Spring- or creek-fed ponds are those where a spring or a creek is the source of water supply to the pond. These ponds are generally found in hilly areas where natural springs are available. The construction of these ponds depends upon the availability of natural springs or creeks.

5.6.1.4 Off-Stream Storage Ponds

Off-stream storage ponds are constructed by the side of streams which flow only seasonally. The idea is to store the water obtained from the seasonal flow in the streams. Suitable arrangements need to be made for conveying the water from the stream to the storage ponds.

5.6.2 Design of Farm Ponds

The design of the farm pond consists of the following parameters

i) Site selection
ii) Determining the capacity of the pond
iii) Design of the embankment
iv) Design of mechanical spillway
v) Design of emergency spillway
vi) Provision for seepage control from the bottom
vii) Rainfall characteristics

The behavior of rainfall is the main factor for surface runoff and infiltration of water into the soil. The higher the rainfall intensity, the greater will be the surface runoff. Again, higher rainfall intensity for shorter spells or lower intensity with longer spells leads to higher infiltration of water into the soil.

5.6.2.1 Site Selection

Farm ponds are simple to construct, easy to maintain, and require little technical knowledge. Selection of site is one of the important steps in the construction of a farm pond. The following points should be kept in mind during the selection of site:

- The site should be at a place where the maximum amount of rainwater can be collected. It is better to construct near any building so that water from the roof of the building can be harvested and easily stored in the poly tank
- The site should be as flat as possible and should not be prone to erosion
- The poly tank should be constructed at least 2 m above the irrigation area so that sufficient head for drip irrigation system can be easily achieved through gravity
- The site should be out of reach from animals and children
- A proper spillway site should be available for the safe disposal of excess water
- The soil should be impervious enough to prevent excessive seepage losses
- Large areas of shallow water should be avoided as it will cause excessive evaporation losses and waterweed to grow

Adequate slope of the sidewall of the tank is a very important factor in the construction of a poly tank. The main purpose of providing a slope in the sidewall is to divert the pressure of water through the sidewall of the tank unless the polythene sheet may be punctured or torn. Normally, the slope of the sidewall is kept at 1:1. In hilly areas, soil depth is comparatively less, so the depth of the tank should not be more than 1.5 m. Construction of tanks with higher depths is tough and expensive.

5.6.2.2 Capacity of the Pond

The capacity of the pond is determined by the size of the catchment and command area in the case of a surface runoff pond. For runoff ponds, the capacity should be determined by the irrigation demand of the command area or runoff yield of a catchment plus provision of evaporation and conveyance losses. The shape of a pond is normally kept as trapezoidal with a side slope of 1:1. The capacity of the pond is determined from a contour survey of the site at which the pond is to be located. The pond capacity is determined with the help of a contour plan of the watershed area where the pond is to be located. From the contour plan of the site, the capacity is computed for different stages using the trapezoidal or Simpson's rule. Simpson's rule gives more accurate values than the trapezoidal formula. The area enclosed by each contour is measured by a piece of equipment called "planimeter." According to trapezoidal rule, the volume V between two contours at an interval H and having areas A1 and A2 is given by

$$V = H/2(A1 + A2)$$

Similarly, by Simpson's rule, volume V is determined by the formula

$$V = H/3(\text{twice the area of odd contour}) + (4 \text{ times the area of even contour})$$
$$+ (\text{Area of first and contours}).$$

The Simpson's rule is also known as the Prismoidal rule. This method is having a limitation that the number of contours should be in odd numbers, i.e., the number of intervals used should be even. A depth capacity curve of the pond is also prepared using different water level depths. This curve is useful in deciding a suitable embankment height with respect to the available capacity of the pond. The effective height of the embankment should be decided on the basis of values of runoff expected from the catchment area and the volume of the water required to be stored. The expected runoff volume should be greater than the designed storage capacity; otherwise, the pond is said to be overdesigned (Singh, 2000).

Example 2: Calculate the volume of water stored in a farm pond, given that the area enclosed by different contours at the site are as follows:

S. No.	Contour Value (m)	Area Enclosed (m²)
1	100	215
2	101	260
3	102	315
4	103	380
5	104	460
6	105	525
7	106	610

Solution: The volume of the farm pond may be calculated by using trapezoidal or Simpson's formula

$$H = \text{Contour interval} = 1.0\text{m}$$

USING TRAPEZOIDAL FORMULA

$$V = H\left\{\frac{\text{Area of 1st contour} + \text{Area of last contour}}{2} + \text{Area of remaining contours}\right\}$$

$$V = 1.0\left\{\frac{215+610}{2} + 260 + 315 + 380 + 460 + 525\right\}$$

$$= 2352.5\text{m}^3$$

USING SIMPSON'S FORMULA

$$V = H/3\{4(\text{Area of even contours}) + 2(\text{Area of odd contours})$$

$$+ \text{Area of first and last contour}\}$$

$$V = 1 \times 100/3\left[4(2.6 + 3.8 + 5.25) + 2(3.15 + 4.6) + 2.15 + 6.1\right]$$

$$V = 1 \times 100/3(70.35) = 2345\text{m}^3$$

5.6.2.3 Design of Embankment

The embankment design of the farm pond consists of the following parameters, which are usually considered for design purposes.

5.6.2.3.1 Foundation

The foundation should be such that it can provide stable support and resistance to the passage of water. Good foundation materials should be used, as they provide both stability and imperviousness to the foundation. A mixture of coarse- and fine-textured soils like gravel-sand-clay mixtures, gravel-sand-silt mixtures, sand-clay mixtures, and sand-silt mixtures will be good foundation materials. Coarse-textured materials like gravel and sand provide good support but it is highly pervious to water. Fine-textured materials such as silts and clays are comparatively impervious to water but they are not very stable materials for foundation. For higher dams, if the foundation materials are not satisfactory, the base width is to be increased so that the load per unit area decreases.

5.6.2.3.2 Cross Section

Cross section of the earthen embankment depends upon the nature of foundation materials as well as the nature of fill materials available at the site. The materials used for embankment construction should be fine and impervious. If it is not available then an impervious core and a cutoff trench should be provided in the embankment to control the seepage. Different embankments are shown in Figure 5.2.

5.6.2.3.3 Side Slope

The side slopes are dependent upon the height of the dam, the nature of the foundation material, and the nature of the fill material. The side slope which is commonly used up to 15 m height for clay soil is 2:5:1 and 2:1 on the upstream and downstream side, respectively, whereas, for sandy loam soils, the side slope of 3:1 and 2.5:1 is usually provided on the upstream and downstream side, respectively. Berms are provided embankment higher than 10 m, on the downstream side of the dam. To protect the embankment from the wave action of water and to avoid overtopping, the upstream slope of the embankment should be flatter than the downstream slope.

5.6.2.3.4 Freeboard

The freeboard is the additional height of the dam provided as a safety factor to prevent overtopping due to wave action and unexpected runoff. It is the vertical distance between the desired water elevation and the elevation of the top of the dam after settlement. Freeboard is generally taken 15% of the depth of flow over the crest. The net freeboard is the sum of the depth of flow, freeboard, and settlement. The net freeboard should be sufficient to prevent the stored water from reaching the top of the embankment due to wave action. Wave height of moderate size embankment is determined using Hawksley's formula;

$F = 1.5\ hw$
$hw = 0.014\ (Dm)^{1/2}$
F = Freeboard in m

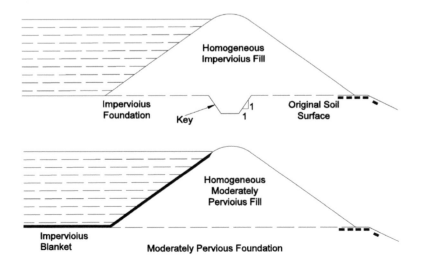

FIGURE 5.2 Different embankments

Application of Plastic in Water Management

where
  hw = wave height in m
  Dm = Fetch length in m

For the design of earthen embankment up to 5 m heights, the depth of flow can be taken between 0.5 m and 0.7 m. The net freeboard as well as wave height freeboard should be determined separately using the above formula.

5.6.2.3.5 Allowance for Settlement

The earthen dam will settle after the construction, and allowances for this settlement should be made. The amount of settlement depends upon the method of construction and foundation material used. Settlement up to 25% can occur when the construction is done by a dragline. The design height of the embankment should be increased by an amount equal to 5% of the depth of flow and freeboard as settlement allowance.

5.6.2.3.6 Stone Pitching

It is a protection provided to the upstream sloping face of the dam. This protection extends from the natural ground surface to the top of the dam excluding top width.

5.6.2.3.7 Riprap

It is also a protection provided to the sloping faces of upstream and downstream of the dam to protect the slopes from waves, men, cattle, rain, etc.

5.6.2.3.8 Cutoff

It is a fairly impervious barrier provided at the foundation in the center of the base of an earth dam. There is a flow of water in the foundation under pressure. The cutoff tries to check this flow of water and increases the path of percolation.

5.6.2.3.9 Design Steps and Computation of Earthwork

The design of the embankment involves the prediction of both the peak rate of runoff and runoff yield. The cross section of the nalla where the embankment is proposed to be constructed is plotted on the map for the computation of earthwork. The sample calculation for earthen embankment design is as under.

Example 3: Design on earthen embankment with the following data

Catchment area	=	20.2 ha
Intensity of the rainfall, I_{tc}	=	17.5 cm/h
RL of natural surface	=	100.00
Reduced level (RL) of Highest flood level (HFL)	=	103.00
Runoff coefficient, C	=	0.30
Soil type	=	Sandy loam
Slope of saturation line	=	4:1

Solution: Peak rate of runoff

$$Q = CIA/36$$

where

  C = Runoff coefficient
  I = Intensity of rainfall for duration equal to time of concentration, cm/h
  A = Catchment area, ha
  Q = [0.30 × 17.5 × 20.2]/36
   = 2.94 m³/s

For rectangular waste weir design,
$$Q = 1.71\, L\, h^{3/2}$$

where

L = Length of waste weir, m
h = Depth of flow over the crest, m
Taking depth of flow over the crest, h = 0.6 m
$L = Q/1.71 h^{3/2} = 2.94/1.71\,(0.6)^{3/2} = 3.70$ m

Adding 15% of freeboard to the depth of flow
$= 0.15 \times 0.60 = 0.09$ m

Freeboard	=	0.60 + 0.09 = 0.69 m
Adding 5% settlement	=	0.05 × 0.69 = 0.035 m
Net freeboard	=	0.69 + 0.035 = 0.725 m = 0.7 m (say)
Wave height, h_w	=	0.014 $(Dm)^{1/2}$
Fetch length, Dm	=	450 m
h_w	=	0.014(450)$^{1/2}$ = 0.29 m
Freeboard, F	=	1.5 hw = 1.5 × 0.29 = 0.43 m

Since the net freeboard is higher than the height of waves, this value can be adopted.

Height of dam, H	=	Storage depth + net freeboard
	=	3.0 + 0.70
	=	3.70 m
Total width of dam	=	H/5 + 1.5
	=	3.7/5 + 1.5
	=	2.24 m = 2.25 m

Upstream and downstream slopes are adopted 3:1 and 2.5:1, respectively (recommended for sandy loam soil)

Bottom width B = top width + upstream slope × height + downstream slope × height
= 2.25 + 3 × 3.7 + 2.5 × 3.7
= 22.60 m

5.6.2.4 Design of Mechanical Spillway

Spillways in the earthen dams are provided to dispose of the excess water and also to serve as a controlled outlet for use of the stored water. In dugout-type ponds, a spillway is not required but the inflow is suitably regulated. For the dam height up to 3 m or catchment area having less than 25 ha, a simple rectangular weir, with stone pitching can be used to dispose of the excess water. A mechanical spillway is provided in the embankment-type pond to let out stored water. Permanent structures are constructed to serve as mechanical spillways for farm ponds. The drop spillway and drop inlet spillway are the two structures that are commonly used. The drop spillway can handle higher discharge whereas the drop inlet spillway provides better control over the water stored in the pond. The drop inlet spillway is also constructed as a simple pipe outlet with the provision of a control valve to regulate the outflow of water. The mechanical spillway is shown in Figure 5.3.

5.6.2.5 Design of Emergency Spillway

To protect the embankment from overtopping due to unexpected inflows into the storage, an emergency spillway is provided. It should be located on one end of the embankment. The bottom elevation of the emergency spillway should be at the maximum expected flood level for the selected frequency of runoff into the pond. The depth of flow over the emergency spillway should not be more than 30

Application of Plastic in Water Management

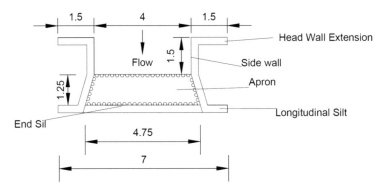

FIGURE 5.3 Mechanical spillway in the farm pond

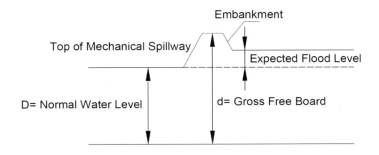

FIGURE 5.4 Emergency spillway

cm. The recommended side slopes are 2:1. An emergency spillway will not be necessary if a surplus weir is provided as the mechanical spillway for the farm pond because the surplus weir will be able to discharge the excess runoff. The drop spillway constructed in the embankment of the pond is termed surplus weir. The location of the emergency spillway in the pond is shown in Figure 5.4.

5.7 TYPES OF POLYTHENE SHEETS

Generally, polythene sheet is available in black, blue, white, and other colors in the market. Its thickness is measured in GSM, micron (μ), and sometimes in gauge. A sheet having a specification of 250 GSM means the weight of 1 m² piece of this polythene sheet will be 250 g or, in other words, 1 kg polythene sheet of 250 GSM specifications will cover 4 m² area. The maximum size of polythene sheet is available in 50 m × 7 m dimensions in the market. A poly tank of a maximum 5 m width can be constructed by this sheet. However, tanks larger than this width can be constructed by sealing two polythene sheets using a heat sealer (joining two sheets at high temperatures). Polythene sheets of the desired width can be obtained directly from the manufacturing company on demand.

5.8 PROCEDURE OF POLY TANK CONSTRUCTION

Construction of poly tanks requires considerations of several factors related to the site and capacity of the tank. A step-by-step procedure of poly tanks construction is provided here.

- The site for constructing a poly tank should be first cleaned so that the size of the tank can be marked on the ground by lime. Once the side of the rectangle/square equal to the side of the top of the poly tank is marked, another rectangle/square will be marked inside this. It

is notable that the side of the inside rectangle/square will be less than the outer rectangle/square by twice the depth of the tank if the sidewall slope is 1:1
- Now, digging will be started straight downward through the sides of the inside rectangle/square up to the depth of the tank. Now, join the sides of the outer rectangle with the bottom side of the inner rectangle by cutting the soil in a sloping manner. Finally, we will get a tank having 1:1 side slope
- All the stones, gravels, roots, and residues of vegetation should be removed from all surfaces of the tank because these may puncture the polythene sheet. A spray of weedicides like Atrazine @ 0.4 g/m^2 is recommended to avoid the growth of any kind of vegetation from the wall of the tank.
- Now, a mixture of clay, dung, and straw is prepared and pasted on the sidewalls and bottom of the tank so that all these surfaces can be smoothened. The ultimate aim is to stop all those chances due to which polythene sheet can be punctured.
- Now the polythene sheet of the required size will be fitted in the excavated tank. The polythene sheet will be well covered by stone/brick and soil from all four upper corners.
- Pitching of boulders, tar felt sheets, and sometimes bricks are recommended for the long life of the polythene but it is not economical from the farmers' point of view. However, nowadays good quality polythene is available in the market, which is resistant to sunlight, air, temperature variation, etc.

Construction of farm ponds in individual farmers' fields or on a community basis for harvesting runoff water when it is in excess and recycling of stored water for irrigation and other purposes when there is a deficiency of water are effective and efficient methods for facing the challenge of water scarcity in rain-fed areas. The role of water-harvesting systems in rain-fed areas is to provide life-saving irrigation to low duty crops in the monsoon season and if possible one or two irrigations to raise another crop in the following dry season. The LDPE-film-lined tanks have been found as one of the best options for water storage in order to maximize water availability for irrigation.

5.9 FARM POND AND ITS BENEFITS

A farm pond is a water body in one part of the farmland, which stores water during rains, and the same is used for life-saving irrigation. There are a number of benefits in having farm ponds both for economic activities and for household uses.

- Provide supplemental irrigation to the standing crops at a critical growth stage
- Reduce waterlogging in high-rainfall and terrain areas
- Harvesting of good quality water and irrigating crops in salt-affected coastal areas
- Raising fish in water bodies when a sufficient quantity of water is available
- Raising fruits and trees in and around the embankment of farm ponds
- Reduction of seepage losses to the maximum extent (95%)
- Harvesting and storing rainwater from early monsoons
- Utilization of harvested rainwater for short-duration crops as well as during the off-season
- The lining of ponds and reservoirs to improve water availability over a longer period of time
- Highly useful in porous soils where water retention in ponds and water-harvesting tanks is minimal
- An economical and effective method of storing water
- Eliminates waterlogging and prevents upward intrusion of salts into stored water
- Prevents soil erosion
- Suitable for effluent ponds and channels to reduce soil and groundwater contamination
- Useful for the purpose of storage of drinking water, for pisciculture, and for providing supplementary irrigation.

Application of Plastic in Water Management

A large number of ponds have been lined with plastics for providing drinking water in the coastal and hilly areas of Gujarat, West Bengal, Karnataka, Himachal Pradesh, and Uttaranchal.

5.10 LINING OF WATER BODIES/FARM PONDS

One of the easiest ways of reducing seepage in water bodies is to provide an impervious lining material. A large number of factors have to be considered before designing an appropriate system for the proper lining of the system. The important geographical factors are topography, rainfall, and the type of soil in the area. The engineering factors are design, alignment, transmission losses, capacity, and structures of the water bodies.

5.10.1 PLASTIC FILM AS LINING MATERIAL

The plastic films have offered tremendous scope as lining material which provides an impervious lining, thus preventing water losses due to seepage (Figure 5.5). These linings use polyvinyl chloride (PVC) and LDPE films. Out of all the types tested so far, LDPE film appears to be the best, whereas PVC lining has several limitations. It cannot be manufactured in wide width and, further, the stability of this film is hampered by the migration of plasticizers, which are essential for extruding the flexible PVC film. In India, where plastic materials are always sold on a weight basis, the PVC film becomes too expensive compared to the LDPE film. Due to its higher specific gravity, the PVC film gives 40% less film for a given weight compared to the LDPE film.

5.11 ROLE OF PLASTIC FILM (AGRIFILM) IN LINING

The use of plastic films (agrifilm) as a lining material was introduced in India as far back as 1959 in canal systems. More than 10,000 km length of canals in different irrigation projects has been lined with LDPE films. This prevents water losses due to percolation and there is tremendous scope for improvement in the years to come. Basically, agrifilm is a tough, wide-width black LDPE film

FIGURE 5.5 LDPE farm pond used for water storage

tailor-made for lining applications. Agrifilm is manufactured by a blown film extrusion technique from virgin Indothene grade 22F A002 and has excellent mechanical properties which meet the ISI specification of LDPE films as per IS: 2508-1984. Agrifilm has got excellent water barrier properties and a very good blend of physical properties.

The LDPE film layer minimizes damage to water bodies due to burrowing animals. For small water bodies like small canals, the construction is also quick and polyethylene film can be laid even in running canals during closure periods of rotation. The breaches which were frequent in these canals either due to rodents or due to soil characteristics have been practically eliminated after lining. There is no effect on canal lining due to alternative wet and dry conditions. Rodents generally boreholes after the closure of canals. But in the case of vertically laid polyethylene films, the nibbling of even parts of a film is minimum. Even if the hole is dug by rodents and the film is pierced, the breach does not get widened easily as the hole gets plugged by the sliding soil.

5.12 FUTURE THRUST AREA FOR AGRIFILM LINING

Freshwater resources are a precious commodity in view of rising demands for the growing population. All steps should be taken to conserve and utilize the water properly and prevent wastage. The lining of canals, ponds, reservoirs, etc. is an important step in this direction. The LDPE can be used with confidence for the lining of canals, distributaries, and ponds. It has been found to be more impermeable and also economical as compared to conventional methods of the lining. It also has a long life if protected by an adequate cover of earth or brick tile lining. Agrifilm being an impervious lining can be effectively utilized in preventing seepage. Agrifilm has brought a new hope in the reduction of lining cost appreciably and at the same time offering a very effective water barrier material. It is easy to handle and apply as a lining material; however, it has to be protected against exposure to sunlight and mechanical damages. Under sunlight, in the presence of UV rays, the strength of the polythene deteriorates very rapidly; therefore, it should be provided with covering material to protect it from sunlight. The use of LDPE film is recommended as a lining material to prevent water losses due to percolation. When the excavation of the pond is complete, the beds as well as sides of the pond have to be leveled and prepared for laying the film (Figure 5.6).

FIGURE 5.6 LDPE farm pond

5.13 LINED PONDS FOR STORAGE OF CANAL WATER

Water management assumes paramount importance to reduce the wastage of water, increase water use efficiency, and ensure equitable distribution. Unscientific and injudicious use of canal water results in heavy conveyance and water application losses and causes waterlogging and salinity in canal command. The situation at the tail of the canal is more severe as the water supply is more erratic. The construction of a water storage reservoir of sufficient capacity is required to maintain the water supply during canal off-periods for higher water productivity. Secondary reservoirs are storage structures (ponds) located in irrigated areas that allow farmers to store part of canal water and use it judiciously. Sometimes, during lean periods or in the rainy season when canal water is not needed for irrigation, water can also be stored in secondary reservoirs for use during critical periods. Similarly, in a period of scarcity, even underground water can also be used along with canal water. The storage of canal water in service reservoirs (ponds) and judicious utilization of stored water provide an excellent opportunity to enhance water productivity and save water to enable bringing more area under cultivation in canal command. Further, the introduction of micro-irrigation in these areas can further improve water productivity by growing high-value crops from the stored precious water. In northern India, where canal water is being used for irrigation, the technology of secondary service reservoirs is gaining popularity and many farmers have constructed plastic-lined ponds for the storage of canal water.

5.14 UNDERGROUND PIPELINE SYSTEM

Plastic pipes used in infrastructure projects are a very long way from the single-use plastic that is causing much controversy worldwide. These pipes remain in the ground performing 24/7 for long periods such as 100 years or more, carrying out the essential tasks of transporting water, wastewater, and sewage safely and efficiently. When they reach their end of life as water and sewer pipe systems, they can be recycled, meaning they have an important role to play in contributing to the circular economy. The underground pipeline system is essentially a low-pressure system, also known as an "open or semi-closed" system. This system is open to the atmosphere, where the operating pressure seldom exceeds 5 m to 6 m. The available level differences of falling topography provide the operating head for the system under gravity for the low-pressure flows. Where large heads are required, an underground pipeline system is used, which is essentially a high-pressure system, also known as a "closed" system. This system is not open to the atmosphere and the operating pressure exceeds 10 m for drip and 20 m for sprinklers. Usually, a gravity head is not sufficient to create such a high pressure; therefore, pumps are used for this kind of system.

5.14.1 DATA REQUIRED FOR UNDERGROUND PIPELINE SYSTEM PLANNING

The following data is required for planning and laying out an underground pipeline system:

i. Topographical map of the area
ii. Subsurface data
iii. Texture and salt component of the soil
iv. Soil characteristics including mechanical properties and shear parameters
v. Permeability of the soil in relation to seepage losses
vi. Rainfall data
vii. Water availability, the subsoil water level in the area, and the quality of the underground water
viii. Possibility of waterlogging and salination
ix. Availability of suitable construction materials
x. Existing drainage and drainage facilities

xi. Existing crop pattern
xii. Existing communication and transportation facilities
xiii. Socioeconomic study and agro-economic survey of the project area
xiv. Adequate investigation should be carried out to collect the data given by digging trial pits and boreholes, where necessary, to ascertain the nature of soil encountered along different alternative alignments

5.14.2 Advantages of Underground Pipeline System

i. As most of a piped distribution system underground, right of way problems are significantly reduced. Because outlet location is not limited by topography, pipe systems are better able to accommodate existing patterns of land ownership with the minimum of disruption
ii. Cross drainage and cross masonry structures can be omitted or minimized
iii. No damage due to heavy rainfall or flood during monsoon
iv. More suitable option for a flood-prone area
v. No hindrance in the movement of farmers and farm equipment
vi. Increase in culturable command area as compared to canals, as the water losses are negligible and acquired land for canal network can also be used for cultivation as piped irrigation network is underground
vii. Better option for undulating fields
viii. Because of shorter transit times for water from the source to field, lower conveyance losses and the smaller volumes of water in the conveyance system, pipe systems can deliver a supply which is more flexible in both duration and timing
ix. The important targets of the modernization of irrigation schemes and digital management will be achieved when water is delivered through piped irrigation network
x. In the case of canals, the marshes and the ponds caused by excessive seepage, in course of time, become the colonies of the mosquito, which give rise to vector-borne diseases and this can be minimized by adopting an underground pipeline system
xi. The increase in project efficiency of the underground pipeline system is about 20% as compared to the canal distribution network
xii. Fertilizers/chemicals can also be mixed with the water
xiii. Quantity of water supplied by piped irrigation network is easily measurable; hence water auditing can be accurately done

5.14.3 Application of Underground Pipeline System

Underground pipeline systems can be used where water is valuable both in terms of the crops which can be grown and limited availability as evidenced by low reservoir capacity or restrictive controls on water abstraction from river or groundwater sources. The following are additional applications of plastic pipe for underground water supply:

1) Where poorly cohesive soils would result in high seepage losses from open canals
2) Where irrigable land cannot be reached by an open canal system due to high ground levels

5.15 LYSIMETER

A lysimeter is a device in which a volume of soil, with or without a crop, is located in a container to isolate it hydrologically from the surrounding. In order to investigate soil moisture depletion under a controlled irrigation/precipitation regime, a lysimeter can be used. HDPE lysimeters can be used widely for monitoring soil-water balance parameters and nutrient leaching (Figure 5.7). It is

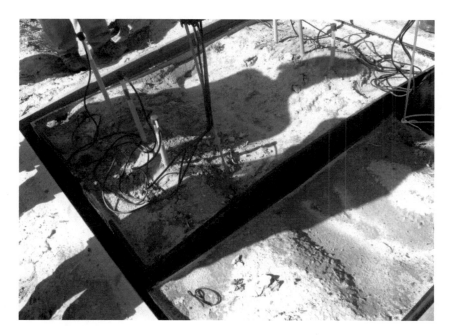

FIGURE 5.7 HDPE lysimeter with sensor for monitoring moisture

usually for the measurement of the accurate water requirement of the crop. For accurate and reliable measurement of evapotranspiration, the lysimeter should be constructed, installed, and operated properly. Singh (1987) discussed the design requirements on lysimeter depth, water control, drainage, area, filling of soil, soil moisture, soil heat flux, and comparability of plant cover which may be helpful in the proper installation of lysimeter. Moisture depletion studies have been conducted at specific intervals throughout the crop period. The measurements involve the amount of precipitation/irrigation applied, the percolated water from the lysimeter, and the soil suction profile at different times. A tipping bucket arrangement has been provided to collect water from the bottom of the lysimeter. Soil moisture content along depth in the lysimeter is required to obtain the soil moisture depletion along the soil profile in the lysimeter.

To reduce cost and secure mobility, a containerized polyethylene (PE-HD) lysimeter can be used for water management. Lysimeters can be broadly categorized as percolation lysimeters (sometimes called drainage lysimeters) and weighing lysimeters (Figure 5.8). Both can serve the purpose of determining the soil-water balance, the vertical percolation flux (drainage), and the chemistry of the percolating water.

Percolation lysimeters may be installed to determine simply the vertical soil-water flux (drainage) or the chemical movement within the soil at a defined boundary. Percolation and weighing lysimeters differ in their measurement methods to determine the vegetation water use and or soil-water evaporation. Percolation lysimeters must be used with a soil-water profile measurement method. Weighing lysimeters permit the mass or volumetric soil-water content change to be determined by weighing the lysimeter and determining its mass change over time. Percolation-lysimeter accuracy of the evaporative water balance is directly related to the precision of the soil-water measurement and its integration through the vegetation root zone.

5.16 APPLICATION OF PLASTIC IN AGRICULTURE DRAINAGE

Irrigation to agricultural fields, no doubt, is regarded as a solution to food security for humanity. But in many areas, it has become a part of the problem because of turning many irrigated

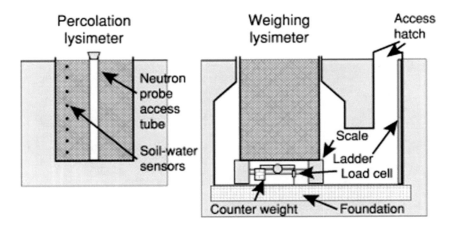

FIGURE 5.8 Percolating and drainage type lysimeter

FIGURE 5.9 Subsurface drainage pipe installation in waterlogged area

farmlands into salty areas and hastens the process of wet desertification. This process has happened all over the world, in developed and developing countries, in the old irrigation commands which initially increased the production, but after a few years, negative effects on productivity of crops were noticed. The waterlogging/salinity is due to seepage and deep percolation losses from canal and field, non-withdrawal of groundwater, constant irrigation, faulty cropping pattern, irrigation by inefficient irrigation method, and closed drainage. The introduction of mechanized pipe drain installation in itself is a major problem, in view of the existing grade of mechanization and knowledge available. The perforated plastic pipe can be used for the installation of a subsurface drainage system to lower down the water from the waterlogged area (Figure 5.9). The planning, design, and construction of the pipe drainage system can serve a pivotal role in subsurface drainage in the future.

Application of Plastic in Water Management

5.16.1 Design Parameters of the Subsurface Drainage System

The system used is a composite system with a lateral corrugated PVC pipe (80–100 mm) and a collector corrugated PVC pipe (200–294 mm). The collector, with laterals connected to it on both sides, discharges into a sump every 50 ha. Here the subsurface drainage water is lifted by a pump and disposed of into a surface drainage system. Filter materials around the pipes are used to prevent the inflow of fine soil particles into the pipes. The materials used are nylon 60 mesh socks for the collectors and polypropylene non-woven fibers (thickness >2.5 mm) and an apparent opening size (O90) > 300 microns.

5.17 PLASTIC MULCH

Plastic mulch can be used for moisture conservation. For moisture conservation, different types of plastic mulch can be used (black mulches, clear or transparent mulches, two-sided color mulches, degradable mulches). The black plastic film does not allow the passage of sunlight onto the soil. Therefore, in the absence of sunlight under the black film, photosynthesis does not take place in the soil. It, therefore, absolutely arrests weed growth. The black plastic mulch helps preserve moisture and control the growth of weeds. It can increase the temperature of the soil. It has been observed that the sunlight will be able to pass through the transparent film and the weeds will rise. Although the black film has been shown to be productive in plains to keep crops cool during summer, transparent film is effective in hilly areas to raise soil temperature during winter in cold weather conditions.

QUESTIONS

1. Describe the farm pond and also discuss its importance.
2. Enlist the types of farm ponds and discuss embankment-type farm ponds.
3. Explain the protection measures of farm ponds.
4. Discuss the design details of the farm pond.
5. Discuss the applications of plastics in agriculture and future prospects of plasticulture in India.
6. Explain design steps of LDPE farm ponds.
7. Write a short note on:
   i) Use of plastic in moisture conservation
   ii) Plastics for loss of water from canals, ponds, and reservoirs
8. Discuss different types of water storage structures.
9. How do we calculate the storage capacity of an LDPE sheet farm pond?
10. Design an LDPE farm pond of 100 m^3 capacity in clay-type soil.
11. Define freeboard and explain the types of freeboard.
12. Describe the application of plastic in water management.
13. How is the plastic used in water conveyance from source to field?
14. Discuss the application of plastic mulch in moisture conservation.
15. Explain the application of plastic in canal lining and water storage.

MULTIPLE CHOICE QUESTIONS

1. In an LDPE farm pond, the sheet thickness to be used may be
   a) 200 microns
   b) 300 microns
   c) 150 microns
   d) None of the above

2. According to the trapezoidal rule, the volume V between two contours at an interval H and having areas A_1 and A_2 is given by
   a) $V = H/2(A_1+A_2)$
   b) $V = H/3(A_1+A_2)$
   c) $V = H/5(A_1+A_2)$
   d) None of the above
3. The lining of water bodies/farm ponds is mainly done for
   a) Reducing seepage
   b) Minimizing evaporation losses
   c) Increasing water use efficiency
   d) None of the above
4. PVC film's specific gravity as compared to LDPE film is
   e) Higher
   f) Lower
   g) Equal
   h) None
5. Air inlet valve in water mains is generally provided at
   a) Summit of the pipe
   b) Upstream of sluice valve
   c) Downstream of sluice valve
   d) Both (a) and (c)
6. Agrifilm lining can be used for
   a) Canal lining
   b) Water storage
   c) Water diversion
   d) All of the above
7. The side slope of the embankment depends on
   a) Fill material
   b) Foundation material
   c) Height
   d) All of the above
8. The capacity of the farm pond is computed using the formula
   a) Rational formula
   b) Trapezoidal rule
   c) Clark's formula
   d) Khosla formula
9. The design of the farm pond is done for the return period of
   a) 25 years
   b) 15 years
   c) 50 years
   d) 35 years
10. The thickness of the plastic sheet used in the lining for pond storage surface varies from
    a) 100 to 200 microns
    b) 100 to 150 microns
    c) 200 to 300 microns
    d) 75 to 100 microns

ANSWER

1	2	3	4	5	6	7	8	9	10
b	a	a	a	a	d	d	b	a	c

REFERENCES

Abu-Awwad, A. M. 1998. Effect of mulch and irrigation water amounts on soil evaporation and transpiration. *Journal of Agronomy and Crop Science*, 18(1), 55–59.

Anikwe, M. A. N., Mbah, C. N., Ezeaku, P. I. and Onyia, V. N. 2007. Tillage and plastic mulch effects on soil properties and growth and yield of cocoyam (*Colocasia esculenta*) on an ultisol in southeastern Nigeria. *Soil and Tillage Research*, 93(2), 264–272.

Clothier, B. E. 1990. Soil water sorptivity and conductivity. *Remote Sensing Reviews*, 5(1), 281–291.

Cooper, P. J. M., Gregory, P. J., Keating, J. D. H. and Brown, S. C. 1987. Effect of fertilizers, variety and location on barley production under rainfed conditions in northern Syria. Soil water dynamics and crop water use. *Field Crops Research*, 16(1), 67–84.

Goel, A. K., Gupta, R. K. and Kumar, R. 2002. Application of plastics for construction of polyhouses for higher agricultural productivity. In *Proceedings of All India Seminar on Advanced Technology for Optimum Agricultural Productivity Organized by Institution of Engineers*, 14th September 2002, Chennai.

Kumar, R., Jat, M. K. and Shankar, V. 2012. Methods to estimate reference crop evapotranspiration - A review. *Water Science and Technology*, 66(3), 525–535.

Meena, N. R. and Meena, K. A. 2019. Micro-irrigation system: A key to water conservation. *Indian Farmer*, 6(10), 670–678.

Meissner, R., Rupp, H. and Haselow, L. 2020. Use of lysimeters for monitoring soil water balance parameters and nutrient leaching. *Climate Change and Soil Interactions*, 171–205.

Men, Q., Li, Y. and Feng, G. P. 2003. Effects of plastic film mulch patterns on soil surface evaporation. *Journal of Irrigation and Drainage*, 22, 17–25.

Mooney, B. P. 2009. The second green revolution? Production of plant-based biodegradable plastics. *Biochemical Journal*, 418(2), 219–232.

Peters, D. B. and Johnson, L. C. 1962. Soil moisture use by soybeans. *Agronomy Journal*, 52(12), 687–689.

Rockstrojm, J. 2003. Water for food and nature in drought-prone tropics: Vapour shift in rain-fed agriculture. *Royal Society Transactions B. Biological Sciences*, 358, 1997–2009.

Salaev, S., Davletov, S. and Eshchanov, R. 2014. Economic feasibility study of canal plastic lining in the Aral Sea Basin. Open Access Peer-Reviewed Chapter. doi: 10.5772/59101.

Singh, R. V. 1987. Design requirements for installation of lysimeters. In *Proceedings of National Symposium on Hydrology Held at National Institute*, Roorkee, 1, VI-14–VI-21.

Singh, R.V. (2000) *Watershed Planning and Management*. Yash Publishing House, Bikaner, India, 459p.

6 Soil Conditioning and Solarization Using Plastics

6.1 INTRODUCTION

Conventional agriculture focuses on the intensive use of fertilizers and pesticides and thus has made a major contribution to meeting global food demands, reducing soil productivity, and improving crop yield and ecosystem health at the expense of soil depletion. A major decrease in soil production potential may be due to its physical constraints, such as surface crusting and hardening, hard pan and compactness of the subsurface, high or slow permeability, hardness extremes, soil water constraints, wind and water erosion, etc. This envisages that soil must be maintained in such a physical condition as to increase crop production in order to allow sufficient crop growth. There is no question that the yield capacity of various crops can be greatly improved if soils are properly maintained for good physical health. Using organic and inorganic soil conditioners and additives to boost agricultural land productivity and increase crop yield can be considered as a means of increasing agricultural production.

Any process that increases soil's ability to boost crop yield or improves soil's efficiency for any purpose is called soil conditioning, whereas a soil conditioner is any product used for soil conditioning. They may also be used as a tool for plant development. Soil conditioning involves the creation and stabilization of soil aggregates that are conducive to seed germination and seedling emergence. For such functions, a good soil stabilizer can reinforce aggregates against breakdown from the impact of raindrops and maintain a high capacity for water infiltration (Gabriels et al., 1981). As described, soil conditioners include many types of organic materials, gypsum, lime, natural deposits, different water-soluble polymers, and cross-linked polymers that hold soil water, living plants, microbes, many products of industrial waste, and others.

6.2 IMPORTANCE AND FUNCTIONS OF SOIL CONDITIONER

The following are the importance and functions of soil conditioners.

- Soil conditioners strengthen the soil's physical, chemical, and biological properties
- They help preserve the soil pH in problematic soils, such as acidic or alkaline soils
- Soil conditioners enhance the water retaining potential, infiltration, percolation, and permeability of water in dry and sandy soils
- Soil conditioners establish a safe soil environment that assists in attracting useful soil microorganisms and earthworms
- The physical properties contributing to improved soil aeration, water conservation, root growth, and soil ecosystems are enhanced by soil conditioners
- Soil conditioners may be used to strengthen or restore soils weakened by insufficient soil management to improve weak soils
- When nutrients are also added, the soil is improved and plants are allowed to grow better, become stronger, and yield more
- Over time, the soil becomes compact and has less air space. The use of a soil conditioner helps reduce the compaction of the soil and the difficult pan problem
- The conditions increase the fertility of the soil and help keep the soil in good condition

6.3 TYPES OF SOIL CONDITIONERS

On the basis of two parameters, soil conditioners can be classified as given below.

(1) History of the materials and origin of the materials
(2) Composition of material

Synthetic or natural materials can occur with regard to their origin. Soil conditioning materials are either organic or inorganic in terms of composition.

6.3.1 Organic Soil Conditioners

Organic soil conditioners are made from biological organisms (e.g., plants and animals). They're used to encourage aggregation, offer a substrate for soil biological activity, improve aeration, lower soil strength, and prevent compaction, crusting, and surface sealing. There are a variety of organic soil conditioners available, but composts are likely the most often used. Crop residues, manures, peat, charcoal, bone meal, blood meal, coffee grounds, compost tea, coir, sewage sludges, FYM, sawdust lignite, humate, and vermiculite are all examples of organic conditioners.

6.3.1.1 Composts

Compost is created when organic wastes are decomposed by microorganisms in the presence of oxygen. Compost as a soil conditioner has a number of advantages, including increased organic C content and microbial activity, a supply of plant nutrients including N, P, K, and Mg, and root reinforcement (Donn *et al.*, 2014). Compost's potential to impact soil microflora by inhibiting several soil-borne diseases such as *Pythium, Phytophthora,* and *Fusarium* spp. is a key aspect. In general, the usage of compost maintains and improves the agricultural soil's stability and fertility.

6.3.1.2 Farm Yard Manure (FYM)

FYM stands for bulky organic manure, which is made up of a naturally degraded mixture of farm animal excrement and urine, as well as litter (bedding material). It is nutrient-dense and improves the physical, chemical, and biological qualities of the soil by adding organic matter. It can be utilized in a variety of soils and crops with a variety of issues.

6.3.1.3 Green Manure

Green manures are crops planted primarily to give nutrients and organic matter to the soil; they are planted in rotation with other crops, plowed, and integrated into the soil to perform the same functions as manures. Green manures aid in the creation of a stable soil structure for plant growth. It improves soil aeration, drainage, and aggregation, increasing microbiological activity and soil fertility (Shinde *et al.*, 2017).

6.3.1.4 Sewage Sludge

Sewage sludge, also known as biosolids, is a type of organic solid that has undergone various treatments to stabilize the organic matter so that it does not attract pests or spread disease (Goss *et al.,* 2013). Biosolids can be applied to agricultural soils because they contain nutrients and organic matter but only under strict regulatory guidelines that minimize heavy metals, weeds, and human and plant infections.

6.3.1.5 Crop Residues

Crop residues are the plant materials that persist after crops have been harvested in an agricultural field or orchard. Stalks and stubble (stems), leaves, and seed pods are among the remnants. Crop leftovers are a valuable source of organic matter that can be used to recycle nutrients and improve the physical, chemical, and biological aspects of soil. Soil integration of crop residues improves nutrient cycling, soil and water conservation, and crop output (Grace et al., 2013; Sharma et al., 2014; and Shinde et al., 2019).

6.3.1.6 Peat Moss

It's a buildup of partially decomposed plants or organic debris. It can be found in peat bogs, moors, and peatlands. In acidic and anaerobic circumstances, peat is generated when plant material does not entirely decompose. It aids in the improvement of soil structure by increasing the capacity to store water.

6.3.1.7 Biochar

Biochar is a solid, fine, granular, black charcoal material made from the slow pyrolysis of biomass, usually from the agricultural or forestry industries. Biochar has a higher potential to absorb cations and retain and exchange nutrients with the soil environment, including microbes and plant roots, because of its larger surface area, negative surface charge, and high charge density. Biochar helps to stabilize biomass and endogenous SOM, improving soil aeration, microbial activity, and nitrogen immobilization, all of which lower emissions of important greenhouse gases such as CH_4, CO_2, and N_2O.

6.3.2 Inorganic Soil Conditioners

Inorganic soil conditioners are quarried, manufactured byproducts, or both, and some are man-made. Water-soluble polymers used to stabilize clay, hydrogel polymers, and synthetic binding agents such as anionic and catalytic polymers, nonviable polymers, and so on are examples of synthetic conditioners.

Soil conditioning utilizing inorganic soil conditioners improves the physical qualities of the soil, allowing for more efficient use of soil and water resources. Soluble inorganic conditioners interact physic-chemically with soil elements, particularly the clay fraction. Thereby, the use of various soil conditioners (VAMA, Krilium, PVA, and Hygromull (a urea-formaldehyde soil conditioner)) improves aggregation, porosity, and hydraulic conductivity, as well as decreasing bulk density, improving porosity, improving infiltration, permeability, and increasing soil profile water (Doyle et al., 1960 and Nimah et al., 1983).

6.3.3 Mineral Soil Conditioners

This conditioner is used to recover soil that has become troublesome. Gypsum, lime, sulfur, crushed rocks, dolomite, and other materials are examples.

6.3.3.1 Gypsum

Gypsum is a fairly soluble source of the vital plant minerals calcium and sulfur, which can help plants grow more quickly. Gypsum has the potential to dislodge exchangeable sodium from high-sodium soil cation exchange sites. It can be used to soften and crumble alkali hardpans, reclaim sodic soil, saline areas, or slippery places, and reclaim sodic soil (Deshpande et al., 2011). It improves infiltration for some puddled soils and supplies calcium to low-exchange-capacity soils.

It also enhances the physical, chemical, and biological aspects of soils, lowering soil erosion and nutrient concentrations (particularly phosphorus) in surface runoff.

6.3.3.2 Lime

Lime is made up of calcium-containing inorganic minerals, mostly oxides, and calcium hydroxide, which typically includes 75–95% $CaCO_3$. It's used to bring very acidic soils back into balance. It acts as a calcium source and helps to increase the pH of the soil. Lime is used most often in extremely acidic soil (pH less than 6.0), where the soil's ability to absorb nutrients, including fertilizer minerals, is impaired.

6.3.3.3 Fly Ash

Thermal power facilities produce fly ash as a byproduct. It's employed as a soil amendment as well as a plant nutrient source. Crop yield and soil fertility can both benefit from fly ash. It improves crop growth and yield in low-fertility soils while also mobilizing macro and micronutrients in the soil. It's utilized to create a new type of soil called "Biosil."

6.3.4 Synthetic Binding Agents

These are the polymers that have been recommended as soil conditioners and are used at much lower rates. The compounds are long-chain, polymeric, organic compounds with a very high molecular weight that bind particles together and create stable aggregates. Organic polymers, primarily polysaccharides (PSD), and polyacrylamides are utilized to improve aggregate stability, preserve fertility, and reduce seal formation (PAM). Low concentrations of these polymers (10–20 kg/ha) were found to be useful in stabilizing and bonding aggregates together at the soil surface and, thereby, sustaining soil fertility in soils with Exchangeable Sodium Percentage (ESP) > 20. It is made up of nonviable polymers as well as anionic and catalytic polymers.

6.3.4.1 Cationic Polymers

Clay absorbs cationic polymers such as polyvinyl chloride (PVC) and polyphenol hydrochloride (PPH) via cation exchange, with calcium acting as a link among clay and polymers. Cationic polymers have strong emulsifying ability.

6.3.4.2 Anionic Polymers

Hydrolyzed polyacrylonitrile (HPAN) and vinyl acetate-maleic acid (VAMA) copolymers are examples of anionic polymers. It is used to keep highly sodic soils from crusting. By forming a sequence of hydrogen bonds, the anion polymeric peripheral complexes link the clay lattice together in an edges-to-edges pattern.

6.4 SOIL SOLARIZATION

Preplant treatment of insecticides, such as the fumigants methyl bromide, chloropicrin, and metam sodium, can reduce soil-borne pests in vegetable and fruit crops. However, because of their toxicity to animals and people, persistent toxicity in plants and soils, the difficulty of soil remediation, and their high cost, these compounds are frequently avoided. Furthermore, when existing environmental legislation is enforced, restrictions on the use of soil-applied pesticides appear to be on the horizon. As a result, reduced-pesticide or non-pesticidal control approaches have become more popular.

Soil solarization is a process in which solar radiation is retained under plastic soil mulch during periods of high ambient temperature, causing temperatures in the upper soil layers to rise to levels that are fatal or sublethal to soil-borne diseases and weeds. Despite the apparent simplicity of the method, research has shown that the effects of soil solarization are caused by a variety of complicated systems that are influenced by a variety of environmental and technical conditions, most notably soil temperatures and the type of plastic films used. A collection of physical, chemical, and biological changes can be recorded to be raised by solarization in the soil, as influencing soil physical and chemical properties and crop production, in addition to soil-borne diseases and pest control.

6.4.1 Mechanisms of Solarization

Although numerous and complex mechanisms for soil solarization effects have been reported in the literature (Katan 1987; Stapleton 1997), the main effects of solarization treatments are found to be primarily related to the thermal action of solarization and the resulting chemical and biological changes in the soil.

6.4.1.1 Thermal Mechanism
Because the solarization process raises soil temperatures to levels that are fatal to many plant diseases and pests, direct thermal inactivation is the most essential and expected mechanism. Heat sensitivity is linked to tiny variations in cell macromolecules, leading to a fatal increase in intramolecular hydrogen, ionic, and disulfide bonds, according to several investigations on the biochemical grounds of organisms' sensitivity to high temperatures (Brock 1978).

6.4.1.2 Chemical Mechanism
Chemical changes in the soil as a result of the heat treatment could be another explanation for the solarization effects (Chen and Katan 1980; Chen et al. 1991). Solarized soil has a higher concentration of soluble mineral nutrients (Katan 1987; Stapleton and DeVay 1995; Stapleton et al. 2002), owing to the death and degradation of soil microbes killed by the heat treatment. In soils with high moisture and organic matter content, toxic ammoniacal nitrogen was found to accumulate due to micro-aerobic conditions caused by the thermal death of nitrifying microorganisms (Hasson et al. 1987), whereas a minimal release of rapidly nitrified nitrogenous compounds was reported in soil with poor organic material when low temperatures raised by ineffective soil solarization and/or a low moisture content increased survival of soil biota and promoted aerobic conditions. In solarized soils, heat-induced breakdown of organic materials may also produce volatile chemicals that are harmful to soil biota (Gamliel et al. 2000).

6.4.1.3 Biological Mechanism
The increased availability of substrate and nutrients following the death of most mesophilic microorganisms resulted in an increased prevalence of microorganisms' antagonists of plant diseases in solarized soil (Stapleton and DeVay 1995). In the presence of heat-induced increased nutrient availability, the improved absorption efficiency of hostile bacteria may advantage them. Further modifications in soil microflora, according to Gupta and Yeates (1997), may be related to decreasing grazing pressure on soil microorganisms by solarization-targeted bacterial grazers and predators. *Bacillus* spp., fluorescent pseudomonads, thermotolerant fungi, and some free-living nematodes were discovered to survive solarization or quickly colonize soil and inhibit pest recolonization, as well as giving superior plant growth (Katan 1987; Stapleton and DeVay 1995). Furthermore, the rise in humid compounds observed as a result of solarization was found to have a favorable effect on luminous pseudomonades (Chen et al. 2000).

6.4.2 Factors Affecting Solarization
Solarization results were shown to be influenced by a variety of factors, while most writers agreed that soil temperature and moisture, environment and weather, and the type and qualities of the mulching layer were all important (Katan et al. 1987; Stapleton and DeVay 1995).

6.4.2.1 Soil Temperature
Due to the above-mentioned critical or fatal accumulation of heat effects beyond a temperature threshold of roughly 37°C for most mesophillic species, the soil temperature is the primary variable for the solar heating effect. Stapleton (1997) found that during solarization, the maximum soil temperatures were found near the soil surface during the day, with temperatures decreasing with depth and at night. Temperatures of more than 50°C were only recorded in the top 5 cm of soil under a clear plastic mulch, but literature reported temperatures of 40–50°C and 36–40°C down to 10–15 cm and 20–30 cm depths, respectively, during summer solarization in warm areas, whereas no lethal or sublethal thermal levels were generally found at deeper soil layers, where temperature

increases by only 3–4°C (Stapleton and DeVay 1983; Greco *et al.* 1985; Chellemi *et al.* 1994). Further suppressive mechanisms, such as the release of volatile toxic chemicals, could explain the nematode population drop seen at 46–91 cm deep, according to Stapleton and DeVay (1986). Solarization-generated thermal levels were found to be substantially greater in closed greenhouses or containerized soil than in open fields (Cartia 1998; Castronuovo *et al.* 2005).

6.4.2.2 Soil Moisture

Moisture in the soil considerably increases heat transfer to microbes, weed seeds, and plants, making it a critical variable in the solarization thermal effect. Furthermore, soil moisture promotes seed cell activity and the proliferation of soil-borne microbes, making them more susceptible to the fatal effects of high temperatures. According to modeling studies, the highest soil temperature was only attained with higher soil water content (Mahrer *et al.* 1984; Naot *et al.*1987), although other authors also claimed a linear relationship between heat capacity and soil water content (Mahrer *et al.* 1984; Naot *et al.* 1987; De Vries 1963; Sesveren *et al.* 2006). In contrast, Al-Karaghouli and Al-Kaysi (2001) found an inverse association between soil maximum temperatures and moisture content, concluding that regular watering during solarization is not required for soil pathogen elimination.

6.4.2.3 Climate and Weather

Climate and weather were also discovered to be important elements in the success of solar heating, as they influence solar radiation levels and, as a result, soil temperatures. Solarization efficiency was observed to be significantly lowered by gloomy and rainy weather by Chellemi *et al.*, 1997, and the best solarizing treatment results were obtained in places with high summer temperatures. Even while summer months in hot climates are undoubtedly the best time for solarization, beneficial applications of this approach against soil-borne diseases, nematodes, bacteria, and weeds have been observed in cooler settings as well (Raio *et al.* 1997; Christensen and Thinggaard 1999; Pinkerton *et al.* 2000; Peachey *et al.* 2001; Tamietti and Valentino 2006). The ability of pathogen heat sensitivity to adapt to diverse climates, resulting in lower thermal thresholds in colder climates, was related to the effectiveness of soil solarization in unfavorable climatic zones.

6.4.2.4 Plastic Film

In the solarizing process, a plastic sheet increases soil temperature by allowing solar energy to flow through while limiting energetic radiative and convective losses (Papadakis *et al.* 2000). The transmission of sun radiation by plastic films was shown to be linked to their radiometric properties, particularly transmissivity, with larger solar transmissivity coefficients resulting in higher temperature rise under the mulch (Scarascia-Mugnozza *et al.* 2004; Vox *et al.* 2005). Other radiometric qualities, such as reflectivity, absorptivity, and emissivity, were also implicated in a plastic film's thermal efficacy (Papadakis *et al.* 2000). The radiometric characteristics of a wide range of plastic films have been carefully investigated in both lab and field studies (Papadakis *et al.* 2000; Heissner *et al.* 2005; Vox *et al.* 2005).

6.4.3 SOLARIZATION RESULTS

6.4.3.1 Increased Soil Temperature

The heating effect of soil solarization is highest at the soil's surface and diminishes as depth increases. The maximum temperature of solarized soil in the field ranges from 108°F to 131°F (42°C to 55°C) at a depth of 2 inches (5 cm) and 90°F to 99°F (32°C to 37°C) at a depth of 18 inches (45 cm). Pest control is normally best in the top 4–12 inches of soil (10–30 cm). Inside greenhouses or by utilizing the second layer of plastic sheeting, higher soil temperatures and deeper soil heating can be attained. Soil solarized in greenhouses can achieve temperatures of 140°F (60°C) at 4 inches

(10 cm) and 127°F (53°C) at 8 inches (20 cm). Soil solarized in black plastic nursery sleeves under a single or double layer of clear plastic can reach temperatures of 158°F (70°C).

6.4.3.2 Improved Soil Physical and Chemical Features

Solarization causes changes in the physical and chemical properties of soil, which aid plant growth and development. It accelerates the breakdown of organic matter in the soil, releasing soluble nutrients like nitrogen (NO_3, $NH4^+$), calcium (Ca^{++}), magnesium (Mg^{++}), potassium (K^+), and fulvic acid, making them more available to plants. Improvements in soil tilth have also been seen as a result of soil aggregation.

6.4.3.3 Control of Pests

Many plant diseases, nematodes, and weed seeds, and seedlings are killed by repeated daily heating during solarization. Many species that can tolerate solarization are also weakened by the heat, making them more vulnerable to heat-resistant fungus and bacteria that operate as natural enemies. Some soil organisms may be killed or weakened as a result of changes in soil chemistry during solarization.

6.4.3.4 Fungi and Bacteria

Much important soil-borne fungal and bacterial plant pathogens are influenced by solarization, including *Verticillium dahliae*, which causes verticillium wilt in many crops; certain *Fusarium* spp., that cause fusarium wilt in some crops; *Phytophthora cinnamomi*, which causes Phytophthora root rot; *Agrobacterium tumefaciens*, which causes crown gall disease; *Clavibacter michiganensis*, which causes tomato canker; and *Streptomyces scabies*, which causes potato scab. Other fungi and bacteria are more difficult to control with solarization, such as certain high-temperature fungi in the genera *Macrophomina*, *Fusarium*, and *Pythium*, and the soil-borne bacterium *Pseudomonas solanacearum*.

6.4.3.5 Nematodes

Many nematode species can be controlled by soil solarization. Because nematodes are extremely mobile and can recolonize soil quickly, soil solarization is not usually as efficient in suppressing nematodes as it is in managing fungal disease and weeds. As a result, nematode management may necessitate yearly therapy. Solarization exerts the most control in the top 12 inches (30 cm) of the soil. Deeper in the soil profile, nematodes may be able to withstand solarization and cause damage to plants with deep root systems. Organic gardeners and home growers will find it extremely handy. Solarization could be an advantageous component of a nematode control strategy. In the San Joaquin Valley, for example, combining solarization with the application of composted chicken manure resulted in good control of root-knot nematodes (Gamliel and Stapleton, 1993).

6.4.3.6 Weeds

Many annual and perennial weeds are controlled by soil solarization. While certain weeds are extremely vulnerable to soil solarization, others are somewhat resistant and require ideal conditions for control (excellent soil moisture, tight-fitting plastic, and intense radiation).

6.4.3.7 Encouragement of Beneficial Soil Organisms

Fortunately, while soil solarization kills many soil pests, it also kills numerous beneficial soil organisms, which can either survive or swiftly recolonize the soil. Mycorrhizal fungi, as well as fungi and bacteria that parasitize plant diseases and help plant growth, are important among these beneficial. Solarized soils may be more resistant to diseases than non-solarized or fumigated soils due to the shift in population in favor of these beneficials.

6.4.3.8 Increased Plant Growth

When plants are cultivated in solarized soil, they frequently develop faster and provide higher- and better-quality yields. This can be ascribed in part to better disease and weed management, but even when the soil that appears to be pest-free is solarized, plant growth rises. A lot of factors could be at play. Minor or unknown pests can be addressed first. Second, increased soluble nutrient levels boost plant growth. Third, solarization has been linked to increased populations of beneficial soil microbes, some of which are known to be biological control agents, such as fluorescent pseudomonades and bacillus bacteria.

6.4.4 IMPROVING SOLARIZATION EFFICACY

Solarization can give effective control of soil-borne pathogens in the field, greenhouse, nursery, and home garden under the right conditions and with adequate application. However, it is often desirable to combine solarization with other appropriate pest management techniques in an integrated pest management approach to improve the overall efficacy of treatment in marginal environmental conditions, with thermotolerant pest organisms or those distributed deeply in the soil, or to minimize treatment duration, to improve the overall efficacy of treatment (Stapleton, 1997). Solarization can be used in conjunction with a variety of different physical, chemical, and biological pest control approaches. This isn't to claim that combining solarization with other techniques always yields better results. Many field trials have indicated that the pesticidal efficacy of solarization or another management method alone could not be improved by combining the treatments under the current conditions (Stapleton and DeVay, 1995). Even in these circumstances, combining solarization with a modest dose of an acceptable pesticide may provide the benefit of a more predictable treatment that commercial users seek.

Although the combination of solarization and a limited dose of 1,3-dichloropropene did not statistically enhance control of the northern root-knot nematode (*Meloidogyne hapla*) over the other treatment by itself, it significantly decreased recoverable numbers of the pest to near-undetectable levels to a soil depth of 46 cm (Stapleton and DeVay, 1995).

Solarization can be coupled with a variety of organic supplements, such as composts, crop residues, green manures, and animal manures, to boost the pesticidal effect of the combined treatments in some cases (Chellemi *et al.*, 1997). By modifying the composition of the resident microbiota or the soil physical environment (biofumigation), incorporation of these organic components alone may operate to lower the number of soil-borne pests in the soil. When these materials are combined with solarization, the biocidal activity of the amendments can be considerably increased. Nonetheless, this seems to be a variable phenomenon, and such effects should not be generalized before preliminary research. Many volatile chemicals released into the soil atmosphere by decomposing organic materials have been proven to be considerably higher when solarized. Researchers have long endeavored to add biological control agents to the soil before, during, or after the solarization organization in order to ensure improved and long-lasting pesticidal efficiency. There have been high hopes for applying appropriate antagonistic and/or plant growth-promoting microbes to solarized soil, either through inundative release or through transplants or any other propagative material, in order to develop a long-term disease-suppressive influence on subsequently planted crops (Katan, 1987; Stapleton and DeVay, 1995). Although this strategy has yet to exhibit a consistent benefit, there have been a few cases of demonstrable benefit. Tjamos and Paplomatas (1988) found that adding the fungus *Talaromyces yavus* to solarized soil that was only heated to sublethal temperatures hindered the survival of *Verticillium dahliae* microsclerotia. Furthermore, it seems that indigenous biota recolonization of solarized soil is just as favorable to following crops as the addition of particular microbes in most trials (Stapleton and DeVay, 1995). For many scholars, this area will undoubtedly remain a source of fascination and experimentation.

6.4.5 CURRENT USAGE

Although soil solarization is now employed on a modest scale as a substitute for synthetic chemical toxicants around the world, its use is likely to grow as methyl bromide is phased out due to its ozone-depleting tendencies. In other areas, however, there are pockets of more extensive uptake for different uses. Solarization, like any other process of soil disinfestations, has advantages and disadvantages. Solarization is dependent on high air temperatures, is most efficacious near the soil surface, somehow doesn't persistently regulate certain heat-tolerant pests (e.g., *Macrophomina phaseolina*), and must be done during the hot months of the year (possibly interfering with biological and chemical control measures) while being simple, safe, and effective within its use restrictions. End users evaluate the practical utility of soil solarization, as with any pest management technique, based on a number of factors, including pesticidal efficacy, effect on crop growth and yield, economic cost/benefit, and peer acceptance. Solarization appears to be most commonly used in greenhouse culture. The flexibility of greenhouse operators to close greenhouses during the hot summer months allows for higher solarization temperatures than are possible with open field treatment. In Japan, for example, more than 5,000 ha of greenhouses were reported to be solarized on a regular basis in 1988. Solarization in greenhouses is also widespread in the other Mediterranean and Near Eastern countries.

Disinfestations of seedbeds, container-based planting media, and cold frames seem to be another usage for which solarization may become widely used, particularly in developing nations. Because individual areas to be treated are low, soil temperature can be vastly expanded, the cost of implementation is low, the value of the plants produced is significant, and the production of disease-free planting stock is essential for producing healthy crops, these are ideal domains for solarization, just as they are in greenhouses (Stapleton, 1997).

Solarization for soil disinfestations in open fields is being introduced in the United States at a modest but steady pace. It has primarily been utilized commercially in locations where summer air temperatures are extremely high and much of the cropland is rotated out of production owing to excessive heat, such as California's central and southern desert valleys. The majority of California growers who are now using solarization in production fields are those who are opposed to the use of methyl bromide or other chemical soil disinfectants, either due to proximity to urban or residential areas, personal preference, or because they are growing for organic markets (Stapleton, 1997). Solarization of producing fields looks to be progressing at a similar rate in other locations with suitable but more tropical climates, such as Florida, US (Chellemi *et al.*, 1997).

Other particular solarization methods that have been verified or just used include removing *Didymella lycopersici* from wooden tomato stakes, utilizing black polyethylene film in open fields or established orchards or wineries (Stapleton and DeVay, 1995), and closing greenhouses in the summer to provide space solarization of aerial equipment. Solarization is widely recognized as important in home gardening and subsistence production, in addition to commercial application. Despite the fact that the majority of these users never employ chemical soil disinfectants, solarization has gained widespread acceptance and is now considered mainstream. Gardeners use it to promote plant health as well as production in these environments (Stapleton, 1997).

QUESTIONS

1. What is soil conditioning?
2. Discuss soil conditioning using plastics.
3. Describe the importance and function of soil conditioners.
4. What is soil solarization? How it can be done using plastic?
5. How can the efficacy of solarization be improved?

6. Enlist different methods of solarization.
7. How plastic is used for the solarization of soil?
8. Describe the importance and functions of soil conditioners.
9. Describe the role of plastic for soil conditioning and solarization.
10. Explain different types of soil conditioners.

MULTIPLE-CHOICE QUESTIONS

1. _____ are used for retaining soil by reducing the rate of evaporation.
   a) Coolants   b) Inhibitors
   c) Mulches   d) Antitranspirants
2. Soil conditioners can be classified on the basis of
   a) History of materials   b) Origin of materials
   c) Composition of materials   d) All of the above
3. Organic conditioners are made of material extracted from
   a) Living things   b) Nonliving things
   c) Both living and nonliving things   d) All of the above
4. Mulching helps in _____
   a) Soil fertility   b) Moisture conservation
   c) Soil structure   d) Soil sterility
5. Inorganic conditioners include
   a) Gypsum   b) Lime
   c) Pyrites   d) All of the above
6. The major effects of mulches, irrespective of their composition, are attributed to:
   a) Suppression of weeds, by restricting the amount of light on the ground surface and thus hindering their germination and growth
   b) Conservation of soil by preventing erosion
   c) Maintenance or improvement of soil structure by eliminating or mitigating the severity of rain and wind action
   d) All of the above
7. Soil solarization is typically carried out when the air temperature reaches ___°C
   a) 25   b) 22
   c) 35   d) 28
8. What is the use of biochar in farming?
   a) Biochar can be used as part of the growing medium in vertical farming
   b) When biochar is part of the growing medium, it promotes the growth of nitrogen-fixing microorganisms
   c) When biochar is part of the growing medium, it enables the growing medium to retain water for a longer time
   d) All of the above
9. The maximum temperature of solarized soil at 2 inches in the field ranges from
   a) 42°C to 55°C   b) 52°C to 65°C
   c) Both a and b   d) None of the above
10. Crown gall disease is caused by
    a) *Verticillium dahliae*   b) *Phytophthora cinnamomi*
    c) *Clavibacter michiganensis*   d) *Agrobacterium tumefaciens*

ANSWERS

1	2	3	4	5	6	7	8	9	10
c	d	a	b	d	d	c	d	a	d

REFERENCES

Al-Karaghouli, A. A. and Al-Kaysi, A. W. 2001. Influence of soil moisture content on soil solarization efficiency. *Renewable Energy*, 24(1), 131–144.
Brock, T. D. 1978. *Thermophilic Microorganisms and Life at High Temperatures*. Springer, New York.
Cartia, G. 1998. Solarization in integrated management systems for greenhouses. In Stapleton, J. J., DeVay, J. E. and Elmore, C. L. (eds.), *Proceedings of the Second International Conference on Soil Solarization and Integrated Management of Soil-Borne Pests*, Aleppo, Syrian Arab Republic, 16–21.
Castronuovo, D., Candido, V., Margiotta, S., Manera, C., Miccolis, V., Basile, M. and D'Addabbo, T. 2005. Potential of a corn starch-based biodegradable plastic film for soil solarization. *Acta Horticulturae (ISHS)*, 698(698), 201–206.
Chellemi, D. O., Olson, S. M. and Mitchell, D. J. 1994. Effects of soil solarization and fumigation on survival of soilborne pathogens of tomato in northern Florida. *Plant Disease*, 78(12), 1167–1172.
Chellemi, D. O., Olson, S. M., Mitchell, D. J., Secker, I. and McSorley, R. 1997. Adaptation of soil solarization to the integrated management of soil borne pests of tomato under humid conditions. *Phytopathology*, 87(3), 250–258.
Chen, Y., Gamliel, A., Stapleton, J. J. and Aviad, T. 1991. *Chemical, Physical, and Microbial Changes Related to Plant Growth in Disinfested Soils*. CRC, Boca Raton, FL, 103–129.
Chen, Y. and Katan, J. 1980. Effect of solar heating of soils by transparent poliethilene mulching on their chemical properties. *Soil Science*, 130(5), 271–277.
Chen, Y., Katan, J., Gamliel, A., Aviad, T. and Schnitzer, M. 2000. Involvement of soluble organic matter in increased plant growth in solarized soils. *Biology and Fertility of Soils*, 32(1), 28–34.
Christensen, L. K. and Thinggaard, K. 1999. Solarization of greenhouse soil for prevention of pythium root rot in organically grown cucumber. *Journal of Plant Pathology*, 81, 137–144.
Deshpande, A.M., Ivanova, I.G., Raykov, V., Xue, Y. and Maringele, L. 2011. Polymerase epsilon is required to maintain replicative senescence. American Society for Microbiology, 31(8), https://doi.org/10.1128/MCB.00144-1
De Vries, D. A. 1963. *Thermal Properties of Soils. Physics of Plant Environment*. North-Holland Publishing Co., Amsterdam, The Netherlands, 210–235.
Donn, S., Wheatley, R. E., McKenzie, B. M., Loades, K. W. and Hallett, P. D. 2014. Improved soil fertility from compost amendment increases root growth and reinforcement of surface soil on slopes. *Ecological Engineering*, 71, 458–465.
Doyle, J. J. and Hamlyn, F. G. 1960. Effect of different cropping systems and of soil conditioners (VAMA) on some soil physical properties and of growth of tomato. *Canadian Journal of Soil Science*, 40(1), 89–98.
Gabriels, D., Maene, L., Lenvain, J. and De Boodt, M. 1981. Possibilities of using soil conditioners for soil erosion control. In Greenland, D. J. and Lal, R. (eds.), *Soil Conservation and Management in the Humid Tropics*. Wiley, Chichester, 99–108.
Gamliel, A., Austerweil, M. and Kritzman, G. 2000. Non-chemical approach to soilborne pest management-organic amendments. *Crop Protection*, 19(8–10), 847–853.
Gamliel, A. and Stapleton, J. 1993. Effect of chicken compost or ammonium phosphate and solarization on pathogen control, rhizosphere organisms, and lettuce growth. *Plant Disease*, 77, 886–891.
Goss, M. J., Tubeileh, A. and Goorahoo, D. 2013. A review of the use of organic amendments and the risk to human health. *Advances in Agronomy*, 120, 275–379.
Grace, K. J., Sharma, K. L., Suma, C. D., Srinivas, K., Mandal, U. K., Raju, B. M. K., Korwar, G. R., Venkateswarlu, B., Kumar, Shalander, MaruthiSankar, G. R., Munnalal, Satish, K. T., Sammi Reddy, K. and Shinde, R. 2013. Effect of long term use of tillage, residues and N levels in sorghum (*Sorghum vulgare*) (L.)-Castor (*Ricinus comminus*) cropping system under rainfed conditions-crop response and economic performance-part I. *Experimental Agriculture*, 42, 1–21.
Greco, N., Brandonisio, A. and Elia, F. 1985. Control of *Ditylenchus dipsaci*, *Heterodera carotae* and *Meloidogyne javanica* by solarization. *Nematologia Mediterranea*, 13, 191–197.
Gupta, S. R. and Yeates, G. W. 1997. *Soil Microfauna as Bioindicators of Soil Health. Biological Indicators of Soil Health*. CABI, Wallingford, UK, 201–234.
Hasson, A. M., Hassaballah, T., Hussain, R. and Abbass, L. 1987. Effect of solar soil sterilization on nitrification in soil. *Journal of Plant Nutrition*, 10(9), 1805–1809.
Heissner, A., Schmidt, S. and von Elsner, B. 2005. Comparison of plastic films with different optical properties for soil covering in horticulture: Test under simulated environmental conditions. *Journal of the Science of Food and Agriculture*, 85(4), 539–548.
Katan, J. 1987. *Soil Solarization. Innovative Approaches to Plant Disease Control*. Wiley, New York, 77–105.

Katan, J., Grinstein, A., Greenberger, A., Yarden, O. and DeVay, J. E. 1987. The first decade (1976–1986) of soil solarization (solar heating): A chronological bibliography. *Phytoparasitica*, 15(3), 229–255.

Mahrer, Y., Naot, O., Rawaitz, E. and Katan, J. 1984. Temperature and moisture regimes in soils mulched with transparent polethylene. *Soil Science Society of America Journal*, 48(2), 362–367.

Naot, O., Mahrer, Y., Avissar, R., Rawitz, E. and Katan, J. 1987.The effect of reirrigation by trickling on polyethylene mulched soils. *Soil Science*, 144, 101–106.

Nimah, M. N., Ryan, J. and Chaudhry, M. A. 1983. Effect of synthetic conditioners on soil water retention, hydraulic conductivity, porosity, and aggregation. *Soil Science Society of America Journal*, 43(4), 742–745.

Papadakis, G., Briassoulis, D., Scarascia-Mugnozza, G., Vox, G., Feuilloley, P. and Stoffers, J. A. 2000. Radiometric and thermal properties of and testing methods for greenhouse covering materials. *Journal of Agricultural Engineering Research*, 77(1), 7–38.

Peachey, R. E., Pinkerton, J. N., Ivors, K. L., Miller, M. L. and Moore, L. W. 2001. Effect of soil solarization, cover crops, and metham on field emergence and survival of buried annual bluegrass (*Poa annua*) seeds. *Weed Technology*, 15(1), 81–88.

Pinkerton, J. N., Ivors, K. L., Miller, M. L. and Moore, L. W. 2000. Effect of soil solarization and cover crops on populations of selected soilborne plant pathogens in Western Oregon. *Plant Disease*, 84(9), 952–960.

Raio, A., Zoina, A. and Moore, L. W. 1997. The effect of solar heating of soil on natural and inoculated agrobacteria. *Plant Pathology*, 46(3), 320–328.

Scarascia-Mugnozza, G., Schettini, E. and Vox, G. 2004. Effects of solar radiation on the radiometric properties of biodegradable films for agricultural applications. *Biosystems Engineering*, 87(4), 479–487.

Sesveren, S., Kaman, H. and Kirda, C. 2006. Effect of tillage and soil water content on thermal properties of solarized soils. In *Proceedings of the International Symposium on Water and Land Management for Sustainable Irrigated Agriculture*. Adana, Turkey, 1–10.

Sharma, K. L., Maruthi Shankar, G. R., Suma, C. D., Grace, K. J., Sharma, S. K., Thakur, H. S., Jain, M. P., Sharma, R. A., Ravindra Chary, G., Srinivas, K., Gajbhiye, P., Venkatravamma, K., Munnalal, Satish Kumar, T., Usha Rani, K., Sammi Reddy, K., Shinde, R. B., Korwar, G. R. and, Venkateswarlu, B. 2014. Effects of conjunctive use of organic and inorganic sources of nutrients on soil quality indicators and soil quality index in sole maize, maize + soybean, and sole soybean cropping systems in hot semi-arid tropical vertisol. *Communications in Soil Science and Plant Analysis*, 45(16), 2118–2140.

Shinde, R., Sarkar, P. K., Bishnoi, S. and Naik, S. K. 2017. Vartmankrishiparidrishya me mridasanrakshan kimahattiavashyaktaevamupay. *Rashtriya Krishi (Hindi)*, 12(1&2), 29–31.

Shinde, R., Sarkar, P. K., Thombare, N. and Naik, S. K. 2019. Soil conservation: Today's need for sustainable development. *Agriculture & Food: e-Newsletter*, 1(5), 175–183.

Stapleton, J. J. 1997. Solarization: An implementable alternative for soil disinfestation. In Canaday, C. (ed.), *Biological and Cultural Tests for Control of Plant Diseases*. APS, St. Paul, MN, 12, 1–6.

Stapleton, J. J. and DeVay, J. E. 1995. Soil solarization: A natural mechanism of integrated pest management. In Reuveni, R. (ed.), *Novel Approaches to Integrated Pest Management*. Lewis, Boca Raton, FL, 309–322.

Stapleton, J. J., Prather, T. S., Mallek, S. B., Ruiz, T. S. and Elmore, C. L. 2002. High temperature solarization for production of weed-free container soils and potting mixes. *Hort Technology*, 12, 541–740.

Tamietti, G. and Valentino, D. 2006. Soil solarization as an ecological method for the control of Fusarium wilt of melon in Italy. *Crop Protection*, 25(4), 389–397.

Tjamos, E. C. and Paplomatas, E. J. 1988. Long-term effect of soil solarization in controlling Verticillium wilt of globe artichokes in Greece. *Plant Pathology*, 37(4), 507–515.

Vox, G., Schettini, E. and Scarascia-Mugnozza, G. 2005. Radiometric properties of biodegradable films for horticultural protected cultivation. *Acta Horticulture (ISHS)*, 691(691), 575–582.

7 Irrigation Scheduling to Enhance Water Use Efficiency

7.1 IRRIGATION

Crop water requirement refers to the amount of water that needs to be supplied for irrigation, while crop evapotranspiration refers to the amount of water that is lost through evapotranspiration (Kumar et al., 2012). The irrigation water requirement basically represents the difference between the crop water requirement and effective precipitation. Crop evapotranspiration can be calculated from climatic data, integrating directly the crop resistance, albedo, and air resistance factors. Irrigation water is applied to replenish the depleted moisture in order to provide sufficient moisture to the plant for optimum plant growth ((Jalota and Arora, 2002; Clemente et al., 2005).

7.2 IRRIGATION SCHEDULING

Crop water use is influenced by different factors related to soil, water, plant, and the environment. As increasingly more production of food is needed with a restricted amount of water supply, and a timely water system with proper depth of irrigation is essential to achieve the greatest advantage from the given amount of water. This has led to the idea of irrigation scheduling. Irrigation scheduling refers to the exact stage of the crop or time during which water should be supplied to the plant. Its aim is to recharge the soil water which has been already taken up by the plant before they are influenced by the deficiency of water. The span between two irrigations should be as extensive as possible without hampering the crop development, yield, nature of yields, and state of the soil. The accessibility of water both in amount and at appropriate period, soil condition, the type of crop grown on it, cost of water, etc., would assist one with choices about the irrigation scheduling.

Irrigation scheduling is fundamental for good management of water and here two basic questions arise, i.e., the amount of water used for irrigation and how frequently to irrigate. How frequently and how to irrigate depends upon the water requirement of the crop. For instance, if the irrigation water need of the crop is 7 mm/day, every day the crop needs a water layer of 7 mm over the entire cropped region. In any case, 7 mm of water need not be provided each day. Drip irrigation systems are designed to meet the irrigation water demand of crops on a daily basis or at an interval of 2 to 3 days. However, in other irrigation systems a larger gap is kept between two irrigations.

Irrigation scheduling is the way toward characterizing the best irrigation water depths and frequencies. It is the process used to decide the right frequency and duration of watering. The objective of irrigation scheduling is to supply sufficient water to completely wet the plant's root zone while limiting overwatering and afterward permitting the soil to dry out in between waterings, to permit air to enter the soil and support root growth but not to the extent that it causes plant stress beyond what is allowed.

In view of the accessibility of water and land cultivated, there are two fundamental approaches in irrigation scheduling.

1. Where water availability is ample as compared to the cultivated land area, the aim should be to achieve maximum production per unit of available land.
2. Where the supply of water is limited as compared to cultivated land area, the objective should be to achieve maximum yield per unit of water used.

For irrigation scheduling, due thought ought to be given to soil water systems; type, variety, and development stage of the crop and climatic conditions. One should gain practical experience before scheduling the irrigation. For instance, paddy needs water when cracks are seen in the soil surface. Light and frequent irrigation is required for crops having shallow roots while crops having deep roots especially in heavy soils usually require more irrigation with large intervals. A light preplanting irrigation assists uniform germination of seed and crop stand. Light and successive water system assists with building up the seedlings better. Planning of the water system should be done at suitable soil temperature since irrigation on hot soil may cause more loss of water through evapotranspiration and irrigation on cool soil may cause some harm to certain crops and may decrease water use effectiveness. So, in hot climatic months, it is smarter to water the field during the night or around evening time and during cold weather conditions, irrigation ought to be applied during the late morning or early afternoon when the soil has begun to heat up.

Irrigation scheduling in many cases is performed dependent on the irrigator's very own experience, the appearance of the plant, observing the neighbor, or at whatever point water is accessible. However, throughout the year, various methods based on soil water checking, observing the plant, and water balance approaches have been created. Each of these methods has a few inadequacies. To overcome these shortcomings, a combination of soil water monitoring and plant status will be the most suitable choice.

7.3 FULL IRRIGATION

It is an irrigation system that provides adequate water to meet the entire irrigation requirement and aims to maximize the production potential of the crop. Excess irrigation reduces crop yields and adversely affects soil physical properties such as soil aeration, soil temperature, and microbial activity. It is justified to use full irrigation when there is no water shortage and the cost is low.

7.4 DEFICIT IRRIGATION

Deficit irrigation meets only a portion of the crops' water needs. If the decline in production costs is greater than the decline in crop value, a reduction in irrigation supply below the full level is economically justified. When there are water shortages or limited irrigation resources, deficit irrigation is used. The crop root zone is not always filled to the desired moisture level when an irrigation system is deficit. When a substantial amount of precipitation falls during the irrigation season, it is possible to only partially fill the root zone, so that some rain flows into the root zone during the irrigation season. Recent concepts in the management of "deficit irrigation" require careful analysis and planning to achieve maximum possible production per unit of water and/or land. Such irrigation replaces only a part of the water that is used by the plant. As an example, it has been observed that partial replenishment of the water used during the period before fruiting does not affect the yield and fruit quality. According to many research studies, crop production could be more efficient in the event of a global water shortage if irrigation systems are appropriately designed. There should, however, be no water shortage during the "critical stages" of the crop's growth.

7.5 IRRIGATION INTERVAL

A time interval between successive irrigations is known as the irrigation interval. Irrigation intervals are usually used at the shortest duration possible in design. In other words, it is the readily available water divided by peak ET. Also, it is given by:

$$i = \frac{(\text{Sfc} - \text{Si})}{\text{ET}} \times D$$

where

    i = Irrigation interval, days
    Sf_c = Moisture content of soil at field capacity
    Si = Depletion of moisture within acceptable limits
    D = Depth of rooting, mm
    ET = Evapotranspiration, mm/month

7.5.1 Factors Affecting Irrigation Interval

How often to irrigate and how much water to use at each irrigation depends on how quickly water is used by the crop, as well as how much water is held in the root zone. There are several factors that determine it: soil texture, soil structure, water penetration, the depth of the effective root zone, the crop grown, and the stage of development of the crop. Organic matter, solid particles, and pores that hold water and air make up all soil. Depending on the texture and structure of the soil, these pores will differ in size and how much water they will hold. In ideal situations, irrigation depth and/or intervals should be adjusted to coincide with the crop's growth stage. The depth of irrigation is kept low at the beginning of the growing season, but applications are frequent. The low ET value and shallow root system of the growing seedling cause this condition. When the crop is in mid-growth level, irrigation depth is increased and irrigation frequency is reduced. Sprinkler and drip irrigation methods make it easy to control irrigation depths. It is, however, difficult to change irrigation depths with surface irrigation methods. It is estimated that in the irrigation following the crop season, the amount of irrigation water used will be reduced by the amount of effective rain which may occur during the crop season. As an alternative, provisions may be made to account for precipitation which is expected during the crop-growing season.

7.6 BENEFITS OF IRRIGATION SCHEDULING

Irrigation scheduling offers a few benefits:

1. It empowers the farmers to plan the water cycle among the different fields to limit crop water pressure and increase yields
2. It diminishes the farmers' expense of water and labor as it helps to limit the number of irrigations
3. It brings down fertilizer and manure costs by holding surface overflow and minimizes permeation (filtering) to groundwater
4. It expands net returns by expanding crop yields and harvest quality
5. It limits waterlogging issues by reducing the seepage necessities
6. With the help of controlled leaching, it helps in controlling root zone saltiness issues
7. It results in additional returns by utilizing the "saved" water to irrigate noncash crops that in any case would not be able to receive water during water stress periods.

7.7 FACTORS AFFECTING IRRIGATION SCHEDULING

During the off-growing season, a farmer's need to irrigate is dependent on several factors. To ensure adequate moisture for planting, irrigation requirements for tillage and salt leaching must be considered. When drought or insufficient water is available during the growing season, the farmer may decide to fill the soil profile to avoid having to irrigate early in the growing season. In addition, farmers may irrigate so weeds will germinate before planting, which can then be removed through cultivation.

    A plant's irrigation requirements depend on the level of moisture around it during planting and germination. A shallow planting depth may require several light irrigations to achieve good stand establishment, while a deeper planting depth may require only one irrigation or possibly none if sufficient moisture is present at planting. The number of irrigations and amount of irrigation water may also be affected by whether and how salts must be pushed below the seedlings. If the soil surface is crusted, one or two irrigations may be necessary in order to soften the soil so that seedlings can emerge uniformly.

After the crop starts to grow and during its vegetative phase, irrigation schedules depend on the rate of crop water use, the source of soil water, the specific crop, and the flexibility of the system. The soil water reservoir accessible to the plant increases as the crop grows and the root system develops. As plants grow, they require higher levels of moisture in the root zone. So, during vegetative development, crop sensitivity, water use, and soil water reservoir should be taken into account when planning the schedule. The maximum and minimum depth of application may also be affected by factors such as soil unevenness, irrigation method, intake characteristics, and water control. Consequently, farmers can increase or decrease irrigation intervals and application depths to avoid deep percolation or runoff without considering how much moisture is required to cause yield reductions.

Increasing irrigation intervals can prevent nutrients from being leached at this point. The soil salinity may require deeper applications for leaching or closer spacing between applications to dilute the salts within the root zone. Additionally, other factors such as weed germination, wind erosion prevention, and soil temperature modification may be relevant.

A well-developed soil water reservoir is present during the mid-season. It's the best time for crops to utilize water because they are particularly sensitive to moisture stress. During this part of the season, irrigation schedules are usually constant as long as the weather is good. However, the amount of effective precipitation could fluctuate with changes in crop water use. The leaching of nutrients and soil salinity are factors to consider. A frequent irrigation schedule or wet conditions can lead to diseases and pest problems. In these cases, irrigation intervals need to be prolonged. Sprinkler irrigation schedules may also be affected by crop cooling and frost protection. When fertilizer, pesticides, or other chemicals are applied through irrigation systems, the timing of application may be most important. During mid-season, many crops reach the stage of fruit formation. A good irrigation schedule during mid-season is critical to good crop production during the most sensitive stage of the season.

A similar set of factors determines irrigation schedules during vegetative and mid-season phases. However, crops are not affected as easily during this time as in the middle of the season. There is a maximum development of soil water in the root zone, water use is decreasing, and the crop is less sensitive to soil moisture depletion. Due to this, irrigations are typically less frequent and deeper. During this period, irrigation schedules can have a significant impact on crop quality. When it comes to scheduling irrigation, availability of water and flexibility of distribution and farm systems are often the most important factors. Taking into account a farmer's preferences, religious obligations, and other cultural influences are also important.

7.8 DIFFICULTIES IN IRRIGATION SCHEDULING AT FARM LEVEL

In spite of the variety of irrigation scheduling methods, farmer adoption is still limited. Understanding what limits and requirements farmers and managers face during irrigation scheduling is essential in selecting the right technique to schedule irrigation. Listed below are some limitations and difficulties associated with the application of irrigation scheduling tools at the farm level.

- Due to limited water resources, irrigation application timing becomes particularly sensitive in situations where there are water shortages. That way, yield reductions can be minimized. Likewise, an understanding of salt tolerance levels is required for scheduling water in saline conditions

- In planning irrigation calendars, it is often difficult to account for rainfall variability. It is important to consider irrigation scheduling options, special requirements, and limitations under variable precipitation patterns
- In order to manage saline waters effectively, water stress indicators are needed, while yield–salinity relationships are very important for the management of inadequate irrigation
- The choice and operation of irrigation scheduling methods should take into account the irrigation efficiency as measured by the adequacy and application efficiency of irrigation, as well as the irrigation method design criteria
- A liberalized cropping system, the pricing of water, and the profitability of irrigation are among the costs and incentives of water-saving irrigation scheduling. It is possible in some cases for technology costs to exceed the budgets of end users
- Many farmers lack proper information on how to operate their tools and implements
- It may be difficult for a farmer to schedule crops according to their needs. Despite having knowledge about hydrology and water budgets, it is not appreciated by some farmers

7.9 IRRIGATION SCHEDULING METHODS

It is possible to estimate the water used by crops in several ways. There is some assumption involved in all of these indirect measurements. Many methods can be used to measure moisture, such as measuring soil moisture, measuring and transpiration. There are several methods for scheduling irrigation in some cases.

Irrigation scheduling can be classified into three broad categories:

- Observations based on personal experience and soil and plant conditions
- The moisture content of the soil
- Estimating evapotranspiration losses

Some of these methods may involve implementing technology, and others may be complex. Each method has its own advantages and disadvantages, and so mixing and matching is often recommended. Listed below are comparisons of various irrigation scheduling methods including how they differ. Information specific to each method can be found in Table 7.1.

7.10 OBSERVATION OF THE PLANTS AND SOILS

Farmers commonly schedule irrigations by observing the crop and assessing the feel and appearance of the soil, as well as changes in plant characteristics, such as changing the color of the plants or curling of the leaves, and eventually wilting of the plants. Rather than examining individual plants, it is often best to observe the crop as a whole in order to detect changes.

1) **Plant Indicator**

    Under wet conditions, plants' roots must be able to draw water from their soil to match the water demand of their environment. Most irrigation scheduling methods for plants use methods based on measuring plant responses to balance the supply and demand of water. It is possible to test whether plants respond favorably to soil-based methods of irrigation scheduling using plant-based methods
    i) **Visual Plant Symptoms**

       Plants are screened via visual signs to determine when to irrigate them. Color, curling, wilting, changes in leaf angle, and curled and rolled leaves are among them. The crop can undergo several different changes due to water stress, such as rapid growth (many young, light-green leaves) to slow growth (fewer young leaves, darker in color, and sometimes dull and grayish). In the case of maize, when plants are under water

TABLE.7.1
Different Methods of Irrigation Scheduling

Method	Measured Parameter	Equipment Needed	Irrigation Criterion	Advantages	Disadvantages
Observation of the plants and soils through personal experience and observation	Feeling the moisture content of the soil	Probe by hand	Moisture content of the soil	With experience, accuracy can be improved. Easy to use; simple	Low accuracy; fieldwork required to collect samples
Monitoring soil moisture with a gravimetric soil moisture sample	Analyze the moisture content of soil through sampling	An auger, a cap, and an oven	Moisture content of the soil	Precision	Labor-intensive, including fieldwork; gap between sampling and results
Moisture monitoring with tensiometers	Moisture tension in the soil.	Vacuum gauge included in the tensiometers	Soil moisture tension	Instantaneous soil moisture reading with good accuracy	Labor to read; maintenance needed; breaks when pressures exceed 0.7 atm
Electrical resistance blocks for soil moisture monitoring	Soil moisture electrical resistance	DC bridges (meter) are blocked by resistance	Moisture tension in the soil	Readings are instant; tensions can be varied; remote readings are possible	Not sensitive to low tension; affected by soil salinity; requires maintenance and field reading
Depending on the model used to predict ET, climate parameters are temperatures, radiation, winds, humidity, and rainfall	Water budgeting	Information from a weather station	Moisture content estimation	With the same equipment, you can schedule many fields with no fieldwork required	Calibrating and adjusting must be done periodically since it is an estimate; calculations require computers
A modified atmosphere	Atmometer gauge	Reference ET	Estimation of moisture content	Direct access to reference ET, easy to use	Calibrating is necessary; it is only an estimate

(Source: Colorado State University. Available at http://www.ext.colostate.edu/pubs/crops/04708.html)

stress they roll their leaves, and in the case of beans, they alter their angle (Figure. 7.1). Observation and experience are key to interpreting crop stress effectively. Otherwise, symptoms may be misleading, and it may be too late by the time they appear.

2) Plant Water Potential

A plant's water potential is a measure of its energy status, which can be compared with the measure of soil water. Physiological and biochemical changes can, therefore, be better reflected in this measure. A pressure bomb or pressure chamber can be used to measure plant or leaf water potential (Figure 7.2). Generally, these methods are used for measuring leaf water potential in situ; dye methods are used in laboratories. Depending

Irrigation Scheduling to Enhance Water Use Efficiency

FIGURE 7.1 Rolling of leaves (in maize) and change in leaf angle (in beans) *(Source: http://www.angrau.ac.in/media/7380/agro201.pdf: accessed on June 3, 2013)*

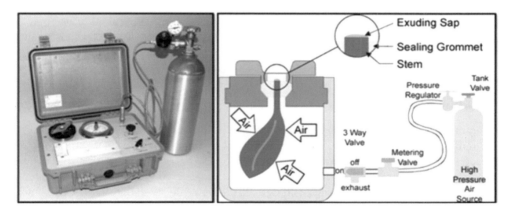

FIGURE 7.2 Pressure chamber apparatus. *(Source: http://www.angrau.ac.in/media/7380/agro201.pdf: accessed on June 3, 2013)*

on the crop, the critical plant water potential varies. A crop in the early phase of growth is at an increased risk of physiological and growth factors adversely affecting its structure and function, so potential values can serve as a guide for irrigation scheduling. In cotton, the critical potential varies between 1.2 and 1.25 MPa throughout plant development, whereas in sunflower, it ranges from 1.0 to 1.2 and 1.4 MPa at various stages of seed formation.

3) **Soil Indicator**

Based on soil indicators, there are several methods. These include feel and appearance, soil moisture monitoring utilizing the gravimetric method, neutron probe, TDRs, soil moisture tension measurements using tensiometers, porous blocks, etc. Using it, one can determine the amount of moisture in the soil. The soil moisture content largely determines the moisture status of the plant. If the soil contains less moisture or the soil is compacted, the roots of the plants will find it difficult to take in the soil moisture. Thus the plant will be under greater stress. To determine how much irrigation should be applied to plants without negatively impacting their growth, the soil moisture content is one of the oldest and simplest methods. One of the easiest ways to determine soil moisture content is by feeling and looking at the soil. The soil is examined visually and by hand. Judgment becomes more accurate with experience.

Scheduling irrigations with this method is quick and easy. It does not require any equipment or technical support. Due to its low-investment requirement, it is popular. Visual observation does, however, have its drawbacks, including that it may not always be accurate and it requires extensive experience in order to be effective. The main reason is that proper assessment of the moisture conditions in the subsoil is very difficult. Also, by the time the symptoms become apparent, most crops have already been deprived of irrigation water for too long, resulting in yield losses, thus, decreasing crop quality and yield. Waiting until symptoms appear is not recommended. It is especially important to apply irrigation water during the early growth stages (the initial and crop development stages).

7.10.1 Methods for Monitoring Soil Moisture

By determining the soil moisture, one can schedule irrigation. Moisture is measured in the soil to detect if there is a shortage of water that will decrease yields or excessive application of water that will lead to waterlogging or leaching of nitrates below the root zone. For effective irrigation water management, soil moisture levels must be monitored. There are several methods for estimating soil moisture and measuring it. Monitoring soil moisture with soil moisture monitoring equipment can provide information about subsoil moisture. Soil moisture monitoring can determine how deep roots are extracting water from, how deep rain or irrigation has penetrated, and when the soil should cease to be irrigated.

7.10.2 Soil Moisture Measuring Devices

The definition of soil moisture varies depending on the discipline. The moisture in the soil is generally determined by the amount of water enclosed between soil particles. It is generally considered that surface soil moisture is the water located in the top 10 cm of the soil, while root zone soil moisture is the water in the top 200 cm of the soil, which is generally referred to as root zone soil moisture. Several soil moisture measurement devices are currently available. The devices include tensiometers, resistance devices (gypsum and granular matrix blocks), neutron probes, and dielectric constant sensors. Many irrigators, however, still use the "feel" method to gauge the moisture content in the root zone of the plants. Some of the instruments used are discussed below.

7.10.3 Tensiometer

A tensiometer measures soil moisture tension at a particular depth in the crop root zone, which is an actual property of soil moisture (Figure 7.3). It is, thus, useful for scheduling irrigations as well. The irrigation is initiated when critical soil moisture tensions are reached. Basically, a tensiometer is a plastic tube packed with water and equipped with a pore on one end and a gauge at the other. With the porous end of the tensiometer in contact with soil, they are installed in the plant zone. As the soil moisture in the root zone dries out, the suction or negative pressure draws the water from the plastic into the soil via the porous tip, creating the same vacuum in the tube as the soil moisture tension. Vacuum gauges display this value. In soil physics, negative pressure (also known as matric potential or tension) is a fundamental concept. Following a negative energy gradient, water moves through the soil, into plants, and finally into the atmosphere. There is more suction at each stage. Based on the volumetric water content associated with soil suction, the volumetric water content of the soils has been determined. The corresponding soil moisture value can, therefore, be calculated by identifying the tensiometric value. Also, the depth at which the tensiometric measurement was taken is just as important as the value of the measurement. Thus, to properly interpret and use SMS devices, both the SMS value and the depth of measurement must be known. The recommended soil moisture tensions for various crops are illustrated in Table 7.2.

Irrigation Scheduling to Enhance Water Use Efficiency

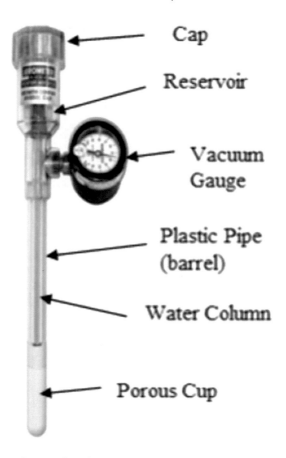

FIGURE 7.3 Tensiometer. *(Source: Google images)*

TABLE. 7.2
Recommended Soil Moisture Tensions for Various Crops

Crop	Tension (centibar)	Crop	Tension (centibar)
Alfalfa seed	80–150	Corn (sweet)	50–80
Pre-bloom	200	Deciduous tree	50–80
Bloom	400–800	Vegetative	40–50
Ripening broccoli	800–600	Ripening grapes	70–80
Cabbage	60–70	Carrot	55–65
Tomato	60–150	Citrus	50–70

7.10.4 Soil Moisture Sensors

For the measurement of soil suction head observations, an advanced system of soil moisture measurement sensors can be used (Figure 7.4). The sensors can measure soil pressure head in a range of 0–199 centibars. It can be installed at a depth of 0.3, 0.6, 0.9, and 1.2 m during crop for the measurement of soil moisture. The sensors are connected with wire lead, which can be connected to the

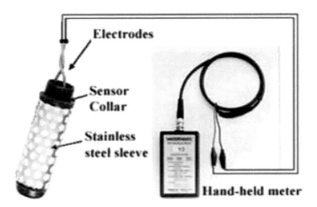

FIGURE 7.4 Soil moisture sensor

Watermark Meter with alligator clips. The moisture meter directly gives the soil pressure head value and is equipped with a temperature adjustment setting. Before installation, the sensors as well as the meter have been checked to ensure that both are functioning properly.

As suggested, the sensors are planted wet before installation, and the wetting and drying cycle is repeated two to three times. The sensors are wetted for 30 minutes, two days prior to installation, and let dry until evening, wet for 30 minutes, let dry overnight, wet again for 30 minutes the next morning, and let dry again until evening. Before the day of installation, the sensors are soaked overnight and installed wet in the morning. This wetting and drying improves the sensor response to the soil moisture. To embed the sensors an access hole is made at each sensing location with the augur. A suitable length of PVC pipe is fitted snugly and over each sensor's collar. The sensor is kept in a vertical position holding the pipe and soil slurry is poured around the sensor to attain firm contact with the soil. At the ground surface, the pipe is capped off, so no surface water can infiltrate the sensor and disturb the sensor reading. The wire left at the top is attached to the pipe and the connecting points are covered with waterproof packing to avoid corrosion and destruction of lead during field operations. To take the moisture suction readings, a portable meter is connected to the sensor wires, which gives the soil suction reading. The degree of soil moisture reduction at various depths shows soil condition at various depths and gives an idea of the irrigation duration needed to rewet the root zone. The values of suction head observed using sensors are converted to the corresponding moisture contents using field calibrated Van Genuchten's relationships.

7.11 RESISTANCE DEVICES

7.11.1 Gypsum Blocks

The use of gypsum blocks for irrigation scheduling has a long history and is based on electrical resistance. Two wires or two concentric wire screens are embedded in a porous material and connected through wire leads. The water content will be monitored by burying one block in the root zone and bringing the leads to the surface of the ground. For additional depths and locations, additional blocks are used. To measure the resistance within the block, wire leads are clipped to a modified ohmmeter. Resistance units have been converted to a numerical scale, generally ranging from 0 to 100. Its two greatest advantages are that it is inexpensive and that it causes very little disturbance to the soil profile. Due to its low cost, irrigators are able to monitor soil moisture more often and in greater depths, as compared to tensiometers.

7.11.2 GRANULAR MATRIX BLOCKS

The main part of a granular matrix block, i.e., resistance block is not made of gypsum but is made of alternative material. Although these blocks may cost two or three times as much as normal gypsum blocks, they last longer, respond more quickly, and have a greater range of sensitivity.

7.11.3 NEUTRON PROBE

Volumetric water content is determined by a neutron probe, an extremely accurate method. Probe tips equipped with neutron emission devices are inserted through an access hole, which is lined with an aluminum or plastic tube to enable repetitive measurements to be made. As the probe tip is lowered through the access tube, successive depth readings are taken. Moisture measurements can be carried out in multiple locations using the probe, which is portable. Neutron probes measure the number of neutrons bouncing back into the soil after they have been emitted. The neutron strikes a molecule of hydrogen in the soil and bounces off in a different direction. There are some neutrons that are deflected but then bounce back in the direction from which they came and that is where the reader on the neutron probe picks them up. Increasing amounts of water in the soil lead to more hydrogen molecules, thus more reflections. Second, due to the presence of radioactive components in the neutron probe, its use is highly regulated. And finally, their accuracy is insufficient at shallow depths (e.g., less than six inches). The neutron probe is shown in Figure 7.5.

7.11.4 GRAVIMETRIC METHOD

Soil water content can be measured using various techniques including evaporation, leaching, chemical reactions, etc. since water is removed and the amount of water removed is measured. Gravimetric measurements with oven drying are among the most widely used methods of determining soil water content. A moist sample is weighed, dried at 105°C for 24 to 48 hours, reweighed, and the water loss is expressed as a percentage of the dried soil mass.

Gravimetric water content calculations –

$$\% \text{ Soil Water} = \frac{\text{weight of wet soil}(g) - \text{weight of dry soil}(g)}{\text{weight of dry soil}(g)} \times 100$$

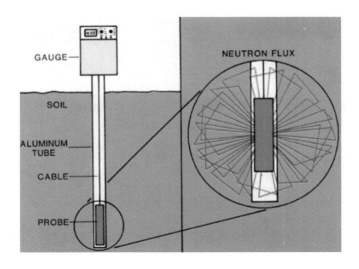

FIGURE 7.5 Neutron probe. *(Source: Google images)*

7.11.5 THERMOCOUPLE PSYCHROMETRY

It is possible to measure soil water potential using thermocouple psychrometry. It can be used to schedule irrigation. In addition to measuring the amount of water in plants, the device can also measure how much water is present in the soil. Because the technique measures water content in the gas phase, it doesn't require a continuous liquid phase to complete the measurements. But the water must be able to evaporate from the sample to the air. For repeated measurements within the same plant or soil, the method uses only a small sample size.

7.11.6 STOMATA RESISTANCE

To maximize crop production and water use efficiency, it is vital to understand plant water relations and predict the onset of water stress. Plant soil atmospheres provide a variety of resistances to fluid flow, with the largest being variations encountered in the diffusive phase of vapor in leaf stomata. Since stomata are characterized by a change in transpiration, they can be used to detect water stress at an early stage.

7.11.7 INFRARED THERMOMETER

For irrigation scheduling in arid regions, infrared thermometers and thermal photographs are new tools that can be used to measure canopy temperature. Temperature can be measured by infrared (IR) thermometers by detecting reflected infrared light that extends beyond visible light's wavelength and corresponds to the surface temperature of an object being measured. In this method, canopy temperature is determined by stomatal closure, which is associated with water stress. Due to the cooling caused by transpiration, the plant temperature increases after the soil water supply is reduced, and the plant becomes stressed.

7.11.8 CLIMATOLOGICAL APPROACH (IW:CUMULATIVE PAN EVAPORATION (CPE) RATIO)

Prihar *et al.* (1974) proposed scheduling irrigation based on the ratio of irrigation water (IW) depth and cumulative moisture from the United States Weather Bureau (USWB). Class A pan evaporimeter, minus precipitation since the last irrigation (CPE). In order for this method to be accurate, the pan evaporimeter and rain gauge must be installed properly and the pan evaporation and rainfall measurements must be accurate. Furthermore, the method is site-specific and dependent on a particular variety of crop. If the IW/CPE ratio is 1.0, crop water is being applied with a volume equal to the amount of water lost by evaporation. Table 7.3 shows a few examples of optimal IW/CPE ratios for important crops.

TABLE. 7.3
Optimal IW/CPE Ratios for Important Crops for Irrigation Scheduling

Crop	Optimum IW/CPE Ratio
Groundnut	0.75 to 1.0 IW/CPE ratio depending on crop developmental stages in Andhra Pradesh, Maharashtra, and West Bengal
Sunflower	0.5 to 1.0 IW/CPE ratio depending on crop developmental stages in Hyderabad and Kanpur
Wheat	1.0 IW/CPE ratio at Ludhiana, Kanpur, and Bikram Ganj
Bengal gram	0.4 IW/CPE ratio at Ludhiana
Mustard	0.4 IW/CPE ratio at Hissar
Maize	0.75 to 1.0 IW/CPE ratio depending on crop developmental stages in Delhi and Hyderabad
Sugarcane	0.5 to 1.0 IW/CPE ratio depending on crop developmental stages in Lucknow

(Source: http://www.angrau.ac.in/media/7380/agro201.pdf: accessed on June 3, 2013)

7.11.9 CANOPY TEMPERATURE

As an indicator of internal water balance in a plant, the canopy temperature could be used to schedule irrigation for crops. A porometer or infrared thermometer can be used to measure it (Figure 7.6). Consequently, for crops such as soybean, oats, barley, wheat, sorghum, and maize, the leaf canopy temperature is a sensitive indicator.

7.11.10 WATER BUDGET METHOD

Using water budgets, we can measure the amount of water lost by evapotranspiration (ET) and the amount that enters soil reservoirs (via effective rain or irrigation). It is a basic tool that can be used to evaluate the occurrence and movement of water through the natural environment.

Water budget methods use the accumulated ET losses from previous irrigations to determine the net irrigation amount. Eventually, the soil profile reaches 100% capacity, and the crops begin to evaporate, thus returning the cycle to its original state. If complete recharge is not possible or desired, the new balance can be determined from net irrigation amount or field observations. This method, however, may not be applicable at locations where crop ET can't be quantified from a water table or other sources.

Data management is required for the water budget. The farmer, therefore, needs to manage soil and crop data, such as calculating crop coefficients, determining field capacity, determining available water, calculating yield thresholds, and determining the starting soil moisture. A farmer needs to monitor the soil moisture inputs (precipitation and irrigation) and outputs (ET) once the starting point is determined. Irrigation should be done prior to reaching the previously established yield threshold depletion level to avoid a fall in yield. In general, farmers set a management allowable depletion level (MAD), which allows the soil to be irrigated once it reaches a certain level of soil depletion. Based on the amount of available water, this could be determined by a percentage. Irrigation scheduling offers the advantage of alerting you when a field is getting close to the MAD, so you can begin to look at it closely. Scheduling irrigation should never be the sole factor determining when to irrigate. Even so, irrigation scheduling will always be able to estimate how much water should be reabsorbed into the soil. Irrigation scheduling of this type has the disadvantage of being complex. A water budget represents water movement following its path into the soil, through the soil, and through a plant. Obtaining an accurate budget requires extensive amounts of data and experience when modeling physical processes.

Thus, by using smart irrigation scheduling, farmers can determine when and how much water their crops need more precisely. Farmers can utilize their water more efficiently either by reducing or maintaining the amount of applied water, while maintaining or improving yields, by using smart irrigation scheduling. These technologies utilize local weather stations for measuring air

FIGURE 7.6 Infra-red thermometer for irrigation scheduling of crops. *(Source: http://www.angrau.ac.in/media/7380/agro201.pdf: accessed on June 3, 2013)*

temperature, humidity, wind speed, rainfall, and soil probes and plant moisture sensing devices that measure water pressure in plant cells. With the help of software, farmers now have easy access to data on field conditions in real time, receive notifications via email and text message, and automate or control irrigation systems remotely.

7.12 GUIDELINES FOR PLANNING IRRIGATION SCHEDULES

Following are the factors that affect irrigation scheduling based on the characteristics of different soil groups and soil depths:

1) **Shallow/Sandy Soil:** The soil moisture retention capacity is low in sandy soils in general and in soils with limited depth (due to the presence of hardpan) close to the soil surface. It is necessary to irrigate these soils frequently and to apply a small quantity of water at each irrigation
2) **Loamy Soils:** Soils with more moisture can store more water than soils with shallow or sandy composition. As a result, irrigation is applied less frequently with a larger depth of water
3) **Clayey Soils:** Clayey soils have a higher moisture-holding capacity than loamy soils. Thus, irrigations are less frequent and the amount of water in irrigation is higher

7.13 NONPEAK IRRIGATION DEPTH ADJUSTMENT

It is best to keep irrigation intervals short and reduce the amount of water used in irrigation during the early growth stages of crops. The young plants may suffer from water shortage if the irrigation interval is kept far apart due to the low ET value as their roots cannot absorb water from the lower root zone layers. During the late season, on the other hand, crops like maize (grain), which is harvested dry, need much less water than other crops. As a result, irrigation is reduced during the late season. During the growing season, irrigation schedules are adjusted based on rainfall. During the crop-growing season, it is sometimes assumed that there is no rainfall. The schedule will need to be adjusted accordingly if rainfall is significant during the period. Usually, irrigation intervals are lengthened or the irrigation depth is reduced in this manner.

7.14 CALCULATION OF APPROXIMATE IRRIGATION SCHEDULES USING A SIMPLE PROCESS

Following are the steps that can be followed to make simple estimations for irrigation scheduling:

1. Estimate the net and gross depths of irrigation (in mm)
2. Determine the water required for irrigation over the whole growing season (mm)
3. Calculate how many irrigations occurred over the growing season
4. Determine irrigation interval (days)

Divide the irrigation water needs during the growing season by the net irrigation depth per application to determine the number of irrigation applications during a crop's growing season. Taking the number of days of the growing season and the number of irrigations, one can calculate the interval between two irrigations. In case of doubt, the irrigation interval should be rounded to the lowest whole figure.

7.15 SOIL WATER CONSTANTS, ROOT DEPTH OF CROPS AND IRRIGATION SCHEDULING

Effective irrigation scheduling requires a comprehensive understanding of the soil water constants and rooting characteristics of crops as they grow.

- **Field capacity:** As a result of surface tension around the soil particle, water remains in the pore spaces for about 48 hours after gravity has drained the soil. Little additional drainage is anticipated thereafter. Field capacity exists when soil can hold the maximum water possible. At this time, soils are able to hold the most water possible. Likewise, soil that is well structured contains enough oxygen to allow roots to live and grow.
- **Wilting point:** Roots work against surface tension to pull water from the soil as they use soil water. In other words, the roots are sucking water from the soil's pore spaces. Normally, plants draw water from the wettest soil first, then as the soil dries they must struggle more to find more water. Until the roots of the plant cannot obtain water from the soil, water is extracted from the soil. It is known as the wilting point for that soil. There is some moisture left in the soil, but the roots cannot extract it due to the tight cling. The plant's growth slows at the wilting point. Without water, plants at a wilting point will not grow and may die. The point at which agricultural crops begin to wilt should never be allowed.
- **Available water (AW):** It is known as the soil moisture available to plants between the field capacity and permanent wilting point. It is possible to express the amount of water in the soil as a percentage or as millimeters per meter. As an example, a soil with 16% water has 160 mm of water per meter of soil. Because soil organic matter and soil structure differ within each texture class, water-holding capacities differ within each class. A well-structured soil and one with more organic matter will have a higher number of pores that are large enough to hold available water. A soil that contains organic matter is more water-retentive and more stable. In a crop root zone, available water is calculated by the following formula:

$$AW = (\text{Field capacity}, \% \text{ by vol.} - \text{Permanent wilting point}, \% \text{ by vol}) \times \text{Root zone depth}(cm)$$

- **Readily available water (RAW):** Readily available water is the volume of soil water per unit surface area between field capacity and critical soil moisture content. RAW is expressed as follows in identical units:

$$RAW = (\text{Field capacity} - \text{Critical moisture content}) \times \text{Depth of root zone of the crop}$$

- Maximum available deficiency (MAD) is defined as both 'maximum available deficiency' and 'management allowable depletion level'. Another concept of importance in irrigation scheduling is the maximum available deficiency (MAD) of soil moisture. It is the ratio of RAW and AW and is expressed as follows:

$$MAD = \frac{RAW}{AW}$$

MAD for most vegetable crops is estimated to be about 0.5, and for other crops (peas, grain crops, fruit plants, sugarcane, and oilseeds) it is 0.65. The exception is potato which has a MAD value of about 0.30. Plants' root depth varies with the type of soil in which they are grown, their physiological characteristics, and their stage of development. Roots can extend freely when they are not bounded by impermeable layers. Most crops, such as lucerne (alfalfa), apples, pears, plums, safflower, and sunflower, can have roots that reach as deep as 180 cm. Several crops develop roots that reach about 90 cm deep, including beans, carrots, sorghum, and wheat. In general, onions, potatoes, radish, and spinach have shallow roots with a maximum root depth of 60 cm. The roots of corn (corn) and cucumber start about 120 cm below the surface. These values represent a fully grown crop with unrestricted soils. Plant roots are restricted from growing near hardpans.

Additionally, the root system of plants gains depth and spreads as they grow and reach full maturity at the full growth stage of the plant. The RAW is calculated as follows:

$$\text{RAW} = (\text{MAD}) \times \text{root depth of crop} \times (\text{Field capacity} - \text{permanent wilting point})$$

It is possible to determine the average root depth of a crop by digging test pits and observing the roots. Yet this method is not very reliable since the root system's most active part is comprised of tiny hairs that are difficult to identify. Moreover, digging up the roots would frequently destroy the hairs. The amount of soil moisture deficits observed in successive layers of the root zone soil is used in agricultural experiment stations to determine root distribution.

7.16 WATER BALANCE APPROACH

Utilizing the water balance approach, irrigation scheduling is based on readily available information about weather, crops, and soils. Based on the soil moisture depletion, we can express soil water balance by:

$$\text{SMD}_i = \text{SMD}_{i-1} + \text{ET}_C + \text{DP}_i - I_i - P_e + \text{GW}_i$$

where
SMD = the amount of soil moisture lost from the root zone; it is defined as the difference between the amount of soil moisture in the root zone at its maximum capacity and the amount of soil moisture in the root zone at the present time
ET_c = evapotranspiration of the crop
DP = deep percolation
I = irrigation amount
P_e = effective rainfall
GW = capillary rise/groundwater contribution
i = time index

Initial depletion of soil moisture can either be assumed at field capacity or be determined using measured moisture content as follows:

$$\text{SMD}_{i-1} = (\theta_{fc} - \theta_{i-1}) \times D_{rz}$$

where
D_{rz} = root zone depth, which increases during the growing season and reaches a maximum during late summer
θ_{fc} = field capacity volumetric moisture content
θ_{i-1} = initial volumetric moisture content

Estimating crop evapotranspiration daily can be done as follows:

$$ET_{ci} = (K_{ci} - K_{si}) \times ET_{Oi}$$

where Et_o = ET for grass reference crops
K_c = crop coefficient (function of the crop type and the growth stage)
K_s = crop stress coefficient (function of the soil moisture available to the crop)

Moisture content affects the crop stress coefficient. It can be calculated as follows for soil moisture depletion greater than readily available water ($\text{SMD}_i >$ RAW):

$$K_{si} = \frac{TAW - SMD_i}{TAW - RAW}$$

For SMD < RAW, $K_{si} = 0$

When soil saturation beyond the level of soil capacity increases due to rain or irrigation exceeds the root zone capacity, deep percolation occurs from the roots of the plant. On the same day after a wet event, soil water content should return to the field capacity. The deep percolation can be obtained as follows:

$$DP_i = P_{ei} + I_i - SMD_{i-1} - ET_{ci} > 0$$

When irrigation and effective rainfall are less than or equal to SMD and ET_c, no deep percolation occurs ($DP_i = 0$).

The capillary rise can be ignored if the water table is below the root zone as we can define effective rainfall as a fixed percentage of rain. Scheduling of irrigation can be done according to a fixed interval, fixed depth, or management allowable depletion (MAD) criteria, and each term of the above equations must be evaluated. In the case of fixed interval irrigation, the estimated irrigation requirement is usually equal to the depletion of soil moisture. In the case of fixed depth irrigation, irrigation is required when soil moisture depletion reaches the irrigation depth.

Irrigation scheduling based on MAD estimates both the day and depth of irrigation in the following manner:

$$AD = TAW \times MAD$$

where
AD = allowable depletion,
MAD = management allowable depletion limit, which is the fraction of TAW that can be safely removed from the soil to meet the daily ET demand on day i.

Here, irrigation is given on the day i, when an allowable depth of soil moisture is depleted. Irrigation depth required equals soil moisture depletion.

MAD should be evaluated according to crop needs and adjusted as needed during the growing season. For shallow-rooted, high-value crops, MAD is usually 25–40%; for deep-rooted, high-value crops, it is 50%; and for low-value, deep-rooted crops, it is 60–65%. In terms of soil texture, the recommended MAD values are:

i) 40% for fine texture (clayey) soils
ii) 50% for medium texture (loamy) soils
iii) 60% for coarse texture (sandy) soils

7.17 CRITERIA FOR SCHEDULING IRRIGATION

Irrigation is scheduled according to the following three criteria:

7.17.1 Soil Moisture as a Guide

A soil water regime concept can be used to schedule (i.e., time) irrigation. Using this concept, water present in a field at full capacity is considered 100% available for crop growth, while water present at a permanent wilting point is considered to be 0%. In order to schedule irrigation, the main criterion for determining the safe limit for soil water depletion must be determined experimentally. Because the allowable soil moisture content for a given crop varies with the agroclimatic condition of the farm, this approach works well in a given climate. It is allowed for irrigation scheduling to reduce available soil moisture to 25–30% for vegetable crops and 50–60% for drought-resistant crops such as barley, sorghum, pearl millet, finger millet, cotton, etc. Consequently, the soil should be allowed to dry to a different level before irrigation based on the crop type. In the case of drought-resistant grain or other crop plants, it could be 1 bar for vegetable crops and 10 bar for drought-resistant grains. Studies conducted in sandy loam soil at Pusa, Bihar, indicate that wheat can be irrigated when 50% of the available soil moisture is depleted. It corresponds to a soil moisture tension of 0.5 bar at a moisture content of about 15%.

7.17.2 Climate as a Guide

In recent years, scientists have discovered that the climate is a major factor in determining the needs of crops and the timing of irrigation. Irrigation schedules cannot be determined only by the soil moisture content without considering climatic factors. As a result, evapotranspiration and pan evaporation were developed as a basis for scheduling irrigation. For irrigation scheduling of crops at a particular region, where soil factors may or may not be constant, only open pan evaporation data can be utilized. The amount of water lost is estimated through evapotranspiration using climatology data, and scheduling of irrigation is done based on a ratio of IW to CPE. IW represents each irrigation depth and CPE represents the cumulative pan evaporation. IW/CPE ratios vary from 0.5 to 0.8 depending on the crop grown in the area and its agroclimatic conditions. Separate relationships are developed for each crop. In Pusa, Bihar, wheat may be irrigated at a ratio of 0.8 IW/CPE. For different crops, similar relationships have been developed. These relationships can be converted for a given region into days after sowing using probability analysis of evaporation data. However, in case of any rainfall, suitable adjustments should be made.

7.17.3 Plant as a Guide

Apart from soil moisture status and climate conditions, plants are a direct indicator of irrigation schedules since they are the main consumers of water. It is possible to plan irrigation according to the physiological stages of growth. Leaf water potential and certain indicator plants can be used to determine plant water status. Experimentally, critical stages of various crops have already been identified. There are four growth stages identified for wheat irrigation, including crown root initiation (CRI), jointing, milking, and flowering. In all stages of crop development, CRI (20–25 days after sowing) is the most important for irrigation, which ensures maximum production per unit of irrigation. After flowering, jointing and milking are the next important stages. For different crops, similar stages have been identified. A technique for scheduling irrigation based on leaf water potential is the latest and most precise way to measure plant water status.

7.18 ASSESSMENT OF THE SCHEDULING CRITERIA FOR SURFACE, SPRINKLER, AND DRIP IRRIGATION

7.18.1 Surface Irrigation Scheduling

Because many surface irrigation systems are inefficient at applying small amounts of water and they are more inflexible, their scheduling criteria differ from those for sprinklers or drip irrigation. Typically, farmers irrigate to fill the soil profile in the higher parts of their fields or at the tail end of their runs. Farmers will find it difficult to plant with less than 8 cm or 10 cm of water unless surface irrigation systems are well designed, leveled, and managed. The use of net irrigation of 2 cm to 4 cm can result in very wasteful and inefficient irrigation practices. So, despite reduced yields due to water stress, a much higher MAD (4 cm to 5 cm) may be economically viable. The greater the MAD, the greater the irrigation efficiency, the lower the leaching of nutrients and the larger the irrigated area. This may significantly increase water yield per unit. Depending on the crop or soil, the economically allowed MAD may only be 2 cm, 3 cm, or 4 cm for sensitive crops or light soils. The application of 3 cm to 6 cm of water uniformly and efficiently may be a challenge without the use of well-leveled basins or graded furrows. The task of determining how a desired net depth of application will be applied is determined after MAD and irrigation intervals are determined for a surface system. The rates of application, the roughness of the surface, the geometry of the furrows, and the depths required for the net application can change as the season goes on.

7.18.2 Sprinkler Irrigation Scheduling

The application of water with sprinkler systems that are well-designed will have a minimum runoff and deep percolation. With proper management, application efficiencies can reach 70% to 90%. Other than rooting depth, water-holding capacities, crop type, infiltration rates, soil surface storage, labor availability, and automation are also important determinants of MAD.

To minimize labor costs and maximize convenience, it is often economically attractive to make the MAD of hand-move systems (and other systems requiring significant labor) as high as possible.

7.18.3 Drip Irrigation Scheduling

There can be fewer irrigation requirements with drip irrigation as opposed to surface or sprinkler irrigation, mainly because there is less evaporation. When the soil surface is bare, plants experience this reduction in water requirements earlier in their growth stage. Water needs are minimal when 50% to 75% of the ground is shaded. ET estimates are frequently reduced when the soil surface area formerly soaked by the drip system is taken into account. This is because most ET estimating methods assume that evaporation is a substantial factor at the beginning of the crop cycle. Furthermore, drip irrigation may result in higher ET because the area is wetted more often than surface or sprinkler irrigation; thus, ET is not necessarily lower. Such adjustments can be made by observing the wetted area, soil moisture levels, and plant response.

In drip irrigation systems, only a portion of the root zone is wetted, compared to other methods. Most systems are designed to operate continuously, or nearly continuously, during peak periods. Drip irrigation implies higher frequency (shorter irrigation intervals) as well as lower MAD (small amounts of water applied daily) than conventional methods. Since there is a limited wetted volume, MAD applies to root zones that are dampened by the drip irrigation system. Scheduling is not complicated. For a wetted volume, MAD is usually set lower than or equal to RAW. The efficiencies of a well-managed system are determined by the amount of water flowing through emitters caused by hydraulic design and manufacturing variations and whether or not some water will be left in the system where it is least used. Well-designed and managed scheduling systems can usually achieve 85% to 90% efficiency without affecting yield noticeably.

7.19 LYSIMETER SET-UP

A Lysimeter is a device in which a volume of soil, with or without crop is located in a container to isolate it hydrologically from the surrounding. For accurate and reliable measurement of evapotranspiration, the Lysimeter should be constructed, installed, and operated properly. Singh (1987) discussed the design requirements on Lysimeter depth, water control, drainage, area, filling of soil, soil moisture, soil heat flux, and comparability of plant cover that may be helpful in proper installation of a Lysimeter. In order to investigate soil moisture depletion under a controlled irrigation/precipitation regime, a Lysimeter can be used. Percolation to the groundwater table from the root zone is being represented by the drainage from the Lysimeter. Figure 7.6 shows the details of the Lysimeter set-up in the field experimental area. The soil suction has been measured using soil moisture measurement sensors embedded at different depths. The measured suction heads have been converted to the corresponding moisture contents using constitutive relationships (Van Genuchten,

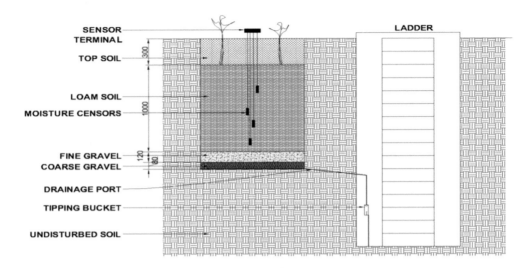

FIGURE 7.7 Lysimeter set-up for crop experiment

1980). The soil and crop pattern inside the Lysimeter is similar to that prevalent in the adjoining field Soil. A Lysimeter (1.5 m deep with a surface area of 1 m²) can be installed in an open field to avoid boundary effects and to simulate actual field conditions. The upper 1.3 m of the Lysimeter is filled with a loam-textured soil, maintaining different layers as per layers hydraulic properties as original field conditions throughout the soil profile, characterized by an organic matter content of 1.1% to 1.2%. The bottom 0.08 m has been filled with very coarse gravel of size more than 3 cm diameter and above it 0.12 m is filled with gravel of about 2 cm in diameter, to allow drainage toward the pipe and avoid clogging. At the bottom, a perforated barrier has been provided to drain off the percolated water uniformly, to the collecting arrangement. The measurements involve, amount of precipitation/irrigation applied, the percolated water from the Lysimeter, and the soil suction profile at different times. The measured suction heads have been converted to the corresponding moisture contents using field-calibrated constitutive relationships given by Van Genuchten (1980). A tipping bucket arrangement has been provided to collect water from the bottom of the Lysimeter. Soil moisture content along depth in the Lysimeter is required to obtain the soil moisture depletion along the soil profile in the Lysimeter. The moisture depletion for different layers (0.3 m each) was measured throughout the crop period. The change in the moisture storage in a soil layer is computed by multiplying the change in the moisture content with the volume of the soil representing that layer. The horizontal area of the Lysimeter is 1 m² and hence area multiplied by soil depth gives the volume of the soil (Figure 7.7).

7.20 MODELING APPROACH FOR IRRIGATION SCHEDULING

7.20.1 Reference Evapotranspiration (ET_0)

The evapotranspiration rate from a reference surface, not short of water, is called the reference crop evapotranspiration or reference evapotranspiration and is denoted as ET_0. The reference surface is a hypothetical grass reference crop with specific characteristics. The use of other denominations such as potential ET is strongly discouraged due to ambiguities in their definitions. The concept of the reference evapotranspiration was introduced to study the evaporative demand of the atmosphere independently of crop type, crop development, and management practices. Consequently, ET_0 is a climatic parameter and can be computed from weather data (Allen et al., 1998). The reference evapotranspiration rate as "the rate of evapotranspiration from the extensive surface of 8 to 15 cm tall, green grass cover of uniform height, actively growing, completely shading the ground and not short of water." Further, Allen *et al.* (1994) defined the grass reference evapotranspiration as

> the rate of evapotranspiration from a hypothetical reference crop with assumed crop height of 0.12 m, a fixed surface resistance of 70 s m^{-1} and an albedo of 0.23, closely resembling the evapotranspiration from an extensive surface of green grass of uniform height, actively growing, completely shading the ground and with adequate water.

The reference surface closely resembles an extensive surface of green, well-watered grass of uniform height, actively growing and completely shading the ground. The fixed surface resistance of 70 s/m implies a moderately dry soil surface resulting from about a weekly irrigation frequency. Estimating procedure for crop evapotranspiration at any given time using a specific approach is to first estimate the reference evapotranspiration from a standard surface and then multiply it with a modified empirical crop coefficient. Many investigators have developed a large number of empirical or semiempirical equations for estimating reference crop evapotranspiration using meteorological data. The commonly used following six reference evapotranspiration models are illustrated in Table 7.4. The local climatic data can be used for reference evapotranspiration estimation and modified crop coefficient value, and reference evapotranspiration has been calculated for each crop. Some of the methods are only valid under specific climatic and agronomic conditions and cannot be applied under conditions different from those under which they were originally developed.

7.20.2 Reference Evapotranspiration (ET$_0$) Estimation Methodologies

The accuracy of different ET$_0$ estimation methods depends on the climatic conditions and availability of data. Data requirements of these methods vary from method to method. Estimation of reference evapotranspiration is a complex phenomenon because it depends on several climatological factors, such as humidity, temperature, wind speed, radiation, types of crops, and their growth

TABLE 7.4
Reference Evapotranspiration Estimation Methods

Sr. No.	Method of ET$_0$ Estimation	Equations Used	Basic Reference	Required Meteorological Data
1.	FAO-24 corrected Penman (c = 1), (F c P-Mon)	$ET_0 = c \left[\dfrac{\Delta}{\Delta+\gamma}(R_n - G) + \dfrac{\gamma}{\Delta+\gamma} 2.7 W_f (e_a - e_d) \right]$	Doorenbos and Pruit (1977)	Net radiation, vapor pressure deficit, and wind velocity
2.	Priestley–Taylor (P-T)	$ET_0 = \alpha \dfrac{\Delta}{\Delta+\gamma}(R_n - G)$	Shuttleworth (1992)	Net radiation, soil heat flux, and vapor pressure deficit
3.	FAO-24 Blaney–Criddle (F B-C)	$ET_0 = a + b \left[p(0.46\overline{T} + 8.13) \right]$	Doorenbos and Pruit (1977)	Annual daytime hours, temperature, and wind velocity
4.	Hargreaves–Samani (H-S)	$ET_0 = 0.0135(KT)(R_a)(TD^{1/2})(TC + 17.8)$ $KT = 0.00185(TD)^2 - 0.0433TD + 0.4023$	Hargreaves and Samani (1982, 1985)	Net radiation, min/max temperature
5.	FAO Pan Evaporation (F E-Pan)	$ET_0 = K_p E_{pan}$	Allen *et al.* (1998)	Pan evaporation
6	Penman–Monteith (P-M)	$ET_0 = \dfrac{0.408\Delta(R_n - G) + \gamma \dfrac{900}{T+273} u_2(e_s - e_a)}{\Delta + \gamma(1 + 0.34 u_2)}$	Allen *et al.* (1998)	Vapor pressure deficit, radiation flux, wind velocity, temperature, and soil heat flux

stages. In addition, ET_0 estimations also depend upon the quality of the meteorological data used. Therefore, it is very difficult to decide on an appropriate ET_0 estimation method among the different available methods for a particular agroclimatic condition. Numerous studies have illustrated that varying performances of the different equations require local calibration (George et al., 2002). The different methods of ET_0 estimation can be grouped into empirical formulations based on radiation (Priestley–Taylor), temperature (SCS Blaney–Criddle and Hargreaves–Samani), combination theory types (Penman–Monteith, FAO-24 Penman (c=1), FAO-24 corrected Penman), and pan evaporation (FAO-24 pan). Six ET_0 estimation methodologies have been chosen to evaluate their comparative performance in estimation of reference evapotranspiration: Priestley–Taylor (radiation); FAO-24 Blaney–Criddle, Hargreaves–Samani (temperature); Penman–Monteith method corrected Penman (c=1) (combination theory type); FAO-24, and FAO pan evaporation (pan evaporation). Different ET_0 models, corresponding equations, parameters used, and the basic references are shown in Table 7.4.

7.20.3 Notation in Reference Evapotranspiration Determination

The notation/symbols used in the reference evapotranspiration models/equations shown in Table 7.5 are given below.

Apart from the site location, different ET_0 models require data on air temperature, humidity, radiation, and wind speed. Allen et al. (1998) have provided the detailed procedure to compute parameters involved in the reference evapotranspiration models. In computation parameters, values of some variables are required to be taken from the standard tables/curves. Most of the parameters involved in the reference evapotranspiration are computed using the mathematical formulas provided by Allen et al. (1998). The equations used for the computation of different parameters involved in the computation of reference evapotranspiration are given below.

a) Wind speeds measured at different heights above the soil surface are different. Wind speed is slowest at the surface and increases with height. For this reason, anemometers are placed at a chosen standard height, i.e., 2 m and denoted as u_2, which can be obtained from the observed wind speed at z m height (u_z), using the equation:

TABLE 7.5
Notation of Parameters Used for Computation of ET_0

ET_0	Reference Evapotranspiration (mm/day)	W_f	Wind Function
α	Priestley–Taylor coefficient ranges from 1.08 to 1.34 depending on the crop and the location	a and b	Regression coefficients
p	Mean daily percentage of annual daytime hours	TD	$T_{max} - T_{min}$ (°C)
TC	Average daily temperature (°C)	T_{max}	Maximum temperature (°C)
T_{min}	Minimum temperature (°C)	u_2	Wind speed at 2 m height (m/s)
e_s	Saturation vapor pressure (kPa)	e_a	Actual vapor pressure (kPa)
$(e_s - e_a)$	Saturation vapor pressure deficit (kPa)	H	Solar radiation at the top of the atmosphere converted to mm of water evaporated
CT	Coefficient of temperature	CH	Coefficient of relative humidity
CU_r	Coefficient of wind	CS	Coefficients of sunshine
CE and CM	Coefficient of elevation and monthly vegetation	K_p	Pan coefficient, value based on local agroclimate
Δ	Slope of vapor pressure curve (kPa/°C)	R_n	Net radiation at the crop surface (MJ/m² day)
G	Soil heat flux density (MJ/m² day)	T	Mean daily air temperature at 2 m height (°C)
γ	Psychrometric constant (kPa/°C)	R_a	Extraterrestrial radiation (mm/day)
E_p	Potential evapotranspiration (mm/day)		

Irrigation Scheduling to Enhance Water Use Efficiency

$$u_2 = \frac{u_z * 1000}{3600} * \frac{4.87}{\ln(67.8z - 5.42)}$$

b) For the calculation of evapotranspiration, the slope of the relationship between saturation vapor pressure and temperature is required. The slope "Δ" of the curve at a given temperature is computed as:

$$\Delta = \frac{4098 * 0.6108 * EXP((17.27 * T_{mean})/(T_{mean} + 273.3))}{(T_{mean} + 273.3)^2}$$

where T_{mean} is the mean daily air temperature in °C, which is defined as the mean daily maximum (T_{max}) and minimum temperatures (T_{min}) rather than as the average of hourly temperature measurements

$$T_{mean} = \frac{T_{max} + T_{min}}{2}$$

c) The psychrometric constant γ (kPa/°C) is given by the following equation:

$$\gamma = \frac{0.00163 * P}{\lambda}$$

where λ is the latent heat of vaporization (MJ/kg) and P is atmospheric pressure (kPa), which is computed as:

$$P = 101.3 * \left[\frac{293 - 0.0065 * z}{293}\right]^{5.26}$$

where z = station elevation above mean sea level (m)

d) The saturation vapor pressure (e_s) is related to air temperature; it can be calculated from the air temperature. The relationship is expressed by:

$$e_s = \frac{e(T_{max}) + e(T_{min})}{2}$$

e) Where humidity data are lacking or are of questionable quality, an estimate of actual vapor pressure, e_a, can be obtained by assuming that dewpoint temperature (T_{dew}) is near the daily minimum temperature (T_{min}). This statement implicitly assumes that at sunrise, when the air temperature is close to T_{min}, the air is nearly saturated with water vapor and the relative humidity is nearly 100%. The vapor pressure at minimum temperature (T_{min}) is computed as:

$$e(T_{min}) = 0.6108 * Exp\left[\frac{17.27 * T_{min}}{T_{min} + 237.3}\right]$$

The vapor pressure at maximum temperature $e(T_{max})$ is computed as:

$$e(T_{max}) = 0.6108 * Exp\left[\frac{17.27 * T_{max}}{T_{max} + 237.3}\right]$$

The actual vapor pressure (e_a) can be computed as:

$$e_a = \left[\frac{e(T_{min})*RH_{max}}{100} + \frac{e(T_{max})*RH_{min}}{100}\right]/2$$

where RH_{min} and RH_{max} are the minimum relative humidity and maximum relative humidity, respectively, expressed in percentage.

f) The extraterrestrial radiation, R_a, for each day can be obtained from the standard table based on the latitude of the place, it can be computed using the equation:

$$R_a = \frac{24(60)}{\pi} G_{sc} d_r \left[\omega_s \sin(\phi)\sin(\delta) + \cos(\phi)\cos(\delta)\sin(\omega_s)\right]$$

where G_{sc}, solar constant, is 0.0820 (MJ/m² min), d_r is inverse relative distance Earth–Sun (equation 5.11), δ is the solar declination (rad), ω_s is sunset hour angle, and lat is latitude (rad).

$$d_r = 1 + 0.0033\cos\left[\frac{2\pi}{365} - J\right]$$

$$\delta = 0.409\sin\left[\frac{2\pi}{365}J - 1.39\right]$$

$$\omega_S = \arccos\left[-\tan(\text{lat}).\tan(\delta)\right]$$

$$\text{lat} = \frac{\text{lat}^0 * \pi}{180}$$

where lat⁰ is the latitude in degrees and J is the number of days in the year between 1 (January 1) and 365 or 366 (December 31), known as Julian day number. The daylight hours N are given by:

$$N = \frac{24}{\pi} \omega_S$$

g) The rate of long-wave energy emission is proportional to the absolute temperature of the surface raised to the fourth power. This relation is expressed quantitatively by the Stefan–Boltzmann law. The net energy flux leaving the earth's surface is, however, less than that emitted and given by the Stefan–Boltzmann law due to the absorption and downward radiation from the sky. Their concentrations should be known when assessing the net outgoing flux. As humidity and cloudiness play an important role, the Stefan–Boltzmann law is corrected by these two factors when estimating the net outgoing flux of long-wave radiation. It is assumed that the concentrations of the other absorbers are constant. This relation is expressed quantitatively by the Stefan–Boltzmann law.

$$R_{nl} = \sigma\left[\frac{T_{max,K}^4 + T_{min,K}^4}{2}\right]\left(0.34 - 0.14\sqrt{e_a}\right)\left(1.35\frac{R_S}{R_{SO}} - 0.35\right)$$

where R_{nl} is net outgoing long-wave radiation (MJ/m² day), $T_{max,K}$ and $T_{min,K}$ are maximum and minimum absolute temperature, respectively, during the 24-hour period (K = °C + 273.16), σ is Stefan–Boltzmann constant (4.903 × 10⁻⁹ MJ/K⁴ m² day), R_s/R_{so} is relative shortwave radiation (limited to 1.0).

The net radiation (R_n) is the difference between the incoming net shortwave radiation (R_{ns}) and the outgoing net long-wave radiation (R_{nl}):

$$R_n = R_{ns} - R_{nl}$$

Complex models are available to describe soil heat flux. Because soil heat flux is small compared to R_n, particularly when the surface is covered by vegetation and calculation time steps are 24 hours or longer, a simple calculation procedure is given by Allen et al. (1998) for long time steps, based on assumption that the soil temperature follows air temperature:

$$G = C_S \frac{T_i - T_{i-1}}{\Delta t} \Delta Z$$

where G is soil heat flux (MJ/m² day), C_s is soil heat capacity (MJ/m³ °C), T_i is the air temperature at time i (°C), T_{i-1} is the air temperature at time i_1 (°C), Δt is the length of time interval (day), and Δz is the effective soil depth (m).

For day and ten-day periods, as the magnitude of the day or ten-day soil heat flux beneath the grass reference surface is relatively small, the value of soil heat flux may be assumed to be insignificant, thus:

$$G_{day} \approx 0$$

h) If the solar radiation, R_s (MJ/m² day) is not measured, it can be calculated using the Angstrom formula which relates solar radiation to extraterrestrial radiation and relative sunshine duration, given as:

$$R_S = \left(a_S + b_S \frac{n}{N}\right) R_a$$

where N is the maximum duration of sunshine or daylight hours (hour), n is the actual duration of sunshine (hour), and n_s/N is relative sunshine duration, as is regression constant, expressing the fraction of extraterrestrial radiation reaching the earth on overcast days ($n = 0$) and $a_s + b_s$ is the fraction of extraterrestrial radiation reaching the earth on clear days ($n = N$).

The Angstrom values a_s and b_s vary depending on atmospheric conditions (humidity, dust) and solar declination (latitude and month). When no actual solar radiation data are available and no calibration has been carried out for improved a_s and b_s parameters, the values $a_s = 0.25$ and $b_s = 0.50$ are recommended. The corresponding equivalent evaporation in mm/day is obtained by multiplying R_s with 0.408. The actual duration of sunshine n is recorded with a Campbell Stokes sunshine recorder at the automatic weather station. The calculation of the clear-sky solar radiation R_{so} (MJ/m² day), when $n = N$, is required for computing net long-wave radiation. When calibrated values for a_s and b_s are available:

$$R_{SO} = (a_S + b_S) R_a$$

When calibrated values for a_s and b_s are not available:

$$R_{SO} = \left(0.75 + 2*10^{-5}*z\right) R_a$$

The net shortwave radiation resulting from the balance between incoming and reflected solar radiation is given by:

$$R_{ns} = (1-\alpha) R_s$$

where R_{ns} is net solar or shortwave radiation (MJ/m² day), α is albedo or canopy reflection coefficient, which is 0.23 for the hypothetical grass reference crop (dimensionless).

7.20.4 CROP COEFFICIENT

For effective and efficient irrigation planning on a regional scale, the crop coefficient (K_c), is required. The K_c value is the ratio of crop evapotranspiration to the reference evapotranspiration. The K_c value represents crop-specific water use and is needed for the accurate estimation of irrigation requirements of particular agroclimatic conditions for different crops. The crop coefficient (K_c) value varies throughout the growing season and its value depends not only on the crop stage but also on the climatic conditions. Variation of crop coefficient (K_c) values during various growth stages is illustrated in Figure 7.8 (Allen et al., 1998).

FAO irrigation and drainage paper-56 has given guidelines for computing crop water requirements. Allen et al. (1998) proposed three values of crop coefficient for three important stages for the development of crop coefficient curves. The crop coefficient for the initial stage is referred to as $K_{c\ ini}$. Similarly, crop coefficients for mid-season and end stages are designated as $K_{c\ mid}$ and $K_{c\ end}$, respectively. Allen et al. (1998) tabulated the values of $K_{c\ ini}$, $K_{c\ mid}$, and $K_{c\ end}$ for different crops under standard growing conditions. In the absence of local calibration of crop coefficients, a procedure has been suggested to modify the FAO reported crop coefficients, for the local climatic conditions, and crop and irrigation practices. Evapotranspiration during the initial stage is predominately in the form of evaporation. Therefore, the frequency with which the soil surface is wetted during the initial period is taken into account. The value of $K_{c\ ini}$ is affected by the evaporating power of the atmosphere, the magnitude of the wetting events, and the time interval between wetting events.

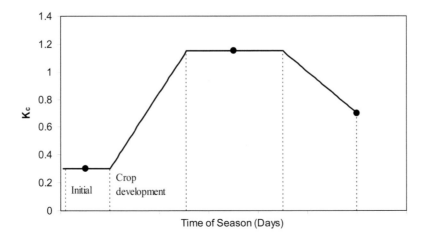

FIGURE 7.8 FAO suggested crop coefficient curve (Allen et al., 1998)

Irrigation Scheduling to Enhance Water Use Efficiency

The value of $K_{c\,mid}$ varies with the climatic conditions and crop height. More arid climates and conditions of greater wind speed will have higher values of $K_{c\,mid}$. More humid climates and conditions of lower wind speed will have lower values of $K_{c\,mid}$. For specific adjustment in climates where the value of minimum relative humidity differs from 45% or where u_2 is larger or smaller than 2.0 m/s.

The FAO recommended crop coefficient values can be modified for the local climatic condition, soil characteristics, and crop according to the procedure given in FAO-56 guidelines.

7.20.5 Modification of the Standard Crop Coefficients

After the selection of the calculation approach, the determination of the lengths for the crop growth stages, and corresponding crop coefficients, a crop coefficient curve can be constructed. The curve represents the changes in the crop coefficient over the length of the growing season. The shape of the curve represents the changes in the vegetation and ground cover during plant development and maturation that affect the ratio of ET_c to ET_0. From the curve, the K_c factor and hence ET_c can be derived for any period within the growing season.

Allen et al. (1998) provided a numerical procedure to compute modified $K_{c\,ini}$ involving computing the impact of the time interval between wetting events, the magnitude of the wetting event, and the evaporation power of the atmosphere. Evapotranspiration during the initial stage for annual crops is predominately in the form of evaporation (Doorenbos and Kassam, 1979). Therefore, accurate estimates for $K_{c\,ini}$ should consider the frequency with which the soil surface is wetted during the initial period. If the soil is frequently wet from irrigation or rain, the evaporation from the soil surface can be considerable and $K_{c\,ini}$ will be large. On the other hand, when the soil surface is dry, evaporation is restricted and the $K_{c\,ini}$ will be small. The value of $K_{c\,ini}$ is affected by the evaporating power of the atmosphere, i.e., ET_0. The value of $K_{c\,ini}$ can vary between 0.1 and 1.15. Hence, the following equation (Allen et al, 1998) is used.

$$K_{c\,ini} = K_{c\,ini(FAO)} + \frac{(I-10)}{(40-10)}\left[K_{c\,ini\,(heavy\,wetting)} - K_{c\,ini\,(light\,wetting)}\right]$$

where I is the average infiltration depth in mm based on the magnitude of the wetting events. Subscripts FAO, heavy wetting, and light wetting refer to the FAO recommended value, $K_{c\,ini}$ derived from the FAO curve corresponding to the heavy wetting, and $K_{c\,ini}$ derived from the FAO curve corresponding to light wetting for the corresponding parameters.

Similarly, the $K_{c\,mid}$ and $K_{c\,end}$ are also modified. The value of $K_{c\,mid}$ varies with the climatic conditions and crop height (Allen et al., 1998). More arid climates and conditions of greater wind speed will have higher values of $K_{c\,mid}$ and $K_{c\,end}$. More humid climates and conditions of lower wind speed will have lower values of $K_{c\,mid}$ and $K_{c\,end}$. For specific adjustment in climates where the value of minimum relative humidity differs from 45% or where u_2 is larger or smaller than 2.0 m/s, $K_{c\,mid}$ and $K_{c\,end}$ values are determined from the following equation:

$$K_{c\,mid/end} = K_{c\,mid/end(FAO)} + \left[0.04(u_2 - 2) - 0.004(RH_{min} - 45)\right]\left(\frac{h}{3}\right)^{0.3}$$

where $K_{c\,mid/end}$ (FAO) is value for corresponding FAO recommended K_c, RH_{min} is mean value for daily minimum relative humidity (%) (20% < RH_{min} < 80%) u_2 is mean value for daily wind speed at 2 m height (m/s) (1 m/s < u_2 < 6 m/s), and h is mean plant height (m) (0.1 m < h < 10 m) during the corresponding crop growth stage. Equations (5.24) and (5.25) are used to modify the FAO recommended K_c values.

During the crop development and late season stage, K_c varies linearly between the K_c at the end of the previous stage ($K_{c\ prev}$) and the K_c at the beginning of the next stage ($K_{c\ next}$), which is K_c end in the case of the late season stage. Allen et al. (1998) provided the method of determination of crop coefficient for an *i*th day in a particular stage as:

$$K_{ci} = K_{c,prev} + \left[\frac{i - \sum(L_{prev})}{L_{stage}}\right](K_{c,next} - K_{c,prev})$$

where *i* is the day number within the growing season, L_{stage} is the length of the stage under consideration (days), K_{ci} crop coefficient on day *i*, and L_{prev} is the sum of the lengths of all previous stages (days).

7.21 FUTURE OF IRRIGATION SCHEDULING – HOW TO TAKE IT FORWARD?

Participation of farmers and irrigation managers is imperative when it comes to formulating, implementing, monitoring, and testing irrigation schedules. There is little benefit to scientifically modifying farming practices that can offset the costs and inputs associated with their use. In many cases, farmers understand when to irrigate, and the refinements offered by scientific scheduling do not compensate for these costs and inputs. According to the conditions of a farm, there are some specific recommendations on how to utilize and further develop various irrigation scheduling techniques. Among the scenarios, low-cost technologies, medium-, and high-technologies conditions under water shortages and normal conditions are discussed. The recommendations for irrigation scheduling under different scenarios are shown in Figure 7.9.

7.22 NUMERICAL PROBLEMS

Example 1: Determine the irrigation interval when the root depth of a crop is 150 cm. With a capacity of 14%, a permissible depletion of 8%, and crop evapotranspiration of 290 mm/month.

Solution: Given

ET = 290 mm/month = 9.6 mm/day
S_{fc} = 0.14
Si = 0.08
D = 150 cm = 1,500 mm

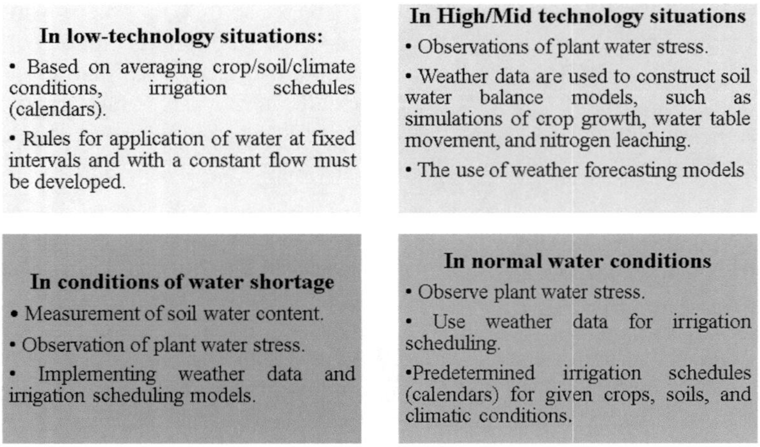

In low-technology situations:
- Based on averaging crop/soil/climate conditions, irrigation schedules (calendars).
- Rules for application of water at fixed intervals and with a constant flow must be developed.

In High/Mid technology situations
- Observations of plant water stress.
- Weather data are used to construct soil water balance models, such as simulations of crop growth, water table movement, and nitrogen leaching.
- The use of weather forecasting models

In conditions of water shortage
- Measurement of soil water content.
- Observation of plant water stress.
- Implementing weather data and irrigation scheduling models.

In normal water conditions
- Observe plant water stress.
- Use weather data for irrigation scheduling.
- Predetermined irrigation schedules (calendars) for given crops, soils, and climatic conditions.

FIGURE 7.9 Recommendations for irrigation scheduling under different scenarios

Irrigation Scheduling to Enhance Water Use Efficiency

To find irrigation interval
Irrigation interval, i, is given as

$$i = \frac{(Sfc - Si)}{ET} \times D$$

$$i = \frac{(0.14 - 0.08)}{9.6} \times 1500$$

$$i = \frac{90}{9.6}$$

$$i = 9.3 \text{ days} \approx 9 \text{ days}$$

Example 2: For the above problem, if there is an effective rainfall of 40 mm during the period under consideration, the irrigation interval will be?

Solution: Given

Effective rainfall (Re) = 40 mm

$$i = \frac{(Sfc - Si)}{ET} \times D + Re$$

$$i = \frac{(0.14 - 0.08)}{290/30} \times 1500 + 40$$

$$i = \frac{130}{9.6}$$

$$i = 13.54 \text{ days} \approx 14 \text{ days}$$

Example 3: A crop requires 9 cm of irrigation water. It was found that the IW/CPE ratio (R) is at 0.75 for irrigation. Determine when to apply water to a crop based on the cumulative pan evaporation (CPE).

Solution: Given

Depth of irrigation water = 9 cm
R = IW/CPE = 0.75
IW = Irrigation requirement = 9 cm

To find CPE

$$R = \frac{IW}{CPE}$$

$$0.75 = \frac{9}{CPE}$$

$$CPE = \frac{9}{0.75}$$

$$CPE = 12 \text{ cm or } 120 \text{ mm}$$

Example 4: Calculate irrigation schedules by means of pan evaporation measurements of Pant Nagar metrological laboratory on the basis of the following points.

Irrigation water to be applied = 6 cm

IW/CPE = 0.9
Date of sowing of crop (November 16, 2002)

Solution: Given

R = IW/CPE = 0.9
IW = Irrigation requirement = 6 cm

$$R = \frac{IW}{CPE}$$

$$CPE = \frac{6}{0.9}$$

CPE = 6.67 cm or 66.7 mm

Month	Date	Pan Evaporation/ day mm	CPE (mm)	Month	Date	Pan Evaporation/ day mm	CPE (mm)
November	16	2.5	4.9	December	1	2.7	41.5
	17	2.4	7.5		2	1.6	44.1
	18	2.6	10.0		3	2.6	48.0
	19	2.5	12.6		4	3.9	53.5
	20	2.6	13.9		5	5.5	58.4
	21	1.3	15.8		6	4.9	63.0
	22	1.9	17.0		7	4.6	65.2
	23	1.2	21.5		8	2.2	66.7
	24	4.5	25.1		9	1.5	68.5
	25	3.6	27.3		10	1.8	70.9
	26	2.2	29.2		11	2.4	76.0
	27	1.9	32.6		12	5.1	77.58
	28	3.4	35.0		13	1.5	78.3
	29	2.4	37.2		14	0.8	79.2
	30	2.2	39.9		15	0.9	80.9

Upon reaching 66.7 mm of CPE, we will irrigate the wheat crop for the first time. CPE reaches 66.7 mm on December 8, 2002, according to the above table. Irrigation will, therefore, be done on 8 December.
In the same way, second irrigation and so on.

Example 5: According to the following soil moisture status and 60% irrigation efficiency, calculate the net and gross amounts of irrigation water required.

Soil Depth (cm)	Soil Moisture at FC (%)	Actual Soil Moisture (%)	Bulk Density (g/cm^3)
0–15	19	9	13
15–30	17	11	14
30–45	17	13	15
45–60	18	15	15

Irrigation Scheduling to Enhance Water Use Efficiency

Solution: Soil moisture deficit in the four soil layers

$$\text{First layer} = \frac{19-9}{100} \times 1.3 \times 15 = 1.95\,\text{cm}$$

$$\text{Second layer} = \frac{17-11}{100} \times 1.4 \times 15 = 1.26\,\text{cm}$$

$$\text{Third layer} = \frac{17-13}{100} \times 1.5 \times 15 = 0.9\,\text{cm}$$

$$\text{Forth layer} = \frac{18-15}{100} \times 1.5 \times 15 = 0.67\,\text{cm}$$

Net quantity of water to be applied = 4.77 cm
Hence, gross water application (GWA), given 60% irrigation efficiency

$$\text{GWA} = \frac{\text{required net irrigation}}{\text{Irrigation efficiency}}$$

$$\text{GWA} = \frac{4.77}{60} \times 100$$

$$\text{GWA} = 7.95\,\text{cm}$$

Example 6: Calculate the soil's field capacity based on the following data.

Depth of root zone = 3 m
Water content existing = 6%
Soil – dry density = 1.6 g/cm³
Applied water to the soil = 515 m³
Evaporation and other water losses = 10%
Plot area = 1,500 m²

Solution: Total water applied = 515 m³
Loss of water = 10%

$$\text{Water used in the soil} = 90\% \times 515$$

$$\text{Water used in the soil} = 463.5$$

$$\text{Water used in the soil} = 463.5\,\text{m}^3 = 463.5 \times 10^6\,\text{cm}^3 = 463.5 \times 10^6$$

$$\text{Total dry weight of soil} = (1500\,\text{m}^2 \times 3\,\text{m}) \times 1.6 \times 10^6 = 7.2 \times 10^9\,\text{gm}$$

$$\%\text{ of water added} = \frac{463.5 \times 10^6}{7.2 \times 10^9} \times 100 = 6.4\%$$

Hence, Net water content = 6% + 6.4% = 12.4%

Example 7: Clay loam soils are being used to grow corn in the Midwest. Considering the plants are already beginning to tassel, you can estimate the available water. Total available water is 3.3 in./ft and D(r) for corn is 5 ft (3–5 ft). Management allowed deficiency (corn) = 0.6.

Solution: Given

AW = 3.3 in./ft
D(r) = 5 ft (3–5 ft)
MAD (corn) = 0.6

$$RAW = MAD \times AW$$

$$RAW = 0.6 \times 3.3 \times 5$$

$$RAW = 9.9 \text{ inches}$$

Example 8: Using a 3.5-ft root zone on loam soil, estimate the available water. When wheat is grown, how much water is readily available? The following data is provided.

Depth of water available = 2.5 in./ft
Volumetric field capacity = 30%
Volumetric permanent wilting point = 15%
Depth of root zone = 3.5 ft
Depth of rooting (wheat) approx. 4 ft (3–5 ft)
Management allowed deficiency (wheat) = 0.56

Solution: Given

AW = 2.5 in./ft
FC (vol) = 30%
PWP (vol) = 15%
D(r) = 3.5 ft
MAD (wheat) = 0.56

$$AW = \frac{(FC-PWP)}{100} \times D$$

$$AW = \frac{(30-15)}{100} \times 3.5$$

$$AW = 0.52 \text{ ft or approx. } 6 \text{ in}$$

QUESTIONS

1. What is irrigation scheduling?
2. Describe the necessity of irrigation scheduling.
3. Enlist different parameters on which irrigation scheduling depends.
4. What is the modeling approach of irrigation scheduling?
5. Define the following terms.
    i) Irrigation efficiency
    ii) Water budgeting
    iii) Water use efficiency

Irrigation Scheduling to Enhance Water Use Efficiency

6. Define or write a short note on the following.
    a) Water budgeting b) Different irrigation schemes c) Parameters used in irrigation scheduling
7. Discuss about irrigation development in India. Also, describe role irrigation engineer in water resources development of India.
8. Define the following.
    a) Evaluation of irrigation systems
    b) Hydraulics of water advance and recession
9. What is irrigation efficiency? How we can improve irrigation efficiency?

MULTIPLE CHOICE QUESTIONS

1. The irrigation of cereal crops is typically done using which of the following methods:
    a) Check method
    b) Furrow method
    c) Sprinkler method
    d) Border method
2. In situations where lands are waterlogged
    a) Saturation of the soil ports occurs within 2 m
    b) Saturation of the soil ports occurs within 55 cm
    c) Saturation of the soil ports occurs within 38 cm
    d) Saturation of the soil ports is up to the crop root zone
3. An irrigation canal generally follows the following alignment.
    a) Ridgeline
    b) Contour line
    c) Valley line
    d) Straight line
4. Crop water consumption given
    a) Is calculated as the volume of water per unit area
    b) Is calculated as the depth of water on irrigated area
    c) May be supplied partly by precipitation and partly by irrigation
    d) All of the above
5. If the water table is relatively high, the irrigation canal will be useless because
    a) There is a large amount of seepage
    b) Cultivated areas will be waterlogging
    c) Uncertain demand for water
    d) All the above
6. Crop ratio is the
    a) Area irrigated between Rabi season and Kharif season
    b) Area irrigated between Kharif season and Rabi season
    c) Irrigated area under perennial crop to total area
    d) Irrigated area under non-perennial crop to that under perennial crop
7. Whenever a crop is raised entirely by rainfall it is referred to as a
    a) Natural crop
    b) Rainy crop
    c) Wet crop
    d) Dry crop
8. Pick up the correct statement from the following:
    a) Gravity water harms crops
    b) Hygroscopic water remains in the soil pores by a chemical bond
    c) Capillary moisture is repressed by surface tension, allowing plants to utilize it
    d) All of the above

9. Preferred location for borrow pits:
   a) Field land on the left side of the canal
   b) Field land on the right side of the canal
   c) Field land on both sides of the canal
   d) The central half width of the canal section
10. The hydraulic structure should be able to withstand
    a) Seepage forces
    b) Hydraulic jump
    c) Hydraulic pressure
    d) All the above
11. Main canals are equipped with cross regulators for
    a) Regulating water supply in the distributaries
    b) Increasing water head upstream when the main canal is running with low supplies
    c) Overflowing excessive flow water
    d) None of these
12. In standing crop, water consumption equals the depth of the water
    a) Transpired by the crop
    b) Evaporated by the crop
    c) Both transpired and evaporated by the crop
    d) Transpired evaporated from adjacent soil used by the crop in trans
13. Effective precipitation for a crop is
    a) The total amount of precipitation minus loss due to evaporation
    b) The total amount of precipitation minus the loss due to infiltration
    c) In the crop period, the total precipitation
    d) Amount of water available in the soil around a crop's root zone
14. Which place is commonly used to irrigate with tank water?
    a) Madhya Pradesh
    b) Himachal Pradesh
    c) Andhra Pradesh
    d) Uttar Pradesh
    Answer: Andhra Pradesh
15. Soil useful moisture is equal to
    a) The difference between field capacity and permanent wilting point within a root zone
    b) Field capacity
    c) Saturation capacity
    d) Moisture content present at permanent wilting point
16. For trickle irrigation (drip irrigation) the field water efficiency is:
    a) 40–55%
    b) 50–85%
    c) 80–90%
    d) 60–80%
17. Subirrigation is used in areas with:
    a) Areas with a low water table
    b) Areas with a high water table
    c) Areas with sloping terrain
    d) Areas with flat terrain
18. Fertigation is a process used in
    a) Sprinkler irrigation
    b) Surface irrigation
    c) Drip irrigation
    d) Center pivot irrigation

19. Which of the following is the incorrect statement:
    a) Free flooding irrigation involves the entry of water from one corner of the field and the spreading of it over the entire field.
    b) In a check irrigation method, the field is divided into smaller compartments, and water is applied in succession to each.
    c) Furrow irrigation consists of injecting water between the rows of plants in the field.
    d) None of these.
20. The depth of the rice root zone is
    a) 55 cm
    b) 40 cm
    c) 85 cm
    d) 90 cm
21. For standing crops in undulating sandy fields, the best method adopted for irrigation is
    a) Sprinkler irrigation
    b) Free flooding
    c) Check method
    d) Furrow method
22. A field's water efficiency can be calculated by:
    a) Water absorbed by crop minus water applied to a field
    b) (Water transpired by crop divided by water applied to a field) × 100%
    c) Water transpired by crop minus water applied to a field
    d) (Water absorbed by crop divided by water applied to a field) × 100%
23. Micro-irrigation is also called:
    a) Nano-irrigation
    b) Petite irrigation
    c) Localized irrigation
    d) Flood irrigation
24. The soil moisture that plants need for growth is
    a) Capillary water
    b) Gravity water
    c) Hygroscopic water
    d) Chemical water
25. Effective precipitation for a crop is the
    a) Stored water available in the soil in the root zone
    b) Total amount of precipitation – the loss due to evaporation
    c) Total amount of precipitation – the loss due to infiltration
    d) Total amount of precipitation during the crop period
26. The highest water use efficiency is obtained in
    a) Sprinkler irrigation
    b) Drip irrigation
    c) Canal irrigation
    d) Fogger irrigation
27. The top of the capillary zone
    a) Is under the water table at every point
    b) Is above the water table at every point
    c) Coincides with the water table at every point
    d) None of these
28. A soil's field capacity is determined by
    a) Capillary tension of the soil
    b) Porosity of the soil
    c) Both a and b
    d) None of the above

29. .... is the ratio between the volume of water delivered to a crop and the area it occupies
    a) Critical depth
    b) Duty
    c) Delta
    d) Base
30. The most common method adopted for irrigation of cereal crops is the
    a) Free flowing method
    b) Check method
    c) Furrow method
    d) Sprinkling method

ANSWER

1	2	3	4	5	6	7	8	9	10	11	12
b	d	d	d	d	b	d	c	d	d	b	d

13	14	15	16	17	18	19	20	21	22	23	24
d	c	a	c	b	c	d	d	a	b	c	a

25	26	27	28	29	30
a	b	b	c	c	c

REFERENCES

Allen, R. G., Pereira, L. S., Raes, D. and Smith, M. 1998. *Crop Evapotranspiration, Guideline for Computing Crop Water Requirements. FAO Irrigation and Drainage*, Paper 56. FAO, Rome, p. 300.

Allen, R. G., Smith, M., Pereira, L. S. and Perrier, A. 1994. An update for the calculation of reference evapotranspiration. *ICID Bulletin*, 43(2), 35–92.

Clemente, R. S., Asadi, M. E. and Dixit, P. N. 2005. Assessment and comparison of three crop growth models under tropical climate conditions. *Journal of Food, Agriculture and Environment*, 3(2), 254–261.

Doorenbos, J. and Kassam, A. H. 1979. *Crop Yield Response to Irrigation*. Irrigation and Drainage Division, FAO, Rome, Paper No. 33.

George, B. A., Reddy, B. R. S., Raghuwanshi, N. S. and Wallender, W. W. 2002. Decision support system for estimating reference evapotranspiration. *Journal of Irrigation and Drainage Engineering*, 128(1), 1–10.

Hargreaves, G. H. and Samani, Z. A. 1982. Estimating potential evapotranspiration. Tech. Note. *Journal of Irrigation and Drainage Engineering*, 108(3), 225–230.

Hargreaves, G. H. and Samani, Z. A. 1985. Reference crop evapotranspiration from temperature. *Applied Engineering in Agriculture*, 1(2), 96–99.

Jalota, S. K. and Arora, V. K. 2002. Model-based assessment of water balance components under different cropping systems in north-west India. *Agricultural Water Management*, 57(1), 75–87.

Kumar, R., Jat, M. K. and Shankar, V. 2012. Methods to estimate reference crop evapotranspiration – A review. *Water Science and Technology*, 66(3), 525–535.

Prihar, S. S., Gajri, P. R. and Narang, R. S. 1974. Scheduling irrigations to wheat, using pan evaporation. *Indian Journal of Agricultural Science*, 44, 567–571.

Singh, R. V. 1987. Design requirements for installation of lysimeters. *Proceedings of National Symposium on Hydrology* held at National Institute, Roorkee, 1, VI-14–VI-21.

Van Genuchten, M. Th. 1980. A closed-form equation for predicting the hydraulic conductivity of unsaturated soils. *Soil Science Society of America Journal*, 44, 892–898.

8 Plastics for Crop Protection

8.1 INTRODUCTION

The protection from hail, wind, snow, or severe rains in fruit-farming and ornamental crops, shade nets for greenhouses, and nets that moderately alter the microenvironment around a crop. Plastic nets are widely used in different agricultural applications. In addition, networks are also used to shield insects and birds from virus vectors, as well as for harvesting and post-harvest operations. Both woven and nonwoven products are defined as nets in the industry. In order to avoid confusion, the following definition of plastic networks is proposed: a plastic net is a substance composed of plastic fibers bound together, forming a regular porous geometric structure in a woven or knitted way and allowing fluids (gases and liquids) to flow through.

The most widely used raw material for farm networks is high-density polyethylene (HDPE). Polypropylene (PP) is also used as a raw material for networks, primarily to treat nonwoven layers. Just a few national standards actually apply specifically to agricultural networks and films. As far as networks are concerned, there is a set of Italian standards covering a wide variety of nets used in agricultural services.

Low-density polyethylene (LDPE) is one of the most widely used greenhouse covers (Dilara and Briassoulis, 2000; Hanafi and Papasolomontos, 1999; Briassoulis and Schettini, 2003; Briassoulis, 2006; Dehbi *et al.*, 2015). In spite of the low cost of these polymeric materials, the economic advantages of plasticulture can be seriously damaged by extreme climatic factors, which decrease the lifetime of the plastic cover (Antignus *et al.*, 1998; Sampers, 2002; Schettini *et al.*, 2011; Schettini and Vox, 2012; Picuno *et al.*, 2012; Dehbi and Mourad, 2016). Despite the continuous efforts made by plastic producers (Stefani *et al.*, 2008, De Salvador *et al.*, 2008), this important characteristic (lifetime) is actually limited to four to five seasons. The specific working conditions, that is, climatic parameters (air temperature, solar radiation, humidity, temperature of the contacting metallic frame, etc.) as well as other local actions (e.g., mechanical constraints, sand wind, contact with agrochemicals and pesticides, etc.), could indeed considerably modify some important technical properties like mechanical strength, radiometric properties, and gas permeability of the exposed plastic films (Briassoulis and Aristopoulou, 2001; Tavares *et al.*, 2003; Schettini *et al.*, 2014).

The greenhouse covers are generally made up of monolayer films of LDPE having 200 µm thickness. Due to the aging and the variation in their radiometric and mechanical properties, trilayer films (Adam *et al.*, 2005) typically with 220-µm thickness made of LDPE, and poly(vinyl acetate) (PVA) with air bubbles entrapped in the middle layer were developed (PROSYN-POLYAN). Due to the air bubbles in the middle layer, this film is more efficient in maintaining the temperature in the greenhouse. However, trilayer films based on LDPE, produced by Agrofilm SA (Algeria) without air bubbles, showed better mechanical properties when compared to the monolayer LDPE film (Dehbi, *et al.*, 2012). Recently, an innovative five-layer film for the greenhouse was produced by Ginegar Plastic Products Ltd.

8.2 LOW TUNNELS

Low tunnels have the same features as greenhouses, except for their complexity and height. Crops that are most commonly cultivated in tunnels are asparagus, strawberries, watermelon, etc. (Figures 8.1–8.2).

FIGURE 8.1 Plastic-covered tunnel with inside plastic soil mulching. (*Source*: https://www.plasticseurope.org/en/about-plastics/agriculture)

FIGURE 8.2 Low tunnels (vegetable grown) (https://www.indiamart.com/proddetail/crop-protection-cover-17967002633.html)

8.3 MULCHES

Mulching or covering the ground with plastic film (usually black, transparent, or white) helps maintain humidity as evaporation is reduced. It also improves thermal conditions for the plant's roots, avoids contact between the plant and the ground, and prevents weeds from growing and competing for water and nutrients (Figure 8.3).

FIGURE 8.3 Plastic mulching *(Source*: https://www.plasticseurope.org/en/about-plastics/agriculture)

8.4 NETHOUSES

Nethouses are naturally ventilated climate control structures used for a variety of applications viz. flowers, cultivation of vegetable crops, herbs/medicinal plants, nursery management (including secondary hardening of tissue culture-raised plants). In nethouses, different environmental conditions like temperature, light intensity, humidity, soil media, irrigation, disease and pest control, fertigation, and other agronomical practices are maintained throughout the season according to the need of the crops grown irrespective of the natural conditions outside.

8.4.1 Structure

In India's first "dome-shaped wire purlin" nethouse was constructed which is a galvanized pipe structure with wires and steel cable closing with mesh. It has galvanized Iron wire and central and perimeter fixing for suspension of the structure during high wind.

8.4.2 Advantages of Nethouses

The naturally ventilated nethouses have different advantages as given below.

i) Has improved water use efficiency: using drip irrigation and irrigating only the root zone of the crop
ii) Regulates variations in environment
iii) Reduces pest infestation: e.g., UV-stabilized covering material like 40/50 mesh insect net
iv) Ergonomic: prefabricated structure makes it easy to assemble and disassemble
v) Indeterminate crop variety: number of fruits per plant is 4–5 times higher as compared to determinate varieties. Seedless parthenocarpic (self-pollinated) varieties can be grown
vi) Increase in yield: yield increases by 5–7 times or even more as compared to crops grown in open fields due to tressling and a controlled environment
vii) Reduced disease attack: reduction in disease-control cost

viii) Eliminates vagarities of the environment: makes cultivation possible in problematic topography, climate conditions, and soil conditions
ix) Improved fertilizer use efficiency: using drip irrigation and fertigation

8.4.3 Net Types

Net types are distinguished by various structural features, such as material form, thread types and dimensions, texture, mesh size, porosity/solidity, and weight; radiometric features, such as color, transmissivity/reflectivity/shading factor; physical features, such as air permeability; and mechanical features, such as tensile tension, strength, and break elongation (Castellano and Russo, 2005). The measurements of nets available typically vary considerably in width and weight. Generally, widths vary from 1 m to 6 m or 12 m to 20 m (depending on the shape of the net) and lengths range from 25 m to 300 m. By adding the width of the required number of nets, larger nets are constructed. Depending on the fiber type, the thread type, and the texture, the color and chemicals used for the first grouping of nets may vary.

8.4.4 Types of Materials

For agricultural networks, polyethylene is mainly used. It is a nontoxic substance that can be used in direct contact with plants; it is completely recyclable, easily convertible, waterproof, and resistant to ultraviolet (UV) radiation agents when stabilized in the correct quantity; also, it has outstanding mechanical properties. Polypropylene is used as a raw material in the manufacturing of nonwoven layers. This type of membrane is used in horticulture and orchards as a direct cover for plants to shield crops from fog, frost, or wind. Nonwoven layers are marked by very low structural integrity and cannot be used as covering for the structural frame. Often used in some groundbreaking net agricultural production are biodegradable starch-based fabrics. At the end of their lifespan, biodegradable goods may be disposed of either in the soil or combined with agricultural ingredients, such as animal and vegetable wastes and as waste in a composting plant to produce carbon-rich compost. Due to their high cost compared to other plastic materials and the decline in their physical and mechanical properties when they are exposed to climate agents, mainly solar radiation, for long periods of time, biodegradable materials are not very common in the market.

8.4.5 Types of Threads and Texture

High-density polyethylene (HDPE) is composed of two large types of threads: spherical monofilaments or flat tapes. The HDPE compound directly extrudes round monofilaments; so in order to obtain flat tapes, the film of the desired thickness and color must first be produced and finally cut. Three main topologies of nets can be defined for traditional agricultural applications on the basis of texture form: flat woven or Italian; English or Leno; knitted or Raschel. The flat woven net is distinguished by a simple orthogonal weaving between weft and warp threads. The weft is the horizontal thread passing between vertical threads and warp, shaping the fabric during the development of the loom. Flat woven nets are light and stable in their shape, but they are relatively stiff and resist deformation. The English woven is a converted flat woven net and is made using the same type of looms. Like the flat woven one it is based on orthogonal weaving between weft and warp threads, but with a double thread in the weft direction, enclosing the warp thread in between. Where a more rigid protective covering is required, during heavy hailstorms, English woven nets are used, for example, for vineyards. Raschel looms develop nets of longitudinal "chains" and transverse knitted strings. In Raschel nets, both threads are connected with each other to prevent the unraveling of threads, for example, as a result of strong wind or hailstorms. Usually, jungles on the borders secure and defend the net, and strengthened buttonholes help the supporting framework to deliver networks more reliably and easily.

8.4.6 Mesh Size, Porosity, Solidity, and Weight

According to the texture, the single threads are connected to one another in such a way that they form a regular porous geometric structure, the mesh. The mesh size is the change in the direction of the warp or weft between two threads. The mesh size is given in mm for both warp threads and weft threads and ranges from 0.2 mm to 3.1 mm for insect nets, 1.7 mm to 7.0 mm for shade nets, 2.5 mm to 4.0 mm for anti-hail nets, and 1.8 mm to 7.0 mm for windbreak nets, while the anti-bird nets are characterized by higher values of 3 cm to 4 cm.

The proportion of the net's open area divided by the net's total area is the porosity of the porous geometric structure. Three techniques can be used to analyze it: radiation balance, solar radiation interception, and material image processing (Cohen and Fuchs, 1999).

8.4.7 Mechanical Properties

The mechanical and physical properties and the durability of a plastic net are strengthened or modified with additives. Additives are also used to increase the permeability of water and polymer fire retardancy and minimize dust accumulation (antistatic additives). They are developed in grains and blended with HDPE, with appropriate proportions, in order to form the compound. The most popular ones are chromatic and UV stabilizers. The stability of the mechanical properties of HDPE depends primarily on its UV radiation resistance, the main cause of HDPE degradation (Kumar and Poehling, 2006; Mourad and Dehbi, 2014) (Table 8.1).

8.4.8 Color

The color of the net is obtained by mixing chromatic additives into HDPE grains prior to the compound's production. The most popular net colors are black, green, or translucent. Transparent networks are used in certain applications in which the shading effect of the net is seen as a negative outcome of the net performance. Black nets are widely used for shading installations where the restriction of incoming solar radiation is advantageous. The black dye is obtained using a "carbon black" additive, which also acts as a UV stabilizer, and therefore the longevity of black thread nets is greater than transparent threads.

8.4.9 Transmissivity, Reflectivity, and Shading Factor

The efficiency of agricultural production and the aesthetic characteristics of the netting system are influenced by the radiometric characteristics of agricultural networks, such as transmissivity, reflectivity, shading factor, or the ability to change the nature of the radiation moving through the net. The construction parameters of the net, along with the structure form, sun positioning, and sky conditions, affect the radiometric efficiency of the permeable structure (Dilara and Briassoulis,

TABLE 8.1
Physical and Mechanical Characteristics of Most Common Agricultural Nets (Castellano et al., 2008)

Agricultural Application	Thread	Shading (%)	UV Res. (kly)	Areic Mass (g/m²)	Br Warp (kN/m)	Br Weft (kN/m)	Plastic Strain (%)	Den
Shading	R, T	25–90	400–800	50–250	4–10	2–15	20–30	450–800
Anti-hail	R, T	10–25	400–800	30–70	4–7	2–4	20–40	500–700
Anti-insects	R	10–20	400–600	70–130	4–5	2–4	20–30	300–450
Windbreaks	R, T	30–70	400–800	60–180	5–15	4–18	20–35	300–450
Ant-birds	R	5–15	300–600	10–30	0.5–2.5	0.5–2.5	20–30	300–450

1998; Hemming *et al.*, 2007). For this purpose, to minimize small-size effects, radiometric properties should be examined on relatively large samples. Nets are fabrics that are nonuniform. Photosynthetically active radiation (PAR, 400–700 nm) transmittance is the most significant radiometric property of covering materials from an agronomic point of view, as PAR is necessary for plant photosynthesis and growth.

A very common commercial parameter which defines a net is the shading factor that describes a net's ability to absorb or reflect a certain portion of solar radiation.

8.4.10 Air Permeability

The power of the net to move the air into it is the permeability of the net. It relies on several variables, such as the air's viscosity and velocity, the dimension and shape of the thread, and the distance between the threads and the texture. Undamaged porous screens are typically used as thermal screens and have a permeability of up to 11 m^2 and insect screens usually have a permeability of less than 8 m^2.

8.5 AGRICULTURAL APPLICATION OF NETS

The main agricultural uses of nets and net-covering systems are protection from meteorological hazards, insects, and small animals; removal of solar radiation; and soil shielding. Nets are also used during harvesting operations to capture the plucked fruits, such as olives, chestnuts, almonds, walnuts, and other small fruits; to package; and to procure cut flowers and dried fruits. Nets installed in the vents of greenhouses or directly on trees are used to protect seeds from insects, birds, and small animals such as rabbits, hares, and mice.

8.5.1 Protection against Meteorological Hazards

One of the most important agricultural uses of permeable plastic coverings is to shield crops from wind, hail, snow, frost, and rain. Windbreak networks are used to minimize the impact of wind on crops, avoid mechanical damage (e.g., breaking of trees, flowers) and biological effects (high evapotranspiration, problems with pollination); improve the quality of products by protecting them from dust, salt, and sand; reduce the wind load on agricultural structures; and minimize the heat loss of animals due to ventilation in open livestock farms. Usually, windbreak nets are attached to a supporting structure consisting of columns or trusses, mounted on a steel, concrete, or wood foundation.

Nets eliminate crop damage associated with hail. In open-field applications, they are frequently used in particular in the production of fruits such as grapes, peaches, apricots, and cherries, where they are mounted or added directly to cultivation with a simple support system. Anti-hail networks are considered a critical protection measure in some cases for greenhouses filled with glass panels, where hail damage can have costly economic effects for materials and crops and can pose a danger to the safety of greenhouse workers.

8.5.2 Reduction of Solar Radiation

Shading nets are designed to reduce solar radiation in order to lower the air temperature inside greenhouses or to decrease the light level for certain shade-loving crops, such as some ornamental plants. The effectiveness of shading systems is dependent on the shading element of the net.

8.5.3 Protection against Insects

Insect-proof nets are widely used in organic farming and are regarded as a pesticide-friendly alternative for environmental and human health. They are used in screen house coverings, with single or double layers, for virus-free production.

8.6 ANTI-HAIL NET

Anti-hail nets may change the microclimate of orchards and hence modify the physicochemical and sensory characteristics of fruits. Anti-hail nets are a part of basic equipment in a modern fruit orchard. They decrease the risks of crop production and thus allow regular and quality fruit harvest. The colors of nets differently obstruct the passing of light through the net, which directly affects some quality parameters of the yield, especially the fruit skin color (Briassoulis, 2007; Dehbi *et al.*, 2017). Anti-hail nets point to the importance of consistently carrying out the technical support measures with the intention of balancing the growing conditions. Another benefit of using the anti-hail nets, as seen in practice, is also the possibility of combining them with the plant support systems. Due to the elimination of production risks, which in some regions poses a distinct problem because of the hail, the placement of anti-hail nets allows for a quicker return of the invested assets in orchard establishment, earlier and more regular yields of high quality, and thus a higher economy of apple production. It has provided a hail-proof solution for safe crop/apple production against hail damage in fruit orchards. An anti-hail net protects crops throughout the year to avoid damage caused by hail, birds, and other attacks. It is suitable for fruits (apples, grapes, and pears in Kashmir) and also crops like vegetables, saplings, and flowers.

The following are the benefits of anti-hail net usage.

i) Helps to protect the crop against hailstorm
ii) Reduce the alternate bearing incidence due to protection of fruit trees
iii) Reduction of sunburn in hot years and protection against birds and insects
iv) Lesser impact of strong wind and continuous market supply of high-quality fruits

8.6.1 Features of Anti-Hail Net

The following are the important properties of the anti-hail net.

- Lightweight and high strength (for crops to be able to take the weight of the net)
- Easy to install (flexible)
- UV stabilized (durable)
- Technically designed for maximum protection from hail
- Designed for covering orchards, vineyards, and greenhouse
- Protects fruit, shrubs, buds, and seeding's from damage caused by hail
- Can be provided in multicolor for better plant photosynthesis

8.6.2 Anti-Hail Net Light Transmittance

Light plays a crucial role when it comes to the intensity of photosynthesis, differentiation of oral buds, and consequently also affects the quality of fruit. It is also a key factor affecting the taste, color, and resilience of fruits. Photosynthetically active radiation (PAR), which is needed for photosynthesis, has a wavelength between 400 mm and 700 nm. Approximately from 85% to 95% PAR radiation is absorbed by a leaf, and the rest either is reflected or penetrates the leaf. The level of light is decreased for approximately 7% (PAR)–20% (UV) with a white net (which allows for the maximum light exposure under a net): 11% (PAR)–28% (UV) with a red-white net, 12% (PAR)–23% (UV) with a green-white net, 13% (PAR) with a light gray net, 15% (PAR)–26% (UV) with a black and green, 16% (PAR)–23% (UV) with a black net and 18% (PAR)–29% (UV) with a red net, measured 50 cm below the net. The level of UV light is decreased by 29% with a black net. With a gray net, the level is decreased by 13% (PAR), whereas with a black net by 18% (PAR).

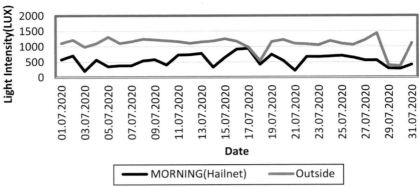

FIGURE 8.4 Variation in light intensity inside/outside an anti-hail net for the month of July

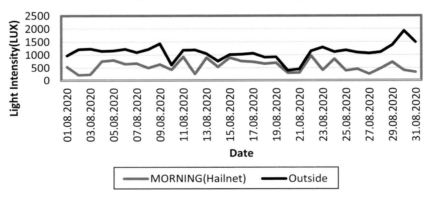

FIGURE 8.5 Variation in light intensity inside/outside an anti-hail net for the month of August

Different types of anti-hail nets obstruct the passing of the light differently and have different life expectancies. The thinner the net and the bigger the meshes, the higher are the swings in transmittance levels. Black anti-hail nets in comparison to white ones have a longer life expectancy, lasting from 15 to 20 years, and blend into the color composition of the landscape more easily but decrease the light transmittance by 20%.

A study was conducted to evaluate the effect of the anti-hail net in apple crops in temperate regions of Kashmir during 2019–2021. A comparison was made between light intensity inside the anti-hail net and outside the hail net. Figures 8.4 and 8.5 indicate the sunshine intensity (lux) which was measured under the anti-hail net and outside the hail net in the apple orchard.

In the case of apples, it has been found that covering apple orchards with hail nets is an effective strategy in reducing the surface temperature of apples and the amount of UV radiation they receive. In areas that will have above optimal temperatures and solar radiation with climate change, the use of hail nets can reduce sunburn losses, improve flavor, and improve consumer desirability of apples. Figure 8.6(a–c) shows the variation in weather parameters viz. temperature, relative humidity, and dew point inside the anti-hail net. The anti-hail net (white) used was HDPE monofilament with an overall dimension of 3,000 m^2 (Figures 8.7–8.10).

Plastics for Crop Protection

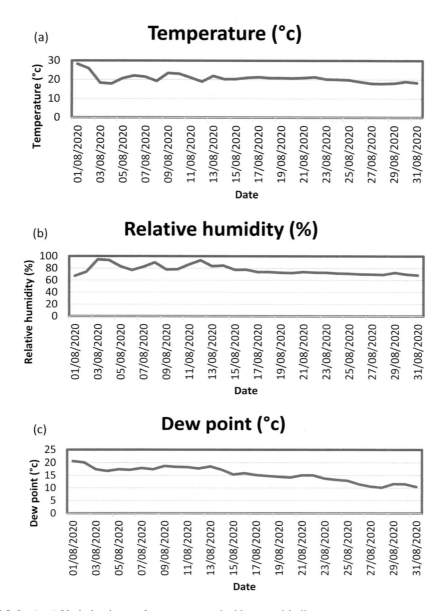

FIGURE 8.6 (a–c) Variation in weather parameters inside an anti-hail net

8.7 ANTI-INSECT NET

Anti-insect nets are made after understanding the pest life cycle and physical properties as each pest has a different body structure and flying habit (pest-control mesh). These are developed considering the parameters such as the diameter of the yarn and spacing between the yarn to prevent insects of different sizes from entering and to provide optimum ventilation. It is made up of HDPE monofilament of mesh 25, 30, 40, 50 with milky white color.

8.7.1 Advantages

The following are the advantages of the anti-insect net.

FIGURE 8.7 Apple plants under transparent anti-hail nets

FIGURE 8.8 Apple plants under the anti-hail nets

FIGURE 8.9 Cherry plants under anti-hail nets at fruiting stage

Plastics for Crop Protection

FIGURE 8.10 Apple plants under anti-hail nets

- Helps to keep out a wide range of pests and insects
- Reduces the need for the application of harmful pesticides
- Reduces downtime due to insect reentry
- Gives improved ventilation

SHORT AND LONG QUESTIONS

1. Explain the role of plastics in crop protection.
2. What is a shade net? What are its types?
3. Discuss the role of shade net in crop protection.
4. Define the following terms:
    a) Mesh size
    b) Porosity
    c) Solidity
    d) Shading factor
5. Explain the physical and mechanical characteristics of agricultural nets.
6. Discuss the role of windbreak and shade net in crop protection.
7. How can nets protect crops against insects?
8. What is plastic mulch? Give its types.
9. How is plastic mulch useful in agriculture?

MULTIPLE-CHOICE QUESTIONS

1. _____ is widely used raw material for farm networks:
    a) LDPE   b) PVC
    c) HDPE   d) None of these
2. Which of the following is used primarily to treat nonwoven layers?
    a) LDPE   b) PVC
    c) PP   d) None of these
3. The width of nets available typically varies from
    a) 1 to 6 m   b) 2 to 9 m
    c) 15 to 30 m d) 50 to 75 m

4. The mesh size for insect nets ranges from
   a) 0.2 to 3.1 mm   b) 2.5 to 4.0 mm
   c) 1.7 to 7.0 mm   d) 1.8 to 7.0 mm
5. The mesh size for shade nets ranges from
   a) 0.2 to 3.1 mm   b) 2.5 to 4.0 mm
   c) 1.7 to 7.0 mm   d) 1.8 to 7.0 mm
6. The mesh size for anti-hail nets ranges from
   a) 0.2 to 3.1 mm   b) 2.5 to 4.0 mm
   c) 1.7 to 7.0 mm   d) 1.8 to 7.0 mm
7. The mesh size for windbreak nets ranges from
   a) 0.2 to 3.1 mm   b) 2.5 to 4.0 mm
   c) 1.7 to 7.0 mm   d) 1.8 to 7.0 mm
8. The efficiency of agricultural production and the aesthetic characteristics of the netting system are influenced by
   a) Transmissivity   b) Reflectivity
   c) Shading factor   d) All of these
9. The most popular net colors are
   a) Black   b) Green
   c) Translucent   d) All of the these
10. _____ eliminate crop damage associated with hail
    a) Shade nets   b) Insect nets
    c) Anti-hail nets   d) None of these

ANSWERS

1	2	3	4	5	6	7	8	9	10
c	c	a	a	c	b	d	d	d	c

REFERENCES

Adam, A., Kouider, S. A., Youssef, B., Hamou, A. and Saiter, J. M. 2005. Studies of polyethylene multi layer films used as greenhouse covers under Saharan climatic conditions. *Polymer Testing*, 24(7), 834–838.

Antignus, Y., Lapidot, M., Hadar, D., Messika, M. and Cohen, C. 1998. Ultraviolet absorbing screens serve as optical barriers to protect greenhouse crops from virus. *Journal of Economic Entomology*, 9, 1140–1405.

Briassoulis, D. 2006. Mechanical behaviour of biodegradable agricultural films under real field conditions. *Polymer Degradation and Stability*, 91(6), 1256–1272.

Briassoulis, D. 2007. Analysis of the mechanical and degradation performances of optimised agricultural biodegradable films. *Polymer Degradation and Stability*, 92(6), 1115–1132.

Briassoulis, D. and Aristopoulou, A. 2001. Adaptation and harmonisation of standard testing methods for mechanical properties of low-density polyethylene (LDPE) films. *Polymer Testing*, 20(6), 615–634.

Briassoulis, D. and Schettini, E. 2003. Analysis and design of low-density polyethylene greenhouse films. *Biosystems Engineering*, 84(3), 303–314.

Castellano, S., Scarascia Mugnozza, G., Russo, G., Briassoulis, D., Mistriotis, A., Hemming, S. and Waaijenberg, D. 2008. Plastic Nets in Agriculture: A general review of types and applications. *Applied Engineering in Agriculture*, 24(6), 799–808.

De Salvador, F. R., Scarascia Mugnozza, G., Vox, G., Schettini, E., Mastrorilli, M. and Bou Jaoudé, M. 2008. Innovative photoselective and photoluminescent plastic films for protected cultivation. *Acta Horticulturae*, 801(801), 115–121.

Dehbi, A., Djakhdane, K. and Mourad, A. H. I. 2012. Impact of degradation of polyethylene films under simulated climatic conditions on their mechanical behavior and thermal stability and lifetime. *American Society of Mechanical Engineers*, 6, 131–135.

Dehbi, A. and Mourad, A. H. I. 2016. Durability of mono-layer versus tri-layers LDPE films used as greenhouse cover: Comparative study. *Arabian Journal of Chemistry*, 9, S282–S289.

Dehbi, A., Mourad, A. H. I., Djakhdane, K. and Hilal-Alnaqbi, A. 2015. Degradation of thermo mechanical performance and lifetime estimation of multilayer greenhouse polyethylene films under simulated climatic conditions. *Polymer Engineering and Science*, 55(2), 287–298.

Dehbi, A., Youssef, B., Chappey, C., Mourad, A. H. I., Picuno, P. and Statuto, D. 2017. Multilayers polyethylene film for crop protection in harsh climatic conditions. *Open Access*, 2017, ID 4205862. doi: 10.1155/2017/4205862

Dilara, P. A. and Briassoulis, D. 1998. Standard testing methods for mechanical properties and degradation of low density polyethylene (LDPE) films used as greenhouse covering materials: A critical evaluation. *Polymer Testing*, 17(8), 549–585.

Dilara, P. A. and Briassoulis, D. 2000. Degradation and stabilization of low-density polyethylene films used as greenhouse covering materials. *Journal of Agricultural Engineering Research*, 76(4), 309–321.

Hanafi, A. and Papasolomontos, A. 1999. Integrated production and protection under protected cultivation in the Mediterranean region. *Biotechnology Advances*, 17(2–3), 183–203.

Hemming, S., Dueck, T., Janse, J. and van Noort, F. 2007. The effect of diffuse light on crops. Paper and Presentation during *ISHS Symposium Greensys 2007 -- High Technology for Greenhouse System Management*, 4–6 October 2007, Naples/Italy. Will be published in *Acta Horticulturae*, 801.

Kumar, P. and Poehling, H. M. 2006. UV-blocking plastic films and nets influence vectors and virus transmission on greenhouse tomatoes in the humid tropics. *Environmental Entomology*, 35(4), 1069–1082.

Mourad, A. H. I. and Dehbi, A. 2014. On use of trilayer low density polyethylene greenhouse cover as substitute for monolayer cover. *Plastics, Rubber and Composites: Macromolecular Engineering*, 43(4), 111–121.

Picuno, P., Sica, C., Laviano, R., Dimitrijević, A. and Scarascia-Mugnozza, G. 2012. Experimental tests and technical characteristics of regenerated films from agricultural plastics. *Polymer Degradation and Stability*, 97(9), 1654–1661.

Sampers, J. 2002. Importance of weathering factors other than UV radiation and temperature in outdoor exposure. *Polymer Degradation and Stability*, 76(3), 455–465.

Schettini, E., De Salvador, F. R., Scarascia-Mugnozza, G. and Vox, G. 2011. Radiometric properties of photoselective and photoluminescent greenhouse plastic films and their effects on peach and cherry tree growth. *Journal of Horticultural Science and Biotechnology*, 86(1), 79–83.

Schettini, E., Stefani, L. and Vox, G. 2014. Interaction between agrochemical contaminants and UV stabilizers for greenhouse EVA plastic films. *Applied Engineering in Agriculture*, 30(2), 229–239.

Schettini, E. and Vox, G. 2012. Effects of agrochemicals on the radiometric properties of different anti-UV stabilized EVA plastic films. *Acta Horticulturae*, 956(956), 515–522.

Stefani, L., Zanon, M., Modesti, M., Ugel, E., Vox, G. and Schettini, E. 2008. Reduction of the environmental impact of plastic films for greenhouse covering by using fluoropolymeric materials. *Acta Horticulturae*, 801, 131–137.

Tavares, A. C., Gulmine, J. V., Lepienski, C. M. and Akcelrud, L. 2003. The effect of accelerated aging on the surface mechanical properties of polyethylene. *Polymer Degradation and Stability*, 81(2), 367–373.

9 Plastics in Drying and Storage

9.1 INTRODUCTION

The Green Revolution came into being with emphasis on high-yielding variety seeds, fertilizers, pesticides, better methods of farming, and postharvest management. It turned India from being deficient in food grains to being self-sufficient. But the increase in agricultural yield has to keep pace with the growing population. To remain self-sufficient in food grains, India needs another green revolution or rather a greener revolution. Innovative agro practices need to be adopted toward transformation of Indian agriculture. Plasticulture is one of the innovative applications of plastics and is the combination of two words plastics and agriculture, which means plastic in agriculture. Per capita consumption of plastic during 2015 in India is 9.7 kg/person which is far below the world average of 45 kg/person. So, the application of plastics in agriculture offers a huge opportunity in modernizing Indian agriculture. Plastics used in packaging solutions help in increasing the shelf life and during collection, storage, and transportation of fruits and vegetables. Plastics can play a major role in energy conservation. They require minimum energy in production and conversion to finished products. They have definite advantages over conventional materials because they have several properties: higher strength/weight ratio, superior electrical properties, superior thermal insulation properties, excellent corrosion resistance, superior flexibility, impermeability to water and resistance to chemicals, and less friction due to smoother surface.

Postharvest management involves a number of unit operations, such as precooling, cleaning, sorting, grading, storing, packaging, etc. The moment a crop is harvested from the ground or picked from its parent plant it starts to deteriorate. Hence, appropriate postharvest management strategies are indispensable to ensure better quality of crops, whether they are sold for fresh consumption or used to prepare processed products. Deterioration of fresh crops mainly causes weight loss, bruising injury, physiological breakdown due to ripening processes, temperature injury, chilling injury, or attack by microorganisms. Therefore, it is essential that crops must be kept in good conditions to acquire excellent quality with maximum shelf life. Various materials, including metals, stainless steel, plastics, woods, fibers, glass, papers, etc. are used in postharvest management operations. However, plastics are found to be the most used materials for the said purpose. A report indicates that the packaging industry in India has seen a strong diffusion of plastics as compared with global standards (Anonymous, 2014).

9.2 DRYING OF CROPS

Drying is a classical method of food preservation, which involves moisture removal through the application of heat. Applied heat raises the vapor pressure in the produce and thus removes moisture in the form of vapors. Drying extends the shelf life of the product and also reduces the weight and volume of the product which in turn facilitates transportation and reduces the storage space. Drying is achieved through the simplest methods like sun drying to the most modern methods like infrared drying, microwave drying, refractive window drying, etc. The application of plastics in some of the important drying methods is discussed below.

9.2.1 Low-Cost Poly House Technology for Drying

Solar drying is a continuous process where moisture content, air, and product temperature change simultaneously along with the two basic inputs to the system, i.e., the solar radiation and the ambient temperature. The drying rate is affected by ambient climatic conditions which include temperature, relative humidity, sunshine hours, available solar radiation, wind velocity, and frequency and duration of rain showers during the drying period. Poly house technology can be used for drying a wide variety of crops depending on the requirements.

The All India Coordinated Research project on Plastic Engineering in Agriculture Structural and Environment Management (PEASEM) scheme has standardized the poly house-type solar dryer with the establishment of different models. A poly house-type solar multi-tunnel dryer having a size of 5 m × 4 m and a floor area of 20 m² was fabricated. Figure 9.1 shows a greenhouse-type solar multitier dryer. The front side height of the dryer is 2.25 m and the back side height of the dryer is 3.25 m. The center height of the dryer is 4 m and the poly house is covered with a transparent UV-stabilized polythene plastic foil of 0.2 m thickness with a transmissivity of 92% for visible radiation which traps the solar radiation in the daytime and enhances the temperature inside to maintain it at the optimum level for drying of fruit and vegetables. The top surfaces of the collector and drying chamber were designed in a curved shape in order to increase the surface area of radiation. The inclination of the top surface of the dryer was kept 30°. The orientation of the dryer was kept in the north-south direction so that radiation from the sun would not be disturbed. The capacity of the dryer ranged from 100 kg to 150 kg of fresh fruit and vegetables depending upon the material and the thickness of the spreading layer.

The mean ambient temperature and relative humidity with the corresponding temperature and relative humidity inside the loaded and unloaded drier at different hours during the period of experimentation. The poly house drying was observed to be quicker than in the open condition. Further, the upper tray has more temperature as compared to that of the lower tray. The inside average temperature of the upper and lower tray of loaded drier was constantly observed to be higher than the ambient by a difference of about 20°C and 17.2°C, respectively. Figure 9.2 shows the view of a low-cost solar poly tunnel vegetable drier. The inside temperature of the upper and lower tray of unloaded drier was constantly observed to be higher than the ambient by a difference of about 22°C & 19.3°C, respectively. The maximum temperature inside was observed to be 59.5 °C (for upper tray) and 47.6°C (for lower tray) in loaded drier and 61.8°C (for upper tray) and 58.9°C (for lower tray) in unloaded drier. This is because the drier is based on the greenhouse effect and it traps the solar radiation resulting in a subsequent increase in temperature. The temperature range conducive for

FIGURE 9.1 Greenhouse-type solar multitier dryer

FIGURE 9.2 View of low-cost solar poly tunnel vegetable drier

TABLE 9.1
Comparative Performance of Solar Poly Tunnel Drier with Open Sun Drying for Different Fruits and Vegetables

Product	Open Sun Drying (days)	Solar Poly Tunnel Dehydration (days)
Tomato	6–7	2.5
Capsicum	5–6	2.5
Cabbage	5–6	2.0
Leafy Vegetable	4–5	1.5
Carrot	5–6	2.5
Apple	9–10	3.0

quick dehydration of fruits and vegetables is recommended as 66–71°C and 60–66°C, respectively. The chimneys drag the moist air within the chamber and remove excess moisture from the vicinity of the product placed on perforated trays. The relative humidity inside the dryer varied between 21% and 74% as compared to outside relative humidity which ranged from 40% to 75%. The temperature inside the dryer was 62–76% higher than the ambient conditions. The cost-benefit ratio of the dryer was 1.5 by adopting solar drying technology. The comparative performance of solar poly tunnel drier with open sun drying for different fruits and vegetables is illustrated in Table 9.1. Figure 9.3 shows the comparative drying of tomatoes.

9.2.2 Poly House Drying

The main objective of poly house dryers is to maximize the utilization of solar radiation. Poly house dryers mainly consist of a drying chamber, an exhaust fan, and a chimney. The roof and walls of such dryers are made of transparent UV-stabilized polythene sheets. Figure 9.4 shows the poly

 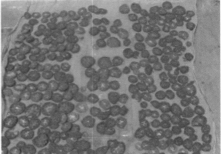

FIGURE 9.3 Comparative drying of tomatoes

FIGURE 9.4 Poly house dryers for drying of crops

house dryers for drying the crops. This sheet has a transmittivity of approximately 92% for visible radiation. It traps solar energy during the daytime and maintains the desired temperature for the drying of produce (Shahi *et al.*, 2011). The UV-stabilized polythene sheet plays a vital role in preventing the deterioration of product in the form of loss in nutritional and organoleptic properties. It allows only short wavelength radiation to penetrate inside and converts them into long wavelengths, when they strike on the surface of the product or floor. Long wavelength radiation cannot move out and thus increase the temperature inside the dryer. Polythene sheets used in the construction of poly house dryers have superior characteristics in terms of transparency, transmittivity, anti-corrosion property, self-adhesive, retraction ratio, tensile properties, tear-resistant, anti-puncture, water-proof, moisture-proof, dust-proof, etc. and hence outnumbered all other materials in poly house drying applications. In some studies, polycarbonate sheets have also been tested for poly house dryers (Janjai *et al.*, 2011). The report indicates that a black surface inside a poly house dryer increases the efficacy of converting light into heat (Shahi *et al.*, 2011). Black polythene sheets of 25–100-micron thickness are used on the floor of the dryer for such purposes.

Poly house dryers are still not very popular among farmers. However, efforts are being made to develop and popularize such dryers. A number of studies have been conducted to develop poly house dryers. Ekechukwu and Norton (1999) designed a natural circulation solar dryer covered with a polythene sheet. It consists of a semicylindrical drying chamber with a cylindrical chimney at one end. The dimensions of the drying chamber were 6.67 m long, 3.0 m wide, and 2.3 m high. Kulanthaisami *et al.* (2009) developed and tested a semicylindrical solar tunnel dryer covered with UV-stabilized semitransparent polythene of 200 microns for drying coconuts. The drying chamber of the solar tunnel dryer was 18 m long and 3.75 m wide and had a drying capacity of 5,000 coconuts. The polythene sheet used in this dryer was opaque to long-wave radiation. Similarly, a poly house dryer suitable for hot and arid regions of northwestern India was developed at ICAR-CIPHET,

Abohar, Punjab, India (Kadam et al., 2011). Dryer was a Quonset shape low-cost poly house having dimensions of 6 m long, 4 m wide, 1.8 m ridge height, and 32.75 m³ volume. It was oriented in the east-west direction. The poly house frame was constructed using bamboo and was covered with a 200-micron-thick UV-stabilized polythene sheet. The poly house dryer was provided with a 9-inch diameter exhaust fan placed opposite the door to remove the moisture accumulated inside. A study reported that the performance of poly house dryers was decidedly dependent on solar radiation, ambient temperature, and relative humidity. The increase in poly house temperature varied from 0.7°C to 19°C, whereas RH inside poly house dryer varied from 16% to 25.70% by an air exchange at a flow rate of 6.1 m³/s.

In a study, Shahi et al. (2011) developed a solar poly house tunnel dryer having 5 m length, 4 m breadth, 3.2 m central height, and side heights of 2.5 m left and 1.5 m right. This dryer consisted of a drying chamber, a small exhaust fan, and a metal duct. A transparent UV-stabilized polythene sheet of 200-micron thickness was used to construct the dryer. The top surface of the dryer was curved in order to increase the area of radiation. The orientation of the dryer was in the north-south direction to achieve maximum penetration of solar radiation in the dryer. The poly house dryer was equipped with a fan of 1,000–1,200 m³/h airflow rate and 1 kW power to achieve forced ventilation. The concrete floor inside the dryer was painted black for better absorption of solar radiation. A glass wool insulation of 2-inch thickness was also provided to the floor in order to reduce heat loss through it. The capacity of the solar tunnel ranged from 1 to 1.5 quintals of fresh fruits and vegetables, depending upon the materials and thickness of the spreading layer. This dryer was found efficient in the drying of different crops in the Kashmir valley. Similar studies have also been conducted by various workers and demonstrated the efficacy of poly house dryers in drying agricultural products efficiently (Janjai et al., 2011; Arjoo et al., 2017).

Attempts have also been made to develop poly house dryers and to evaluate their performance in farmers' fields. Along this line, a simple semicircular solar tunnel poly house dryer was developed in the field of a farmer at Pali, Rajasthan, India, having a dryer floor area of 5 m × 3.75 m. The orientation of the dryer was in the east-west direction. The framed structure was covered with a UV-stabilized polythene sheet (200 microns) having a ridge height of 2.25 m. The dryer was found suitable for drying various agricultural products on a large scale under a controlled environment. It was large enough to permit a person to enter into it and carry out operations such as loading and unloading the crops to be dried.

It can be stated that by using apposite poly house dryers, farmers can easily and properly manage their high-moisture perishable crops and thus reduce the postharvest losses noticeably. Figure 9.5 shows a poly house dryer in the field of a farmer.

Some researchers designed the solar tunnel dryer for drying 250 kg/h fish in batch mode to reduce the moisture content to 16% (wb) from initial moisture of 84% (wb). Greenhouse-type solar tunnel dryer for industrial drying of selected species of fish Croaker, Anchovy and Ribbon in the western coastal town Veraval (20° 53′ N, 73° 26′ E), Gujarat, India and was installed at Jose and Brothers Fish Industry. The single-span arc-type GI pipe frames were used to construct the dryer (collector area of 150.9 m²) and covered with a single-layer 200-micron-thick UV-stabilized polyethylene sheet. The test results showed that developed dryer can reduce the moisture content of salt-treated fish up to 42.85% to 66.66% (db) between 8 h and 16 h, whereas in the case of unsalted fish, moisture content was reduced up to 17.64–25% (db) in 24 h to 32 h of drying depending on a variety of fish and initial moisture content. The quality analysis showed that fish drying is better in the low-temperature range 0.96693.

9.2.3 Refractive Window Drying

Poly house drying is nearly impractical without application of suitable plastics. Similar is the case of refractive window drying. Refractive window drying is a novel drying technique that converts foods in the form of liquids or slurries into flakes or powders within a typical residence time of 3–5

FIGURE 9.5 Poly house dryer in the field of a farmer

min (Nindo and Tang, 2007). Generally, conventional drying methods involve high temperatures and/or longer durations which result in the degradation of heat-sensitive bioactive compounds (Pavan et al., 2012). This limitation is addressed by refractive window drying. In this drying system, heat energy is transferred from hot water ($\geq 95°C$) circulated beneath a plastic conveyor belt (Mylar™) and is used to dry a thin layer of liquid or slurried food product spread on the belt surface. On the end portions of this belt, cold water is circulated to assist the detachment of dried material by scraping (Nindo et al., 2003). The most important component of the refractive window drying system is the conveyor belt made of a food-grade plastic sheet. In general, drying is completed only when the heat transfer takes place from a hot medium (air or hot surface/liquid) to the product to be dried. In this system, heat is refracted from hot water to the product through a plastic conveyor belt. Hence, the characteristics of plastic belts decide the drying efficiency as well as end-product quality.

In the refractive window drying system, the product is dried as a thin film and cold air is circulated over the product layer to achieve its cooling which results in dried products exhibiting excellent color, vitamins, and antioxidant retention in comparison to other conventional drying methods (Caparino et al., 2012). Reports indicate that products dried using the refractive window drying system are comparable in quality to freeze-dried products. However, refractive-window drying is advantageous over freeze-drying as it is faster, energy-efficient, and cheaper (Nindo and Tang, 2007).

Postharvest management is essential to minimize the postharvest losses of agricultural crops. Various postharvest unit operations are practiced in order to achieve safe handling, shelf life enhancement, and value addition of agricultural crops. Numerous indigenous and engineered materials are used in different postharvest management operations. However, in recent times, plastics have outnumbered all other materials due to their bountiful advantages. From simple to the most complicated postharvest operations are carried out using various types of plastics. Major applications of plastics are seen in packaging, drying, storage, and transportation of crops. It appears that postharvest operations like packaging are almost impossible without plastics. However, a major drawback in the application of plastics in the postharvest management of crops is their adverse effect on the environment. Efforts are being made to develop and commercialize biodegradable plastics. Such plastics would be a boon to the postharvest management field in near future.

9.2.4 Open Sun Drying

Open sun drying is one of the traditional methods of drying. It involves spreading the harvested crops under the sun in open yards. This is still one of the most common drying methods used in India and other developing countries. It is a continuous process where moisture content, air temperature, and product temperature change simultaneously. In this process, the rate of drying is largely affected by air temperature, relative humidity, sunshine hours, solar radiation, wind velocity, duration of rain showers during the drying period, etc. Open sun drying is predominantly accomplished using plastics. Owing to the easy availability, low cost, and inertness of plastics, farmers use them for sun drying of harvested and threshed grains, fruits, vegetables, etc. Drying is completed even in the farms using plastics. There is no need for any furnished floor for the drying of crops on polythene. Tarpaulins made of high-density polythene (300–500 g/m^2) are used for drying purposes. Although open sun drying is economical and simple, it has drawbacks such as no control over drying rate, nonuniform drying, and chances of deterioration due to the exposure of products to rain, birds, dust, storm, rodents, insects, and pests, which results in poor quality of dried products. The direct exposure of products to the sun's UV radiation may also reduce the nutrients such as vitamins, carotenoids, etc. in the dried product. Likewise, UV radiation in the sun's rays changes the organoleptic properties, such as texture, color, and flavor of food materials (Sangamithra *et al.*, 2015). Therefore, indirect-type sun drying is found advantageous over the open sun drying method. Poly house dryers may be considered indirect-type solar dryers.

9.3 UNIT OPERATION AFTER HARVESTING

9.3.1 Field Handling of Crops

Field handling is the first postharvest unit operation carried out immediately after harvesting. It involves three major operations, namely collection, packaging, and transportation of crops. Crops are collected and subsequently transported, with or without packaging, to the desired estimation. Harvested crops are collected in field containers, such as bags, plastic buckets, plastic crates, woven baskets, gunny bags, etc. However, presently in India, plastic crates have replaced all other containers due to their strength, portability, corrosion resistance, etc. They are smooth with no sharp edges or projections to damage the produce. They are quite sturdy and hence do not bend out of shape when lifted or tipped. Packaging of harvested crops directly into packages in the field immediately after harvest reduces the damage caused by multiple handling.

9.3.2 Removal of Field Heat

Temperature management immediately after harvesting is the most essential step in extending the shelf life of crops. Freshly harvested fruits and vegetables contain significant amounts of field heat which needs to be removed prior to packaging, transportation, or storage. Recently, farmers have started using shade nets having 30–50% shading intensity to provide a cooler environment compared to ambient air temperatures. These shade nets are mainly made of UV-stabilized plastics (Ghosh and Ghosh, 2009).

9.3.3 Field Curing

Curing of roots, tuber, and bulb crops such as potatoes, sweet potatoes, and onions is an important practice if these crops are to be stored for a substantial period of time. Curing is accomplished by holding the produce at high temperatures and high relative humidity for several days after harvesting. The main objective of curing is to heal the wounds and form a new protective layer of cells. Curing becomes difficult in regions where harvesting coincides with the rainy/moist season.

Unexpected rains also damage the bulbs during curing. Under such circumstances, poly tunnels, and polythene sheet covers were found very useful in curing crops. The crops can also be cured after packing into 15–25 kg net sacks (Kitinoja and Adel, 2002)

9.3.4 Grading and Sorting

Grading and sorting are two important unit operations to increase the market value of a produce. Specially designed graders and sorters are used to accomplish these unit operations. Although various materials are used in the fabrication of these pieces of equipment, the plastics are also being used due to their low weight, sturdiness, corrosion resistance, etc.

9.3.5 Conveying

Conveying involves physically moving or lifting the produce in bulk from one location to another. Belt conveyors, screw conveyors, or elevators are used for this purpose. Strong or liquid structures may transfer fresh produce. Examples of sturdy structures used to shift either loose or packed materials include metal chain links, leather, plastic, or canvas belts, and wood, metal, or rubber rollers. For water-tolerant crops such as apples, carrots, and tomatoes, hydro-handling by flumes may be used. Engineered plastics are extensively used in conveyors and bucket elevators.

9.3.6 Storage

The shelf life of the product depends on its moisture content at the time of packaging and the rate of moisture gain during storage which is also called sorption isotherm study. Significant volumes of fresh and processed food materials are stored before ultimate consumption. From very simple to the most modern storage structures are utilized for the storage of agricultural crops. Clamps, root cellars, evaporative cooling structures, bunkers, cover and plinth (CAP), godowns, ventilated structures, and such types of storage structures are being used conventionally in India (Kale et al., 2016; Kale and Nath, 2018). However, in recent times, with the introduction of mechanical refrigeration systems, automatic conveyors, and elevators, the storage structures have seen a paradigm shift. Presently, cold storage, modified and controlled atmospheric storage, hypobaric storage, vertical metal silos, hermetic storage, etc. are used for bulk storage of various crops. It can be seen that plastic is extensively used in almost all types of storage structures as construction material. It can be noticed that out of all these structures, CAP storage is impossible without the use of suitable plastics. In India, CAP is still used to store very large volumes of wheat and paddy. In fact, CAP storage is a necessity in India as Indian grain production increased faster than storage capacity (Bhardwaj, 2015). The storage of food grains under large polythene has been practiced in India as well as in other countries for a long time. In CAP storage, outdoor stacks of bagged grains are covered with a water-proof polythene sheet. The advantage of CAP storage is its low cost. It is considered that the cost of CAP storage is only one-fourth of the cost of godown storage. However, CAP storage is vulnerable to wind damage and needs to be inspected frequently to detect the damage. The system requires careful management if severe losses are to be avoided. Careful quality control is achieved with regular sampling. For CAP construction, a plinth with hooks for the ropes lashing the stack is constructed on a suitable site. Dunnage is provided and the covers are made of black polythene sheet of 250 μm thickness shaped to suit the stack. The covers are held down by nets and nylon lashing. Condensation is prevented by placing a layer of paddy husk-filled sacks on the top of the stack under polythene. For typical 150-ton CAP storage, the commonly constructed size is 8.55 m × 6.30 m for 3,000 bags each of 50 kg capacity. It is generally provided on a raised platform where grains are protected from rats and the dampness of the ground. The grain bags are stacked in a standard size wooden dunnage. The stacks are covered with 250–350 μm low-density polyethylene (LDPE) sheets from the top and all four sides. Wheat grains are generally stored in such CAP storage for 6–12 months. It is the most economical storage structure and is widely used by the Food Corporation of India for bagged grains.

9.3.7 Transportation of Crops

Transportation is an inevitable postharvest operation in the journey of crops from farm to fork. Crops are transported many times after harvesting till consumption. Transportation occurs within the field, field to the packhouse, field to cold storage, packhouse to cold storage, field to the warehouse, cold storage to distant markets, markets to consumers, etc. Freshly harvested grains are generally transported in bulk/gunny bags using trolleys, whereas fresh fruits and vegetables are transported using rigid crates, sacks, wooden boxes, etc. One of the most important containers used for transporting fruits and vegetables is the returnable/reusable molded plastic crates. These reusable boxes are molded from high-density polythene (HDPE) and are widely used for transporting crops in many countries. They are strong, rigid, smooth, easily cleaned, and made to stack and nest when empty in order to conserve space. One major drawback is that these crates deteriorate rapidly when exposed to sunlight unless treated with a UV inhibitor.

9.4 PACKAGING FRESH AND PROCESSED CROPS

Packaging is one of the most important steps in the long and complicated journey of fresh and processed crops. Different types of packaging systems, such as flexible bags, crates, baskets, cartons, bulk bins, palletized containers, etc., are used for handling, transportation, and marketing of fresh and processed crops (Boyette *et al.*, 1996).

9.4.1 Classification of Packaging Systems

Packaging systems can be classified as below:

1. Flexible sacks: made of plastics and nets
2. Wooden crates
3. Cartons (fiberboard boxes)
4. Plastic crates
5. Pallet boxes and shipping containers
6. Baskets made of woven strips of leaves, bamboo, plastic, etc.

9.4.2 Plastic Bags

The key containers for packing fresh fruits and vegetables are polythene bags. Plastic sacks are cheaper and have very low bagging and sealing costs. They are clear, thus allowing easy stuffing inspection and easy acceptance of high-quality graphics. Plastic films are available in a wide variety of thicknesses and grades, and the greenhouse gases within the bag can be designed to control them. To ensure the proper balance of oxygen, sulfur dioxide, and carbon dioxide, the film content is sufficiently permeable for the vapors of water inside the bag. Various patches, in addition to engineered synthetic films, valves that connect to low cost, and ordinary plastic film bags have been developed. These instruments are temperature-responsive and monitor the greenhouse gas blending.

9.4.3 Shrink-Wrap

Shrink-wrapping is one of the most recent developments in the packaging of freshly harvested fruits and vegetables. In this type of packaging, individual fruits or vegetable items are wrapped with polythene film. Shrink-wrapping has been used successfully to package kinnow, capsicum, cabbage, cauliflower, potatoes, apples, cucumbers, bitter gourd, and various other fruits and vegetables. Shrink-wrapping with plastic can reduce weight loss and shrinkage, protect the produce from disease, reduce mechanical damage, and provide a good surface for stick-on labels (Dhall et al., 2012).

9.4.4 Rigid Plastic Packages

Rigid plastic packages include clamshells. Packages having top and bottom with heat formed from one or two pieces of plastic are known as clamshells. Clamshells, which are inexpensive and versatile, provide excellent protection to the produce, and present a very pleasing consumer package. They are often used with smaller packs of high-value crops, such as small fruits, berries, mushrooms, etc., or food items that are easily damaged by crushing. They are also used extensively in the packaging of minimally processed, precut, and prepared salads. Molded polystyrene containers have been found as a substitute for corrugated fiberboard. At present these containers are not cost-effective, but as environmental pressures grow, they may be more common. Similarly, heavy-molded polystyrene pallet bins have been used as a substitute for wooden pallet bins. Although at present their cost is almost double that of wooden bins, they are durable, are easier to clean, recyclable, do not decay when wet, do not harbor disease, may be nested, and are made collapsible (Boyette et al., 1996).

9.4.5 Biodegradable Films

Environmental pressure is increasing continuously due to human interruptions. Under such circumstances, disposal and recyclability of packaging of all kinds of materials are becoming very important issues. It has been found that common polyethylene takes about 200–400 years to break down in a landfill. However, the addition of 6% starch reduces this time to about 20 years or less. Hence, nowadays, the packaging material companies are developing starch-based polyethylene substitutes that break down in a landfill as fast as ordinary paper. This move toward biodegradable or recyclable plastic packaging materials may be driven by cost in the long term but by legislation in the near term. Some authorities have proposed a total ban on plastics.

9.4.6 Modified Atmospheric Packaging

Modified atmospheric packaging (MAP) is almost impracticable without using suitable plastics. MAP of fresh fruits and vegetables is based on modifying the concentrations of O_2 and CO_2 in the atmosphere that are generated inside the package (Mangaraj et al., 2009). It is desirable that the natural interaction occurring between respiration of the product and packaging generates an atmosphere with low levels of O_2 and/or a high concentration of CO_2. The growth of harmful microorganisms is thereby reduced and the shelf life of the product is extended. In a modified atmosphere packaging, gases of the internal and the external ambient atmosphere try to equilibrate by permeation through the package walls at a rate dependent on the differential pressures between the gases of headspace and those of the ambient atmosphere.

In this context, the barrier to gases and water vapor provided by the packaging material must be considered. Thus, it can be stated that the success of the MAP largely depends upon the barrier (packaging) material used. These packages are made of plastic films with relatively high gas permeability (Figure 9.6). Packaging films with a wide range of physical properties are used in MAP. There are several groupings in MAP films, such as in the plural, vinyl polymers, styrene polymers, polyamides, polyesters, and other polymers. Polypropylene (PP) is part of the polyolefin group and is used largely in MAP, in both forms: continuous and perforated. Although various types of plastic films for packaging are available, relatively few have been used to pack fresh fruits and vegetables and even fewer have permeability for gas that makes them suitable for MAP. It has been

Plastics in Drying and Storage

FIGURE 9.6 Plastic crate (*Source*: Green Processing company, 2021)

FIGURE 9.7 Plastic fish icebox fish bins and big fishing iceboxes (*Source*: Changzhou Treering Plastics Co. Ltd.)

recommended that the permeability for CO_2 should be three to five times the permeability for O_2. Many polymers used to formulate packaging films are within this criterion (Figure 9.7).

Problem 1: 200 kg of tomato at 22% moisture content 90% (wb) is dried 10% (wb). Calculate the amount of moisture removed in drying.

Solution:
Amount of initial moisture content = 0.90 × 200 = 180 kg
Dry matter in tomato = 200 − 180 = 20 kg
Final moisture content in 10% wb tomato

$$\text{Total weight of tomato} = \text{weight of water} + \text{dry matter}$$

$$mc = \frac{\text{weight of water } w_w}{\text{total weight } w_t}$$

$$0.10 = \frac{w_w}{w_w + 20}$$

$$w_w = 2.22 \, \text{kg}$$

Total water removed = 180 − 2.22 = 177.78 kg

QUESTIONS

1. Explain the role of plastics in drying and storage.
2. Discuss the packaging of fresh and processed crops.
3. Give the classification of packaging systems.
4. What are plastic bags? What are their functions?
5. Define biodegradable films.
6. Discuss the modified atmospheric packaging.
7. What is drying? Discuss the types of drying.
8. Write a short note on poly house drying.

MULTIPLE-CHOICE QUESTIONS

1. Postharvest management involves
   a) Precooling   b) Sorting
   c) Grading   d) All of these
2. Shade nets made of _____ plastics are used by farmers to provide a cooler environment to remove field heat prior to packaging.
   a) UV-stabilized   b) PVC
   c) PP   d) None of these
3. The shading intensity of shade nets used by farmers to provide a cooler environment to remove field heat prior to packaging varies from
   a) 30% to 50%   b) 50% to 60%
   c) 70% to 80%   d) 80% to 90%
4. The main objective of curing is
   a) to heal the wounds   b) form a new protective layer of cells
   c) Both a and b   d) None of these
5. _____ are important unit operations to increase the market value of produce.
   a) Grading and sorting   b) Curing and conveying
   c) Both a and b   d) None of these
6. _____ are used for conveying the produce in bulk from one location to another.
   a) Belt conveyors   b) Screw conveyors
   c) Elevators   d) All of these
7. Tarpaulins made of _____ are used for drying purposes
   a) High-density polythene   b) Polypropylene
   c) Low-density polyethene   d) None of these
8. The main objective of poly house dryers is to _____ the utilization of solar radiations.
   a) Minimize   b) Maximize
   c) Limit   d) All of these
9. Refractive window drying converts food in the form of liquids or slurries into flakes or powders within a typical residence time of _____.
   a) 15–20 min   b) 25–30 min
   c) 5–10 min   d) 3–5 min

10. Drying is completed only when heat transfer takes place from
    a) Product to be dried to hot medium
    b) No heat transfer takes place
    c) Hot medium to the product to be dried
    d) None of these

ANSWERS

1	2	3	4	5	6	7	8	9	10
d	a	a	c	c	d	a	b	d	c

REFERENCES

Anonymous. 2014. Potential of plastics industry in northern India with special focus on plasticulture and food processing. A report on Plastics Industry, FICCI, New Delhi.

Arjoo, Y. and Yadav, Y. K. 2017. Performance evaluation of a solar tunnel dryer for around the year use. *Current Agriculture Research Journal*, 5(3), 414–421. doi: 10.12944/CARJ.5.3.22

Bhardwaj, S. 2015. Recent advances in cover and plinth (CAP) and on-farm storage. *International Journal of Farm Sciences*, 5(2), 259–264.

Boyette, C. D., Quimby, P. C., Caesar, Jr., A. J., Birdsall, J. L., Connick, W. J., Daigle, Jr., D. J., Jackson, M. A., Egley, G. H. and Abbas, H. K. 1996. Adjuvants, formulations, and spraying systems for improvement of mycoherbicides. *Weed Technology*, 10, 637–644.

Caparino, O. A., Tang, J., Nindo, C. I., Sablani, S. S., Powers, J. R. and Fellman, J. K. 2012. Effect of drying methods on the physical properties and microstructures of mango (Philippine 'Carabao'var.) powder. *Journal of Food Engineering*, 111(1), 135–148.

Dhall, R. K., Sharma, S. R. and Mahajan, B. V. C. 2012. Effect of shrink wrap packaging for maintaining quality of cucumber during storage. *Journal of Food Science and Technology*, 49(4), 495–499.

Ekechukwu, O. V. and Norton, B. 1999. Review of solar-energy drying systems II: An overview of solar drying technology. *Energy Conversion and Management*, 40(6), 615–655.

Ghosh, A. and Ghosh, A. 2009. *Greenhouse Technology: Future Concept of Horticulture*. New Delhi: Kalyni Publishers.

Janjai, S., Intawee, P., Kaewkiew, J., Sritus, C. and Khamvongsa, V. 2011. A large-scale solar greenhouse dryer using polycarbonate cover: Modeling and testing in a tropical environment of Lao People's Democratic Republic. *Renewable Energy*, 36(3), 1053–1062.

Kadam, D. M., Goyal, R. K., Singh, K. K. and Gupta, M. K. 2011. Thin layer convective drying of mint leaves. *Journal of Medicinal Plants Research*, 5(2), 164–170.

Kale, S. J. and Nath, P. 2018. Kinetics of quality changes in tomatoes stored in evaporative cooled room in hot region. *International Journal of Current Microbiology and Applied Sciences*, 7(6), 1104–1112.

Kale, S. J., Nath, P., Jalgaonkar, K. R. and Mahawar, M. K. 2016. Low cost storage structures for fruits and vegetables handling in Indian conditions. *Indian Journal of Horticulture*, 6(3), 376–379.

Kitinoja, L. and Kader, A. A. 2002. *Small-Scale Postharvest Handling Practices: A Manual for Horticultural Crops*. University of California, Davis, Postharvest Technology Research and Information Center.

Kulanthaisami, S., Subramanian, P., Mahendiran, R., Venkatachalam, P. and Sampathrajan, A. 2009. Drying characteristics of coconut in solar tunnel dryer. *Madras Agricultural Journal*, 96(1/6), 265–269.

Mangaraj, S., Goswami, T. K. and Mahajan, P. V. 2009. Applications of plastic films for modified atmosphere packaging of fruits and vegetables: A review. *Food Engineering Reviews*, 1(2), 133–158.

Nindo, C. I., Sun, T., Wang, S. W., Tang, J. and Powers, J. R. 2003. Evaluation of drying technologies for retention of physical quality and antioxidants in asparagus (*Asparagus officinalis*, L.). *LWT— Food Science and Technology*, 36, 507–516.

Nindo, C. I. and Tang, J. 2007. Refractance window dehydration technology: A novel contact drying method. *Drying Technology*, 25(1), 37–48.

Pavan, M. A., Schmidt, S. J. and Feng, H. 2012. Water sorption behavior and thermal analysis of freeze-dried, refractance window-dried and hot-air dried açaí (Euterpe oleracea Martius) juice. *LWT-Food Science and Technology*, 48(1), 75–81.

Sangamithra, A., Sivakumar, V., Kannan, K. and John, S. G. 2015. Foam-mat drying of muskmelon. *International Journal of Food Engineering*, 11(1), 127–137.

Shahi, N. C., Khan, J. N., Lohani, U. C., Singh, A. and Kumar, A. 2011. Development of polyhouse type solar dryer for Kashmir valley. *Journal of Food Science and Technology*, 48(3), 290–295.

10 Plastics in Aquaculture

10.1 INTRODUCTION

Application of plastics in aquaculture plays an important role to increase its production. The fish are caught and eaten from ancient times, but recent innovation is the culture of fish in captivity. In China, the first mention of aquaculture was in 500 BC and the Arthasastra of Kautilya, one of the oldest Indian epics, suggests that the activity of fish culture in India dates back to 300 BC. Fish culture was recorded in the eastern part, especially in the Bengal areas, before the advent of scientific culture. "Trapping and holding" of fish seed followed the conventional method and lifted it to table size, thereby marking the beginning of aquaculture in India (FAO, 1999; Kushreshths and Awasthi, 2000; Nayak et al., 2010). In these parts of the world, such a scheme persisted until recently. With the establishment of the Central Inland Fisheries Research Institute (CIFRI) in Barrackpore in 1945, organized aquaculture research was initiated. In 1977, under CIFRI at Kauslyaganga, Bhubaneswar, which also housed the Pond Culture Division, the Indian Council of Agricultural Research (ICAR) placed great emphasis on aquaculture research and training and founded the Freshwater Aquaculture Research and Training Center (FARTC). At the same time, the Trainers Training Center (TTC) and Krishi Vigyan Kendra (KVK) were also established there. As the Central Institute of Freshwater Aquaculture (CIFA), this center eventually became an autonomous institution in 1987.

Many inventions, packages, practices, and gadgets have been developed by focused aquaculture research. It has been learned over time that the potential of plastic use in aquaculture is high and, therefore, systematic research has been undertaken in CIFA in this regard. The All India Organized Research Project on the Use of Plastics in Agriculture (APA) was commissioned by CIFA as a center to take care of the aspects of aquaculture during 1988-89. In certain fields of agricultural activities, plastics are applied and have produced promising outcomes. Similarly, it is strongly committed to developing the aquaculture sector through applications in many aspects, from pond management to usage kit management. Over the past two decades, with numerous studies and findings, working on plastic applications in aquaculture has passed through several stages (Plast India 2009; Shastry 1989). Such discoveries need to be made available to young generations in order to use them in these fields for further growth.

The fishing industry has also reacted favorably to the challenges. Research into breeding and cultivation of fish began in the country in the 1950s, but real technical progress began in the early 1970s when the seed production and cultivation of fish began to expand. The sector has developed in leaps and bounds since then, sustaining a growth rate of about 6.2% over the last 25 years. Aquaculture innovations are emerging at a rapid rate, and the implementation of research and development in other fields is thus welcome in the industry. Most of these advances are the use of plastics in agriculture. The use of plastics in agriculture/aquaculture in the country has increased rapidly and has acquired considerable importance in the conversion of materials and resources, thus contributing significantly to an increase in agricultural production.

Plastics are probably the most flexible of all recognized materials. To satisfy the very unique performance criteria of the end use, they may be customized to be synthetic. Due to their durability, plastics easily replace traditional materials such as wood, glass, metals, paper, etc. in different segments, such as agriculture, drainage, water conservation, aquaculture, packaging, etc. India has seen a significant increase in plastic use.

Medium-density polyethylene (MDPE), high-density polyethylene (HDPE), polyvinyl chloride (PVC), polypropylene (PP), polystyrene (PS), polyamide (nylon), polycarbonate (PC), acrylic (PMMA), fiber-reinforced plastics (FRP), etc., are the plastics most widely utilized in aquaculture.

In India, the use of plastics in aquaculture and fishing is very small. Plastics are commonly used for the packaging of raw fish, the manufacturing of crafts and gear, etc. In 1988, ICAR initiated an All India Organized Research Project on the "Application of Plastics in Agriculture" for the successful use of plastics in agriculture in India. A Co-operating Centre on the Use of Plastics in Aquaculture was set up at the Central Institute of Freshwater Aquaculture, Bhubaneswar, as part of this program. This Centre has produced numerous plastic wares for aquaculture purposes since its inception. For two decades, research into the use of plastics for aquaculture has developed several products that have been tested for their applicability in research stations as well as in the fields of farmers. Many of these gadgets are currently at various stages of growth, assessment, and acceptance. A short description of these requests is provided in the segment below.

10.2 USE OF PLASTICS IN AQUACULTURE

In India, the use of plastics in aquaculture and fishing is relatively small. Plastics have been commonly used for the packaging of raw fish, the manufacturing of crafts and gear, etc. For two decades, the use of plastics for aquaculture has created several items that have been tested for their applicability at research stations as well as in the fields of farmers. Plastic has been used very significantly in the field of aquaculture. The major uses of plastics in aquaculture are nylon fishnets, framed plastic pipe and nylon net cages, pen culture in water-locked areas, etc.

10.2.1 Fishnet

A net that is used for catching fish is called a fishing net or a fishnet. Fishing nets are typically constructed of the mesh created by knotting a comparatively thin yarn. Modern nets are commonly constructed of nylon-like artificial polyamides. Since antiquity, a variety of fishing nets have been made. For thousands of years, local net fishing has been practiced using nets constructed from locally available materials (Figure 10.1).

10.2.1.1 Types of Fishnets

The various types of fishnets that are commonly used are given below:

- Hand net
- Cast net
- Chinese nets
- Gillnet
- Drift net
- Stake net

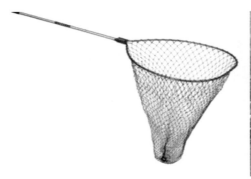 

FIGURE 10.1 Fishnet

Plastics in Aquaculture

10.2.2 Cages

Rigid PVC pipes and FRP floats are used to design and fabricate floating plastic cages. 3.0 m × 3.0 m × 1.5 m cages of 6 numbers each are set in parallel in two rows with a wooden walkway between them. FRP floats (45 cm in diameter and 32.5 cm in height) are connected in every corner of all the chambers to keep the complete structure above the surface of the water. Rectangular cages with two chambers measuring 13.0 m × 7.0 m are made of GI pipe structures backed by HDPE drum floats. These cages float in ponds with a depth greater than 5.0 m for the rearing of multiple fish species. The size of cages varies according to the requirement, cultural type, water body depth, water body size, wind speed, water current financial availability, experience, materials, etc. In addition to the high production capacity in the method, the ease of stocking, processing, feeding and control of biological parameters makes it very adaptable to aquaculture growers.

10.2.3 Seed-Rearing Tanks

FRP translucent tanks with 1.52 m diameter and 1.0 m height of built and produced with 1 mm thick FRP translucent sheets are used for high-density fish seed rearing. For rearing of spawn to fry, fry to fingerlings, and fingerlings to juveniles with high survival and development compared to that of standard culture process, systems of recirculatory water flow, aeration, and biofiltration are integrated. The FRP pools are ideal for aquaculture purposes due to their specific features of simple handling, abrasion resistance, corrosion resistance, temperature resistance, and ease of incorporation into any form of fittings needed to improve the performance of the rearing systems.

10.2.4 Pens

Pens are made with bamboo/wooden poles protected by plastic netting (HDPE) materials at the outlet of the farm channel or at the shallow depth of the ponds, streams, etc. In the pen, fish are stored and fed with conventional feeding methods. Around 85–90% of survival is observed in this sort of society. The pens are also used for the management of brood stock, mono/polyculture activities, etc.

10.2.5 Trays Used for Packing

In general, trays used for packaging are overwrapped with a protective film, sometimes shrinking with PE wrapping. The shrinking of the film is done by the use of hot air or hot water. Stretch wrapping is also used with heat-sensitive items. Manually (very often in the supermarket) or by computer, the film is spread over the product. Foils used as wraps or bags for food tray packaging must be puncture-resistant, expandable, and impervious to gases such as oxygen.

10.2.6 Films Used for Packing

The packaging industry uses hundreds of different films. These can be divided generally into two groups:

- Basic films that consist of a single film layer
- Laminate consisting of two or more simple films glued together by heat or by adhesives or bound together

For the packaging of frozen goods, plastics such as polyethylene film or ethylene copolymer and vinyl acetate are also commonly used. Using premade containers, polyethylene packs can be produced manually. To seal the bags that are hand-filled, an impulsion or bar sealer is used.

10.2.7 CATFISH HATCHERY

The catfish hatchery system consists of the elements below. The egg incubation unit consists of 16 HDPE circular tube numbers (40 cm diameter and 10 cm height) each fitted with a tap-associated PVC inlet pipe (15 mm diameter). Every tub is given a PVC pipe outlet (6 mm diameter) and the entire device is held on a metallic frame. Facilities are also given via the siphoning device for the excess release of water. Two numbers of FRP rectangular tanks form the post-larval rearing unit (2.97 m × 0.54 m × 0.30 m each). Each unit is equipped with arrangements for receiving water via PVC pipes (25 mm diameter). The egg incubation unit, larval rearing, and post-larval rearing units are interconnected in series and have full-flow valve water reception facilities from a common source.

10.2.8 FISH FEEDER/FEED DISPENSER

In various production systems, fish feed is the single most significant factor. A fish farm's economy depends heavily on the efficiency with which the fish use the food supply. Depending upon the form of feed used and the strength of other management activities, it accounts for more than 70% of the overall operating cost. Different conventional feeding systems are in vogue in India, but the performance of those feeding systems is not scientifically assessed. The quantity of feed dispersed by simple water turbidity determination is modified mainly by farmers. Feed wastage is higher in the conventional feeding system due to primitive feeding methods and also requires more recurrent expenditure due to labor involvement. Keeping in mind, demand and automated feeding systems are implemented that not only economize the process of culture but also reduce the loss of feed, maintaining the system's environmental composition (Mohapatra et al., 2009).

10.3 POLY HOUSE PONDS

Application of greenhouse pond environment accelerates the fish productivity (Figure 10.2). The water temperature is one of the most important parameters for fish growth. Therefore, fish culture inside a poly house is used for the production potential and growth in the greenhouse and open pond environment during low-temperate periods. It is a profitable business and the farming community finds great profitability in adopting this technology. The fish growth is mainly influenced by the water quality parameters, stocking density, and availability of natural and artificial feed. Dissolved oxygen (DO), water temperature, and pH are important water quality parameters. DO below 4 ppm is detrimental to the fish growth. Hence, a provision for the aeration of water is very important. The optimum temperature is 25–29°C; the pH of the water should remain slightly alkaline (7.5–8.5) to minimize the disease problems. The other factors, such as water salinity and turbidity, are also taken into account. Frequent water exchange improves water quality and controls the disease to a great extent, resulting in reduced expenditure on disease control measures. At least 50% water has to be exchanged weekly.

During winter periods, the growth of fish is least, as there is an acute fall in the water and air temperatures. Fish do not accept feed during low air temperatures. Hence, the best growing period for fish is summer during which intensive feeding can be done to fish stock and maximum growth can be achieved to make it profitable. Application of poly house-covered ponds may be helpful in improving the growth of fish even in harsh winter periods.

The main factor for fish growth is water temperature. In low-temperature regions, metabolic activity is greatly reduced during the winter season, thereby affecting fish development (Bandyopadhyay et al., 2000).

10.4 MOST COMMONLY USED PLASTICS IN AQUACULTURE

- PVC: pipes and fittings, aeration pipeline, hosepipes and fittings, pumps, cage floats, cage frame, drums, prawn shelter, fish handling crates, jerry cans, etc.

FIGURE 10.2 Farm pond inside poly house

- HDPE: floats for cages, twines and cords, net weaving, monofilament for the development of nets and hapas, storage tanks, water supply pipes and fittings, aeration, drainage, water-holding ponds, tubs, buckets, trays, basins and various aquaculture equipment components, and laboratory goods
- PP: Twines and rope, crates, tubs, buckets, trays, basins, and laboratory wares
- Nylon: Twine and ropes, fishnets, components for implements
- LDPE: Pond lining, greenhouse canopy cover, fish seed transportation carry bags
- FRP or Composite: Boats, floats, plastic gadgets, live fish transportation tank, carp hatchery, fish feeding device, magur hatchery
- Acrylic (PMMA): Glass jar hatchery, small container

10.5 HOW TO SELECT THE PLASTICS

To establish whether the material is thermoplastics or thermosetting, examine it and perform a cutting test. It's most likely thermoplastics if a shave can be removed with a knife. It's presumably thermosetting if the material is hard and won't pare but instead flakes or powders. The following points should be considered to select the plastic in aquaculture.

- Paring PMMA and polystyrene is tough
- For identification, an electric soldering iron is a useful instrument. Press it against the unknown samples before it becomes red hot. If the sample sinks, it is thermoplastics; if it does not sink, it is thermosetting
- Water floatation test; nylon, polycarbonate, PF, UF, and other sinks. LDPE, HDPE, or PP floats

TABLE 10.1
Specific Gravity Test in Water

Floats on Water (SG<1)	Specific Gravity	Sinks in Water (Floats in HYPO)	Specific Gravity	Sinks in HYPO	Specific Gravity
Polypropylene	0.90	Polyethylene oxide	1.06	CAB	1.20
LDPE	0.92	Polystyrene	1.07	PC	1.20
HDPE	0.96	SAN	1.10	PMMA	1.20
Ethylene vinyl acetate copolymer	0.93	ABS	1.10	Polysulfone	1.24
		Nylon	1.10	PP	1.30
				CA	1.30
				Cellulose Nitrate	1.37
				PVC	1.40
				Polyamides	1.40
				Acetal resin	1.40
				PTFE	2.75
				PF	1.42
				MF	1.48

- Keep an eye on the color: brown or black: PF, most likely. Colors that are light or pastel: UF or MF, most likely
- On a hard surface, perform a drop test: PC, PS, SAN, or ABS metallic ring, CA, PVC, nylon, PTFE, PMMA, and acetal resin have a dull sound
- Test from the scratch: Low sheen scratched easily: Scratches can be made on high-gloss surfaces: HDPE is most likely. Scratch-resistant high gloss: PP, most likely.

10.5.1 Specific Gravity Test in Water

The specific gravity test in water is illustrated in Table 10.1.

10.6 PLASTIC PACKAGING FOR FRESHWATER FISH PROCESSING AND PRODUCTS

Each technical procedure requires quality assurance, and adequate packing materials and processes are critical for freshwater fish. If these standards are not met, all processing efforts may be for naught, resulting in significant financial losses. Packaging should protect the product from contamination and spoiling while also providing the following benefits:

- Extend the shelf life of a product
- Facilitate distribution and display
- Give the product greater consumer appeal
- Facilitate the display of information on the product

10.6.1 Fresh Fish Packaging

Freshwater fish that is supplied live to a consumer or a processing factory has a high degree of reliance on proper handling during transportation and, once processed, requires appropriate packaging. Live fish can be carried in insulated containers with lids over short distances. Fish can also be transported in regular trucks, but the water in the containers must be aerated and cooled by portable devices over lengthy distances. Fish boxes constructed of suitable materials should be used to keep the quality of fresh fish during shipping. The six requirements listed below should be kept in mind when selecting fish boxes:

- Suitable size for the range of fish to be handled or the product to be put into them
- Convenient size for manual handling or lifting by mechanical equipment
- Stackable such that the weight of the containers on top rests on the containers underneath and not on the fish
- Constructed of impervious nonstaining materials
- Easy to clean
- Drainage provision for melted ice

High-density polyethylene is commonly used in fish boxes. Although this has numerous benefits, such as longevity, lightness, and simplicity of washing, it also has drawbacks, such as a high price and the fact that they are nonreturnable. As a result, disposable fish boxes with a capacity of roughly 25 kg (fish and ice) are more commonly used. Fiberboard cartons, waxed and waterproof boxes are examples. Insulated boxes, such as boards constructed of molded polystyrene, are favored for shipment by lorries without a cooling system. The latter is frequently used to transport chilled and frozen fish and fish products to wholesale and retail stores. Each layer of fillets should be thinly packed and separated from one another in the case of fillets.

10.6.2 Evaluation of Polyethylene, Polypropylene, and Laminated Polypropylene Packaging Material in Fish Retailing

CIFA, Bhubaneswar, tested PE, PP, and laminated PP packaging materials. Spice mix Nisin and pro-biotic cultures such *L. casei, P. pentosaseus, L. bulgaricus*, and *S. thermophilius* were used to treat rohu and catla fish pieces. All of the treatments improved the fish patties' keeping quality for up to 12 days at 52°C and 1 month at −20°C. Even greater keeping quality was achieved when fish flesh was blended with 20% fish gel of the same species, resulting in 15 days of chilling and 2 months of freezing. When it came to improving the keeping qualities of vacuum packaged products and modified environment packages, the findings were similar.

The packaging materials PE, PP, and laminated PP were tested. Vacuum packing and MAP were found to be best served by laminated PP packaging materials. Both the ready-to-eat and ready-to-cook fish patties demonstrated improved keeping quality in chilled and frozen settings. Even after three months of frozen storage, the product's quality remained unchanged.

10.6.3 Evaluation of Polyethylene

Indian big carps (rohu, catla, and mrigal) weighing 1–2 kg were packaged in 70 cm polyethylene carry bags with a length of 25 cm (200 microns). The carry bags have windows to make them easier to handle. Fresh entire fish were wrapped individually in carry bags following thorough washing. This packaging gave the goods a sanitary appearance and was useful for fish chilling and chilled display. After harvesting to whole fish retailing, chilled fresh whole fish in carry bags retained quality for 36 h. On the carry bag, there was information on the date of harvest, fish species, weight, and use directions, among other things. This strategy was proven to be suitable for the marketing of fresh entire fish.

10.6.4 Evaluation of Polypropylene Containers with Lid for Retail Marketing of Fish Cutup Parts during Chilled and Frozen Storage

Fish cutup portions, fingers, chunks, and nuggets were hygienically wrapped in circular polypropylene rigid containers with lids having a 500-micron thickness and were either refrigerated or frozen

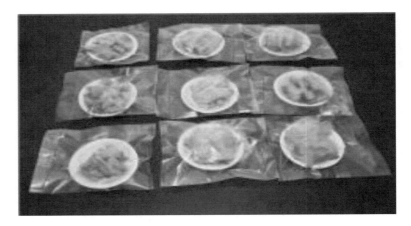

FIGURE 10.3 Packaging of fish cutup parts

FIGURE 10.4 Whole fish packaging

and exhibited for retailing. The quality of chilled and frozen items was maintained for seven days and three months, respectively.

10.6.5 EVALUATION OF PET BOTTLES FOR PACKAGING OF FISH PICKLE

The pickle (Murrel and Anabas) was tested in PET bottles with capacities of 500 ml and 250 ml to see how long it could preserve its quality. Under chilling and ambient storage, the pickled fish product's keeping quality remained at its peak for six months. After 15 days of manufacture, the pH of the final product was discovered to be in the range of 3.4 to 3.6, and it had stabilized. Wide mouth PET bottles have been shown to be a good alternative to glass bottles for preserving pickle quality for up to a year (Figures 10.3–10.4).

10.7 CONTRIBUTION OF AQUACULTURE TO MARINE LITTER

Plastics have an important role in fisheries and aquaculture. Regrettably, some of the items utilized in fisheries and aquaculture may end up as marine garbage. When user products and plastic objects reach the end of their useful lives, they must be removed from the environment and reclaimed, repurposed, or disposed of in the appropriate waste streams.

Plastic materials are exposed to direct UV light, wave action, abrasion, and temperature variations as a result of their continuing use. Embrittlement and disintegration can be exacerbated by these circumstances. The objects will no longer be fit for purpose if they break. Furthermore, plastics are chosen over traditional natural fibers because the latter degrade more quickly in the environment; however, the currently employed synthetic polymers do not disintegrate as quickly and will take a long time to degrade.

As a result, losses from fisheries and aquaculture are commonly reported in surveys of marine litter on beaches (Browne *et al.*, 2015; Nelms *et al.*, 2017; Slip and Burton,1991), hovering on water bodies (Cózar *et al.*, 2014; Thiel *et al.*, 2003), positioned on the seafloor (Cózar *et al.*, 2014; Thiel *et al.*, 2003), and floating on the surface.

There are no worldwide estimates of how much plastic garbage the fisheries and aquaculture industries generate. The Republic of Korea's first national assessment of fisheries and aquaculture debris input to marine habitats (Jang *et al.*, 2014) found that annual input from lost fishing gears was 44,081 tons, with 2,374 tons of waste tossed overboard from fishing.

Shipping and fishing activities have been blamed for the majority of the debris accumulated around the North Sea coasts of Belgium, Denmark, France, Germany, the Netherlands, Norway, and the United Kingdom (Galgani *et al.*, 2000; Vauck and Schrey, 1987; Unger and Harrison, 2016; Williams, Tudor and Randerson, 2003).

Some marine debris can be traced back to scouring and commercial netting operations, and tagged pots and bait boxes can be used to identify individual fisheries and home ports in some circumstances.

10.7.1 Abandoned, Lost, or Otherwise Discarded Fishing Gears

Abandoned, lost, or otherwise discarded fishing gears (ALDFG) are the largest source of plastic trash in the marine environment in the fisheries and aquaculture sectors, yet there is significant heterogeneity in its distribution and abundance (Figure 10.5). Some fishing systems often catch species near to the seafloor, which can cause problems for fishing gear owing to irregularities such as rocks, fissures, and crevices, which can entangle fishing gear and cause breakages and loss. In most

FIGURE 10.5 Crabs caught in the abandoned, lost, or discarded fish gear

cases, anomalies on the seafloor will affect trawls, dredges, and pots. Gillnets, trawls, handlines, and longlines are all important parts of the ALDFG.

Enforcement on fishermen to abandon gears (e.g., illegal fishing or illegal gears), operational pressure (e.g., use of too much gear in limited time periods), environmental conditions (e.g., weather, seabed irregularities), lack of/inaccessible/expensive onshore gear, and waste disposal facilities can all contribute to the loss of fishing gears (Macfadyen, Huntington and Cappell, 2009; Gilman *et al.*, 2016).

ALDFG might keep fishing or get hooked up in other fishing gear. Because the density of plastics (such as HDPE and PS) is lower than that of seawater, most fishing trash will float; however, entangled and entrapped gears are more likely to continue floating on the seafloor until they are removed. The overall debris on the seabed of the Mediterranean Sea and Northeast Atlantic was composed primarily of plastics (41%) and ALDFG (34%), but ALDFG represented more than 75% in some locations, such as north of the Faroe Islands, the Norwegian continental shelf, and several seamounts and ocean ridges. A detailed investigation of floating macro-debris (>200 mm diameter) found that fishing was responsible for 20% of the number and 70% of the weight, namely floats and buoys (Eriksen *et al.*, 2014). The presence of ALDFG in the marine environment can have a considerable influence on commercial fishing and the shellfish sector, making it a major issue for fisheries and marine conservation.

ALDFG can result in ghost fishing, stock depletion, nontarget species capture, conservation concerns, and risks to other vessels, and it is expensive to remove (Arthur *et al.*, 2014; Bilkovic *et al.*, 2016; Derraik, 2002; Laist, 1987; Wilcox *et al.*, 2015). Derelict crab pots in Chesapeake Bay, Virginia, United States, are one example. During four consecutive winters (2008-2012), marine debris was collected, with blue crab pots being the most abundant.

10.7.2 Plastic Debris from Aquaculture

Plastics, like ALDFG, can be released from aquaculture and mariculture facilities and end up in the ocean as marine debris. Anchor rope wear and tear, storms, accidents, and disagreements with other maritime users all contribute to the loss of aquaculture buildings. Severe weather can wreak havoc on aquaculture structures, resulting in significant amounts of marine debris in some cases (Lee *et al.*, 2015).

Derelict gear from the pearl oyster aquaculture in French Polynesia (Andréfout, Thomas, and Lo, 2014) and shellfish aquaculture contamination of Puget Sound are two examples of plastics waste from aquaculture (Bendell, 2015). The seafloor and beaches surrounding these two aquaculture facilities have been recorded as having lost or discarded debris from the operations. On shellfish beds, PVC tubes, net caps, plastic bands, zip ties, oyster bags, and nets have been observed. The scope and ecological impact are unknown, but continuous concerns have raised the possibility of navigational risks, ecological disturbances, animal hazards, and hazards for boat traffic and fishermen. Finally, the breakdown of these objects may result in the development of microplastics, which may have additional environmental consequences.

QUESTIONS

1. How are plastics used in aquaculture?
2. How does aquaculture contribute to marine pollution?
3. Draw a diagram of a fishnet.
4. Enumerate commonly used plastics in aquaculture. What are their uses?
5. Define the following terms:
    1) Cages
    2) Pens
    3) Catfish hatchery
    4) Fish feeder

MULTIPLE-CHOICE QUESTIONS

1. Most commonly used plastics in aquaculture are:
   a) PVC   b) Nylon
   c) HDPE   d) All of these
2. _____ are used to design and fabricate floating plastic cages:
   a) Rigid PVC pipes   b) FRP floats
   c) Both a and b   d) None of these
3. Diameter of translucent tanks used for high-density fish seed rearing is
   a) 1.52 m   b) 7 m
   c) 2.98 m   d) 3.75 m
4. The height of translucent tanks used for high-density fish seed rearing is
   a) 1 m   b) 2 m
   c) 10 m   d) None of these
5. Translucent tanks used for high-density fish seed rearing are made of _____mm thick FRP translucent sheets
   a) 1   b) 2
   c) 5   d) None of these
6. The egg incubation unit of the catfish hatchery system consists of _____ HDPE circular tubes
   a) 200 b) 20
   c) 160 d) 16
7. The dimensions of two FRP rectangular tanks that form post-larval rearing are
   a) 2.97 m × 0.54 × m 0.30 m each   b) 2.67 m × 1.54 m × 0.30 m each
   c) 3.97 m × 0.54 m × 0.90 m each   d) None of these
8. The fish growth inside poly house ponds is influenced by
   a) Water quality parameters   b) Stocking density
   c) None of these   d) All of these
9. The main factor for fish growth inside a poly house pond is
   a) Temperature   b) Relative humidity
   c) Dew point   d) All of these
10. For packing frozen foods _____ are commonly used.
    a) Polyethylene film   b) Vinyl acetate
    c) Ethylene copolymer d) All of these

ANSWERS

1	2	3	4	5	6	7	8	9	10
d	c	a	a	a	d	a	d	a	d

REFERENCES

Andréfouët, S., Thomas, Y. and Lo, C. 2014. Amount and type of derelict gear from the declining black pearl oyster aquaculture in Ahe atoll lagoon, French Polynesia. *Marine Pollution Bulletin*, 83(1), 224–230.

Arthur, C., Sutton-Grier, A. E., Murphy, P. and Bamford, H. 2014. Out of sight but not out of mind: Harmful effects of derelict traps in selected US coastal waters. *Marine Pollution Bulletin*, 86(1), 19–28.

Bandyopadhyay, M. K., Tripathi, S. D., Aravindakshan, P. K., Singh, S. K., Sarkar, B., Majhi, D. and Ayyappan, S. 2000. Fish culture in polyhouse ponds-A new approach for increasing fish production in low temperature areas. Abstract. *The Fifth Indian Fisheries Forum*, 17–20 January, 2000, CIFA Bhubaneswar, India, p. 7.

Bendell, L. I. 2015. Favored use of anti-predator netting (APN) applied for the farming of clams leads to little benefits to industry while increasing near shore impacts and plastics pollution. *Marine Pollution Bulletin*, 91(1), 22–28.

Bilkovic, D. M., Slacum Jr., H. W., Havens, K. J., Zaveta, D., Jeffrey, C. F. G., Scheld, A. M., Stanhope, D., Angstadt, K. and Evans, J. D. 2016. Ecological and economic effects of derelict fishing gear in the Chesapeake Bay 2015/2016. *Final Assessment Report*. Revision 2. Virginia Institute of Marine Science, College of William & Mary.

Browne, M. A., Chapman, M. G., Thompson, R. C., Amaral-Zettler, L. A., Jambeck, J. and Mallos, N. J. 2015. Spatial and temporal patterns of stranded intertidal marine debris: Is there a picture of global change? *Environmental Science & Technology*, 49(12), 7082–7094.

Cózar, A., Echevarría, F., González-Gordillo, J. I., Irigoien, X., Úbeda, B., HernándezLeón, S., Palma, Á. T., Navarro, S., García-de-Lomas, J., Ruiz, A. and Fernández-dePuelles, M. L. 2014. Plastic debris in the open ocean. *Proceedings of the National Academy of Sciences of the United States of America*, 111(28), 10239–10244.

Derraik, J. G. 2002. The pollution of the marine environment by plastic debris: A review. *Marine Pollution Bulletin*, 44(9), 842–852.

Eriksen, M., Lebreton, L. C., Carson, H. S., Thiel, M., Moore, C. J., Borerro, J. C., Galgani, F., Ryan, P. G. and Reisser, J. 2014. Plastic pollution in the world's oceans: More than 5 trillion plastic pieces weighing over 250,000 tons afloat at sea. *PloS One*, 9(12), e111913. doi: 10.1371/journal.pone.0111913

FAO. 1999. Rural aquaculture in India. RAP publication 1999/21. Food and Agriculture Organization. Regional Office for Asia and Pacific, Thailand, Bankok.

Galgani, F., Leaute, J. P., Moguedet, P., Souplet, A., Verin, Y., Carpentier, A., Goraguer, H., Latrouite, D., Andral, B., Cadiou, Y. and Mahe, J. C. 2000. Litter on the sea floor along European coasts. *Marine Pollution Bulletin*, 40(6), 516–527.

Gilman, E., Chopin, F., Suuronen, P. and Kuemlangan, B. 2016. *Abandoned, Lost or Otherwise Discarded Gillnets and Trammel Nets: Methods to Estimate Ghost Fishing Mortality, and the Status of Regional Monitoring and Management*. FAO Fisheries and Aquaculture Technical Paper No. 600. Rome.

Jang, Y. C., Lee, J., Hong, S., Mok, J. Y., Kim, K. S., Lee, Y. J., Choi, H. W., Kang, H. and Lee, S. 2014. Estimation of the annual flow and stock of marine debris in South Korea for management purposes. *Marine Pollution Bulletin*, 86(1), 505–511.

Kushreshtha, A. K. and Awasthi, S. K. 2000. The present status of the Indian plastics industry and its future developments: Popular Plastics and Packaging. *Plastindia Exhibition Special Issue*, 2000, 29–60.

Laist, D. W. 1987. Overview of the biological effects of lost and discarded plastic debris in the marine environment. *Marine Pollution Bulletin*, 18(6), 319–326.

Lee, J., Lee, J. S., Jang, Y. C., Hong, S. Y., Shim, W. J., Song, Y. K., Hong, S. H., Jang, M., Han, G. M., Kang, D. and Hong, S. 2015. Distribution and size relationships of plastic marine debris on beaches in South Korea. *Archives of Environmental Contamination and Toxicology*, 69(3), 288–298.

Macfadyen, G., Huntington, T. and Cappell, R. 2009. *Abandoned, Lost or Otherwise Discarded Fishing Gear*. UNEP Regional Seas Reports and Studies, No. 185. FAO Fisheries and Aquaculture Technical Paper, No. 523. UNEP/FAO, Rome, 115p.

Mohapatra, B. C., Sarkar, B., Sharma, K. K. and Majhi, D. 2009. Development and testing of demand feeder for carp feeding in outdoor culture system. *Agricultural Engineering International: the CIGR Ejournal*, XI, 2009.

Nayak, S. K., Yadav, S. N. and Mohanty, S. 2010. *Fundamentals of Plastics Testing*. Springer (India) Private Limited, New Delhi, 1–421.

Nelms, S. E., Coombes, C., Foster, L. C., Galloway, T. S., Godley, B. J., Lindeque, P. K. and Witt, M. J. 2017. Marine anthropogenic litter on British beaches: A 10-year nationwide assessment using citizen science data. *Science of the Total Environment*, 579, 1399–1409.

Plast India. 2009. *7th International Plastics Exhibition and Conference*. Pragati Maidan, New Delhi, February 4–9, 2009.

Shastry, H. V. 1989. *Identification and Selection of Plastics for Product Development*. Training Manual, CIPET, Bhubaneswar Centre, 1989.

Slip, D. J. and Burton, H. R. 1991. Accumulation of fishing debris, plastic litter, and other artefacts, on Heard and Macquarie Islands in the Southern Ocean. *Environment Conservation*, 18(03), 249–254.

Thiel, M., Hinojosa, I., Vásquez, N. and Macaya, E. 2003. Floating marine debris in coastal waters of the SE-Pacific (Chile). *Marine Pollution Bulletin*, 46(2), 224–231.

Unger, A. and Harrison, N. 2016. Fisheries as a source of marine debris on beaches in the United Kingdom. *Marine Pollution Bulletin*, 107(1), 52–58.

Vauck, G. J. M. and Schrey, E. 1987. Litter pollution from ships in the German Bight. *Marine Pollution Bulletin*, 18(6), 316–319.

Wilcox, C., Heathcote, G., Goldberg, J., Gunn, R., Peel, D. and Hardesty, B. D. 2015. Understanding the sources and effects of abandoned, lost, and discarded fishing gear on marine turtles in northern Australia. *Conservation Biology*, 29(1), 198–206.

Williams, A. T., Tudor, D. T. and Randerson, P. 2003. Beach litter sourcing in the Bristol channel and Wales, U.K. *Water, Air, & Soil Pollution*, 143, 387–408.

11 Plastics in Animal Husbandry

11.1 INTRODUCTION

Plastic is the most useful synthetic man-made substance, made up of elements extracted from fossil fuel. It can be used in almost every part of human life starting from domestic to manufacturing industry (Varda *et al.,* 2014). Among all, plastics are also used in every field of agriculture, aquaculture, and animal husbandry.

Plastic material can play an important role in storing animal grains and straw during winter. The plastic films used for storing silage are resistant and it is easy to transport and store the material for longer periods. The silage storage in wrapped bales is a very popular method in many countries, because it offers advantages over hay processing, such as more flexible harvest times, less dependency on the weather, and greater versatility in ration formulation. The application of plastic in different sectors is shown in Figure 11.1. Usually, there are two types of bale-wrapping systems: (1) bale is wrapped as a stand-alone or silo unit completely enclosed or (2) in-line in which bales are placed end to end with PE-film wrap applied around the circumferential surface of large-round bales. Silage from piles of grass and sugar beet pulp is used in addition to bale-wrapping to store and ferment the product after harvesting. After harvest, sugar beets may temporarily be put in a silage pile covered by plastic to protect the beets from rain. Compared to Northern Europe, the use of silage plastics is higher than in Southern Europe, while in Southern Europe, more plastics are used in greenhouses, tunnels, and irrigation pipes. The literature search did not give rise to data in this application on the tons of plastics used or the quantity of silage protected worldwide or for Europe. Silage accounted for approximately 3% of Italy's total agricultural plastic use in 2005 (Ashbell and Weinberg 1992; Scarascia *et al.*, 2011).

11.2 POLYMERS AND THEIR PRODUCTS DURING PREHARVEST AND POSTHARVEST

Low-density polyethylene (LDPE) and polypropylene are the two most common polymers used in the production of plastic products used in livestock farms. LDPE is used to make nearly 70–85% of agricultural plastic products, while polypropylene is used for some other applications, such as nets to tie round bales and rope. Table 11.1 illustrated the most common plastic products used in livestock farms, as well as the polymers used to make them. The majority of the plastic used in livestock farms is LDPE film, which is used to conserve livestock feed. When the plastic is no longer usable, it must be disposed of outside the farms, contributing to an expense to the farm. It was estimated 0.3 kg/ton of dry matter (DM) for bunker silos, 2.1 kg/ton of DM for tube line-wrapped bales, 2.5–2.7 kg/ton of DM for individual round bales wrapped with 4 layers, and 2.0 kg/ton of DM for bagged silos on Canadian farms. Plastic can be used in animal production. Plastic can be used for the following purposes.

1. Preproduction: plastic helps in the production of their food
    a. Mulching
    b. Fertilizer
    c. Pesticide and insecticide
2. Postproduction: to store animal food after the production
    a. Storage

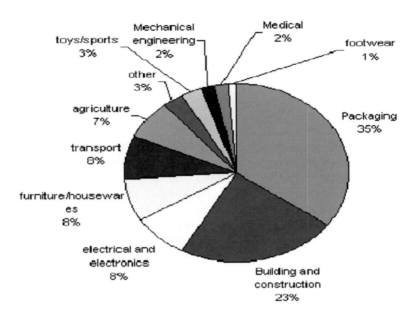

FIGURE 11.1 Plastics used in different sectors (*Source*: Varda et. al. 2014)

TABLE 11.1
Polymers and Their Applications on Livestock Farms

Use of Film	Types of Polymers
Plastic film to cover horizontal silos and stretch film to wrap bales	LDPE, linear LDPE, and coextruded PE-EVOH
Mulch films for cereal crops	LDPE and LLDPE
Plastic bags for grain storage	LDPE

11.3 APPLICATION OF PLASTICS IN FARMS

The films used for the conservation of silage accounted for 80% and 93% of the plastic purchased yearly by meat and dairy farms, respectively, and only 37% by pig farms. The majority of the farms generate corn silage (with a mean DM content varying from 30% to 38%), stored in horizontal silos covered with LDPE ranging in thickness from 100 μm to 250 μm. A bunker silo of 40 m long, 10 m wide, and 2.5 m high weighing around 650 tons of silage requires a 50 m × 12 m plastic film weighing approximately 110 kg of plastic cover. As a result, the plastic consumption for 1 ton of silage is 0.17 kg or approximately 0.53 kg/ton of DM. The utilization of plastic will rise to 0.28 kg/ton of silage approximately for 0.87 kg/ton DM if the wall of the silo is lined with a plastic film layer. All of the farms surveyed produced hay bales tied with LDPE nets. The average net consumption is approximately 0.20 kg/bale. During certain seasons (spring and fall), more than 80% of the farms produced covered bales, using individual wrapping with 4 to 6 layers of 25-m-LDPE stretch film. In relation to the bale diameter (1.2 m to 1.8 m) and the number of layers applied, the average film consumption ranged from 0.7 kg/bale to 2.1 kg/bale.

Plastic is now used in farms for forage, pesticide, and fertilizer containers. The fertilizers were stored in either large bags (PP fabric) containing 500–600 kg or small bags containing 500–600 kg (LDPE-sealed film), 0–100–200–300–400–500. It is utilized in farms that produce milk from cow and buffalo ranches. Pig farms are used to raise pigs. The plastic purchased for feed conservation (kg/farm per year) is shown in Figure 11.2.

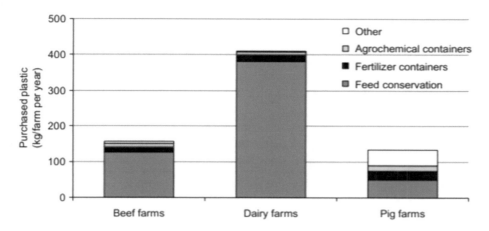

FIGURE 11.2 Yearly plastic procurement by livestock farms in northern Italy (*Source*: Borreani, and Tabacco, 2017)

11.4 PLASTICS IN ANIMAL PRODUCTION

The average plastic consumption per ton of fertilizer varied depending on the type of the bag used. The agrochemical storage containers, which are mostly constructed of HDPE, use only a small percentage of the plastic used in farms. In brief, the plastic films used to cover silages, the stretch films used to wrap bales, and the net used to knot bales have been found to account for the majority of the plastic utilized in farms. Depending on the herd size, the plastic films used to cover bunker silages account for 33% to 45% of the total amount of plastic purchased by cattle farmers for fodder conservation. Because of its increased flexibility, bale wrapping is commonly used in dairy and meat farms, and stretch film accounts for almost one-third of these farms' total plastic use. The net used to tie haylage, hay, silage, or straw bales can account for up to 40% of a farm's total plastic consumption. The total amount of plastic acquired annually by livestock farms in northern Italy ranged from 0 kg to over 700 kg of film to cover silages, 40 kg to over 800 kg of net to tie bales, and up to 500 kg of stretch film on farms that employed wrapping as their primary forage conservation technique. Plastic use per unit of land area is at 13 kg/ha on average, with peaks around 25 kg/ha per year. These values were lesser than those found on horticultural farms that employ greenhouses or mulching, where annual plastic consumption varies from 1,200 kg/ha to 1,700 kg/ha for greenhouse covers and 300 kg/ha to 700 kg/ha for mulching. When plastic usage is compared to the number of dairy cows, an average of 10.3 kg of plastic is used per cow per year, with 3.9 kg used to cover bunker silages. Levitan *et al.* (2005) conducted a survey on plastic use in dairy farms in New York State's Central Leatherstocking-Upper Catskill Region and found that each cow utilizes 3.4 kg of plastic film per year.

A survey was conducted on 961 Irish farms with the purpose of quantifying the primary and postprimary use of PE- and PP-based products used for packing fertilizer, mulching maize, tying silage bales, and sealing silage. The plastic used per unit of farm surface was 4.7 kg/ha of new film to cover pit silages.

11.5 CONSERVATION OF FODDER

Food storage means the storage and diligent preservation of some amounts of fodder that are unchanged during chemical reactions or physical conversions of the fodder when overly available for future uses. The important points for the conservation of fodder are given below.

- To preserve feed when it is available in excess
- Maintain the optimum nutritional value of fodder

- Shifting available food from the present to the future
- Moving the feed from one location to another
- To assist pasture management

11.6 SILAGE

Silage is the green material produced by the controlled fermentation of the high moisture content retained by the green fodder crop (Figure 11.3). Silage is a type of fodder made from green foliage crops which have been preserved by acidification and achieved through fermentation (FAO, 1994).

11.6.1 SILAGE BAGS

India is the largest milk-producing country globally. However, the average milk yield of cattle in India is abysmally low (50% of the global average) on account of current fodder management practices. A recent report published by the Planning Commission indicates the deficit of green fodder is to the extent of 35%. Green and nutritious "forage" (maize, sorghum, etc.) can be conserved through a natural "pickling" process in a sealed airtight container; this is known as "ensiled forage" or "silage." Silage can be prepared when surplus green fodder is available and can be used during the lean period.

There are numerous advantages of converting green fodder to silage:

- Silage is a substitute for green fodder during the lean period
- Silage ensures improved quality and digestibility of fodder to the livestock
- It helps to increase milk production and helps sustain higher milk production during the lean period

11.6.2 ADVANTAGES OF "SILAGE BAGS"

Silage bags made from Repol Polypropylene are revolutionizing "fodder management" vis-à-vis traditional methods of "mud silos," "underground silos," and "concrete structures." Repol "silage bags" are variants of large sacks (known as flexible intermediate bulk containers) with a polyethylene

FIGURE 11.3 Silage

liner. Silage made in Repol silage bags can easily be stored for longer periods of time without affecting its nutritional value. The following are the characteristics of silage material.

- Being a lighter material, transportation is possible unlike traditional "silos"
- The bags are readily available in different sizes – 100 kg to 1,000 kg
- These bags are tough with high tear- and puncture-resistant properties

Silage preparation in Repol Polypropylene bags involves the following characteristics.

- Selection of green fodder having proper maturity and desired moisture level (58–68%)
- Chopping of fodder into desired sizes
- Addition of diluted molasses to the chopped fodder for proper fermentation
- Selection of the desired size of flexible intermediate bulk container (FIBC) bag with "liner"
- Filling the chopped fodder with intermittent ramming from top to evacuate the air
- Closing the liner and storing the silage for a specific time, and the silage is ready for feeding the livestock later
- When the weather does not allow haymaking, the silage will be prepared from green fodder
- Silage can be prepared from plants with thick stems and is typically not very suitable for haymaking, such as sorghum and maize
- Weeds can also be used for silage making along with primary fodder crops. The processing of silage destroys most weed seeds
- It is incredibly palatable

11.7 SELECTION OF CROPS FOR SILAGE MAKING

It is easier to ensile forages that have the following characteristics.

- High fermentable sugar level
- Low protein level
- Limited buffering capacity
- At the time of ensiling, 35% of dry matter should be available
- Pasture grasses: elephant grass (Napier grass), guinea grass, Rhodes grass, Sudan grass, Setaria, ruzi grass, etc.
- Straws: rice straw, wheat straw, soybean chaff, peanut hulls, etc.

11.8 SILO

Animal fodder and feeds, fertilizers, seeds, vegetables, milk and milk products, agricultural machinery, and other items must be stored in various storage structures. A silo is a farm facility that stores and processes animal fodder in a controlled environment. To promote partial fermentation, the hay is cut and put in an airtight condition in the silo. The stored fodder is known as silage. Silage can be made from any green crop with moderately tough stalk, e.g., grass, sugarcane, legumes, etc. Silage is more nutritive than dried stalk. But when compared to green fodder, there is some loss of nutritive value in silage, because the sugar is converted into lactic acid giving a sour taste. Losses are also occurring due to surface spoilage, fermentation, and seepage. Loading of tower silos is difficult. It needs a mechanical loader or a large capacity blower for elevating the cut fodder. The wall should be smooth, circular, and strong enough to avoid cracking due to lateral pressure. Hence, a heavy reinforcement is a must. The only advantage is if the water table is very close to the ground level, tower silos are preferred. The cost of constructing these silos is comparatively much higher than that of horizontal types.

11.8.1 Site for Construction of Silo

Following are the important points to be considered when constructing silos.

- The site where the silo is to be built should be elevated from other locations
- The site must be clear from the blockage of water
- Silo pit walls must be leakproof
- It should be at least 6 ft. away from the shed formed for animals
- Construct the silo to the south side of the animal shed if possible

11.9 KINDS OF SILOS

11.9.1 Stack Silo

It is the simplest type of silo. A plastic sheet of 0.1 mm thick is placed over the ground and it is covered completely by equally chopped silage materials on the sheet. There is a need for proper tread pressure and full sealing (Figure 11.4).

11.9.2 Bunker Silo

A bunker silo is usually built on the ground, but there are other construction methods to build a silo using the ground configuration (slope 3–5%) or a semiunderground type, which is half below the ground level. There is a need for sidewalls made of wood and concrete, and plastic sheets are preferably used to seal the interior. Correct tread pressure must be applied, and complete sealing is necessary. Supports are required to avoid the external collapse of the sidewalls. The front width should be such that to prevent aerobic deterioration, the total amount of silage per day with a thickness of 20–30 cm can be removed. The bunker silo is shown in Figure 11.5.

11.9.3 Pit/Trench Silo

Pit silo is a circular deep well. It is completely lined on the sides and bottom to avoid water inflow inside the silo. It can be formed of bricks, stone, or lime, or cement-based concrete. A 22.5 m wall

FIGURE 11.4 Plastic sheet silos

FIGURE 11.5 Bunker silo

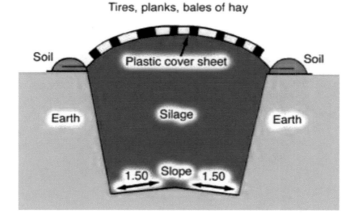

FIGURE 11.6 Trench-type silo

along the length of a pit silo which has 15 m depth has been provided with plaster to the entire inner surface. To protect the silage from the sun and rain, a simple roof can be created over the silo. In general, a cow is typically fed 1.4 kg of silage per 45 kg of body weight each day (1.4 kg/45 kg of body weight/day or 3 kg per 100 kg of body weight/day. The amount of silage feed each day is used to determine the silo's diameter. The rate of elimination of silage should be 10cm per day. The diameter is limited to 6 m, and the depth is 2–3 times that of the diameter.

A trench silo can be constructed by simply digging the earth, but in order to avoid damage, it is easier to put plastic sheets inside. For a long period of time, a trench silo whose interior is paved with concrete may be used (Figure 11.6).

11.9.4 Plastic Bag Silo

Plastic bags of 0.1 mm thick are used in this form and filled with chopped raw materials, compressed to the maximum extent necessary to remove the internal air, and then sealed fully. The plastic bag silo is shown in Figure 11.7.

FIGURE 11.7 Plastic bag silo

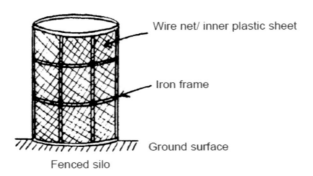

FIGURE 11.8 Fenced silo (framed silo)

11.9.5 FENCED SILO (FRAMED SILO)

The frame is made of bamboo, wood, iron, etc. The shape of the cross section can be circular or rectangular and the interior is covered with plastic sheets. The fenced silo (framed silo) is shown in Figure 11.8.

11.9.6 TOWER SILO

The storage silos of this type are cylindrical structures usually 10–90 ft. (3–27 m) in diameter and 30–275 ft. (10–90 m) in height. Tower silos that hold silage are normally unloaded from the top of the mound (Figure 11.9).

This application, developed to store grains for animal feed and straw during the winter, is further proof of the value of plastics. Plastic films used to store silage are resistant and the content can be easily transported and stored for long periods (Figure 11.10).

11.10 SEALING METHODS

Sealing methods affect the extent of aerobic deterioration in horizontal silos. In horizontal silos, during the storage period, a spoiled layer is formed below the sealing sheet, known as "surface

Plastics in Animal Husbandry

FIGURE 11.9 Tower silo

FIGURE 11.10 Plastics used for silo

waste." There is also some evidence that invisible oxidation losses occur throughout the whole mass of silage during the storage period. A large quantity of the silage mass (about 25%) can be within the top 1 m depending on silo size and depth. The most common material used to seal horizontal silos is plastic film. The principal function of the film is to seal the forage and allow anaerobic conditions to establish.

11.10.1 Unsealed Silos

These are silos that haven't been sealed, and a bunker silo's correct cover is just as crucial as its proper harvesting and filling processes. Previous research has shown that if horizontal silos are not coated with plastic film, the quality and recovery of silage are harmed. When bunker and stack silos are not sealed, DM and nutrients are lost, according to a study (Berger and Bolsen, 2006). The top 0.90 m of silage from 127 horizontal silos was examined at three locations across the silo face from 1990 to 1993. The depths of sampling from the surface were 0–0.45 m (depth 1) and 0.45–0.90 m (depth 2). The silos can be sealed with a single PE film (from 100 m to 150 m thick) of black or white-on-black, which was fixed with tires, sidewall discs, or soil. Bunkers and stacks that were not sealed suffered more losses. Silage in the unsealed silos' periphery area had pH values ranging from 4.75 to 8.55, which was indicative of ruined silage. The organic matter losses in the topmost layer (top 0.45 m) were reduced when a plastic sheet was added. Silage sealing also helped to decrease spoiling.

In an exposed silo, the aerobic deterioration is initially limited to the top 15–30 cm. The reason for this is that at the upper layer, aerobic microbial activity is high enough to remove all of the oxygen entering the crop by diffusion or convection. The pace of microbial activity decreases as the rapidly degradable components of the crop in the top layer are exhausted, allowing oxygen to flow deeper into the silo and induce deterioration at that level (Muck, 1998).

11.10.2 Lining Bunker Walls with Plastic

Using plastic to line the bunker walls, a substantial portion of the silage kept in horizontal silos is exposed to air and susceptible to spoiling, particularly near the walls (at the silo's shoulders), because silages are difficult to seal correctly. According to one study, silage DM losses near the surface of bunker silos are largest (76%) along the silo wall and lowest (16%) near the core. As a result, the connection of the cover to the bunker silos is still a concern. Before attaching the main sheet, place an additional film of 1–2 m deep (depending on the size of the silo) between the wall and the forage and then over the forage. As a result of this extra effort, the quality of the silage along the wall is comparable to that throughout the silo (Honig, 1991, Lima et al., 2016).

11.10.3 Plastic Film to Cover Silage

Three important functions must be fulfilled by a plastic film that is used to cover silage. First and foremost, the film should resist precipitation as well as damage caused by weather and animal attack. Second, the film should be UV-resistant to withstand continuous sunshine exposure. Finally, the silo film's duty is to ensure anaerobic conditions in the silage.

11.10.3.1 Plastic Film Color and Thickness

Oxygen permeability into the silage is greatly dependent on the temperature of the plastic, therefore the quantity of air infiltration and subsequent aerobic losses are affected by the color of the sheet. Only a few studies on the thermal effects of covers on the upper have been published. These surface layers are extremely prone to poor fermentation because of the inadequate packing density and proximity to the plastic film. Furthermore, the high temperature creates a microclimate in the upper layer that substantially promotes the proliferation of harmful microbes (yeasts, molds, and aerobic bacteria). The largest DM losses and yeast counts when corn silages were sealed with black PE, which is consistent with their findings. During storage, the black sheet likewise shows a greater temperature than the white-on-black film. The effects of color on the temperature of film surfaces were investigated in a study by different researchers. Temperature peaks for the black film were up to 16°C higher in the early hours than for the white film, according to the authors. The maximum temperatures were recorded at midday, as expected, with the black and green colored films exhibiting fairly comparable thermal characteristics. The same was true in the evenings. Savoie (1988) created a model to calculate the costs of plastic and respiration losses due to air penetration through the film. The following parameters were taken into account while determining the ideal thickness: storage length, silage density and DM content, film permeability, and the relative value of plastic and silage. When fully sealed, polyethylene silage bags of various thicknesses (100 m, 150 m, and 200 m) did not create significant changes in losses in 130 days, averaging 0.2% loss per month (Savoie, 1988). Modeling of several film thicknesses on a stack silo revealed that 100 m was the most cost-effective for 3 months of storage, 150 m for 7 months, and 200 m for 12 months. It's worth noting that thicker films are more puncture- and tear-resistant than thin ones.

11.10.3.2 Oxygen Permeability of Plastic Film

In silage, the air is the most common cause of deterioration. Polyethylene isn't completely impervious to oxygen diffusion; therefore, it won't completely prevent oxygen from entering. As a result, there is widespread consensus that sheets with low oxygen permeability should be pursued. When a coextruded PE polyamide film was created for covering horizontal silos in the early 2000s, it became the first generation of barrier films (Borreani et al., 2007). It had a thickness of 125 m and was made up of 2 PE outer layers and a polyamide middle layer. However, this idea revealed numerous flaws, such as rigidity and fragility, which resulted in reduced use in farm settings.

Oxygen barrier (OB) films composed of PE and ethylene-vinyl alcohol (EVOH) have just recently become accessible. Ethylene-vinyl alcohol has the best barrier properties as well as good mechanical properties as puncture resistance, tear resistance, and stretchability (Borreani et al., 2011). In the European and American markets, there are two types of OB films to choose from. The first is a white-on-black sheet made by sandwiching an EVOH layer between two layers of PE during the production process. The second is a thin film (45-m-thick PE + EVOH) that must be covered with a tarp or a second layer of PE when used in real-world situations. Because it is not UV-stabilized, this treatment is required. Originally, the thin OB film was used in conjunction with a tarp to protect against UV light and physical damage. This form of UV shielding, on the other hand, is prohibitively expensive for farmers with limited resources. To solve this difficulty, a method was developed that mixes the thin film with a standard PE sheet. An experiment was conducted to see if this approach of covering maize silage in bunker silos was beneficial.

The following two systems were evaluated: the first method involved draping a sheet of 45-m-thick OB film along the length of the sidewall before filling with approximately 2 m of extra. The extra film was pulled over the wall after filling, and a sheet of PE was laid on top. A typical sheet of 180-m-thick PE film was used in the second system. Over the course of two years, eight commercial bunker silos were divided in half longitudinally, with one half covered with OB and the other with the normal system. When compared to the core, the oxygen barrier method generated well-fermented silages that were similar to the middle part of the silo (core), whereas the PE system produced less lactic acid and had higher pH and mold counts.

The predicted milk yield for the PE system was 116 kg/ton lower than the core, despite the fact that the OB system and core had equal milk yields (1,258 kg/ton and 1,294 kg/ton, respectively). Even though the OB films cost more than the PE layer, these results and those obtained by Borreani and Tabacco (2014) revealed a net economic advantage when the OB films are utilized due to both reduced nutrient losses and labor time necessary to clean the upper layer.

The plastic cover is being weighed. The plastic film kept in place with recycled car tires is commonly used to prevent degradation in horizontal silos. Because of their inexpensive cost and widespread availability, tires have been widely used. When the number of tires per square meter increased, there may be a drop in temperature and better protein availability of hay crop silage, according to Ruppel (1993). Higher tire density (30 tires/10 m^2) and sandbags along the shoulders resulted in fewer silage losses in the top layer, according to a study by Ruppel (1993). The findings of a research project on several silage sealing techniques were presented. Borreani and Tabacco (2011) presented the findings of a study on several silage sealing techniques. A single white-on-black sheet was used to cover a farm bunker silo. Tires were used to cover half of the width of the sheet (25 kg/m^2), and gravel was used to cover the other half (200 kg/m^2). The silo was opened for summer consumption; however, the feed-out rate was modest (12 cm/day).

The temperature in the corn silage's outer portions was affected by the variation in the sealing method, according to the findings. The silage that was covered with tires reached a maximum temperature of above 40°C, whereas the silage that was covered in gravel did not. The amount of soil applied to the PE plastic cover has an impact on the quality of the silage. Griswold et al. (2010) investigated the efficiency of numerous sealing techniques followed in Brazil for reducing top-layer losses. Covering a black plastic sheet with dirt (100 kg/m^2) reduced losses, which was linked to lower pH and ash content, as well as lower yeast counts. Most farmers, however, are adamant about not covering the horizon. Also, they are hesitant to cover horizontal silos with soil, especially if the silo is huge, because they do not believe the labor and expenditures required in doing so are appropriate and cost-effective. Furthermore, when unloading, the soil utilized as a cover can contaminate the silage. Alternative covering tactics were studied to decrease aerobic deterioration in the peripheral sections of corn silage in a warm climate which include (1) black PE film (control), (2) black PE film plus sugarcane bagasse (10 kg/m^2) over the sheet, and (3) black PE film plus soil (30 kg/m^2) over the sheet (Borreani, and Tabacco. 2007). Temperatures were unaffected by treatments during the early stages of storage, but after around 80 days of fermentation, the temperature in the controlled silage began to rise but not in the others. Because the gas transfer rate is reduced by the presence of soil or sugarcane bagasse over the sheet, this can be attributed to the influence of oxygen permeability of the film throughout a lengthy storage time. These findings also imply that the material covering the film minimizes wind-induced billowing, which has an impact on the volume of the air sucked into the silo. It is crucial to stress the importance of keeping the plastic cover weighed down during the storage and feed-out periods. Air can penetrate the peripheral areas of a silo up to 1 m or more beyond the feed-out face during unloading, especially when the sealing cover is not weighed down or is only weighed with tires, implying that daily removal rates should be higher than 30 cm/day in these situations to avoid extended aerobic spoilage (Bernardes, 2016).

Example 1: Design a trench silo for a small farm having 120 buffaloes weighing 650 kg each and having to be fed at the rate of 4 kg/100 kg of its weight. The silage is fed 150 days a year.

SOLUTION: GIVEN

Number of buffaloes =120
Weight of each animal = 650 kg
Feeding rate of silage = 4 kg/100 kg of animal body weight
⇒ [(100 kg animal fed 4 kg of silage
650 kg animal's required silage = (4/100) × 650 = 26 kg)]
Number of days the animal is fed = 150 days

ASSUMPTION

Depth of silo = d = 2.5 m
Length of silage feed per day = 15 cm = 0.15 m
Side slope = ½, i.e., tan θ = 0.5
1 m^3 of silage density = 650 kg/m^3
W_t of silage fed per day = feeding rate of silage × weight of each animal × number of animals

$$= \frac{4}{100} \times 650 \times 120 = 3120 \text{ kg/day}$$

Density = weight/volume
Volume = weight/density
Volume of silage required per day = 3,120/650 = 4.8 m^3/day
Volume according to design
Volume = length × breadth × depth = area × depth
Therefore, the volume of silage stored per day in trench silo = ½(w + d + w) × d × l

$$\text{Volume of silage stored per day in trench silo} = \frac{1}{2}(w + d + w) \times d \times 0.15 \qquad (2)$$

From equation (1) and (2)
Volume of silage required per day = volume of silage stored per day in trench silo
½(w + d + w) × d × 0.15 = 4.8
½(2w + d) × d × 0.15 = 4.8
½(2w + 2.5) × 2.5 × 0.15 = 4.8

2w + 2.5 = 25.6
w = 11.55
Bottom width = w = **11.55 m**
Top width = w + d = **14.05 m**
Length of silo = (length of silage withdrawn/day) × no of days
= 0.15 × 150 = 22.5 m
Considering 20% losses, the length of silo as a safety factor = (22.5 + 22.5 × (20/100)) = **27 m**

11.11 PLASTIC FOR LIVESTOCK SHELTER

Poly houses are intended to let in much-needed light and warmth. Not only does this increase food conversion rates but it also provides a healthier, dryer environment that is less prone to bacteria and fungi. Poly houses are becoming more popular for calf housing, poultry sheds, dairy herd housing, beef cows, goats, pigs, and alpaca housing (Northern Polytunnels, 2017). Cattles need to be protected from extreme changes in climate and predator attacks. Therefore, suitable shelter or housing that matches climatic conditions and a suitable protection system need to be provided. For protein sources, sheep and goats are used in the diet of the human population. Also, they are utilized for wool production. Animal growth, production, and reproduction are the ultimate concern of the farmer. Plastic-based animal sheds are utilized to give comfort to the sheep and goat family. It can be detached and set up elsewhere as per the shelter requirement during the migration of nomadic tribes. The following are the engineering design parameters that should be considered.

- Space requirements
- Wind direction
- Ventilation assessment
- Natural light direction
- Thermal insulation
- Overall structural stability
- Waste management and disposal hygiene factor
- Durability

Animals' conditions can be checked as per the following parameters.

1. Normal physiological parameters
   - Respiration rate
   - Heart rate
   - Temperature
2. Hematological parameters
   - Hemoglobin levels
   - Total erythrocytes count (TEC)
   - Total leukocytes count (TLC)
   - Differential leukocytes count (DLC)
   - Sugar levels
   - Total protein content
3. Stress hormone parameters.
   - Cortisol levels
   - T3, T4 (thyroid investigation)
4. Growth parameters
   - Birth weight
   - Weight of animals in different months

11.11.1 Plastic Use in Animal Shelter

For the safe design of any structure, it is essential to have knowledge of various types of loads and their worst combinations to which it may be subjected during its life span (Figures 11.11–11.13). Following are the various types of loads that should be calculated.

1. Dead load
2. Live load
3. Wind load

FIGURE 11.11 Plastic shelter for animals

FIGURE 11.12 Plastic shelter for sheep

Plastics in Animal Husbandry

FIGURE 11.13 Sheep inside a poly house

4. Seismic load
5. Snow load
6. Earth pressure
7. Water current load
8. Impact load
9. Temperature and erection effects

11.11.2 Advantages of Poly House as Livestock Shelter

The following are the advantages of plastic-based animal shelters.

a) Short-wave radiation from sunshine promotes warmer and drier housing conditions for livestock, which reduces bedding expenses
b) Increases natural light, lowering energy use
c) A warmer climate boosts the food conversion ratio, cutting feed costs and lowering the risk of twin lamb illness
d) Content livestock – improved welfare and lowered lamb mortality
e) Poly houses are less expensive per square meter than traditional structures
f) The natural light keeps bacteria, fungi, molds, and odors to a minimum and provides livestock with natural vitamin-rich sunlight as well as warmth.

QUESTIONS

1. Describe the application of plastics in animal husbandry.
2. How plastic is used in the conservation of fodder?
3. What is silage and what are its advantages?
4. How to select the site for the construction of the silo?
5. Discuss different types of silos.
6. Describe the animal shelter using plastics and its advantages.
7. Explain the design steps of the animal shelter.

MULTIPLE-CHOICE QUESTIONS

1. Plastics are used in animal husbandry for
   a) Storing animal grains
   b) Storing straw during winter
   c) Storing silage
   d) All of the above
2. Silage is used
   a) To storage in wrapped bales
   b) It is advantageous over hay processing
   c) It is less dependent on the weather
   d) All of the above
3. What is the purpose of conservation of fodder
   a) To preserve feed when it is available in excess
   b) To maintain the optimum nutritional value of fodder
   c) To make available the food throughout the year
   d) All of the above
4. What is silage?
   a) Silage is the green material produced by the controlled fermentation of the high moisture content retained by the green fodder crop
   b) It is dry material that conserves moisture
   c) It is dry material which increases temperature
   d) None
5. Selection of crops for silage making is based on
   a) High fermentable sugar level
   b) Limited buffering capacity
   c) Pasture grasses
   d) All of the above
6. Different kinds of silos are
   a) Stack silo
   b) Bunker silo
   c) Pit/Trench silo
   d) All of the above

ANSWERS

1. d   2 d   3. d   4. a   5. d   6. d

REFERENCES

Ashbell, G. and Weinberg, Z. G. 1992. Top silage losses in horizontal silos. *Canadian Journal of Chemical Engineering*, 34, 171–175.

Berger, L. L. and Bolsen, K. K. 2006. Sealing strategies for bunker silos and drive-over piles. In *Proceedings of Silage for Dairy Farms: Growing, Harvesting, Storing, and Feeding. NRAES Publ* (Vol. 181, pp. 266–283). Brno, Czech Republic.

Bernardes, T. F. 2016. Advances in silage sealing. *Advances in Silage Production and Utilization*, 1, 53–62.

Bisaglia, C., Tabacco, E. and Borreani, G. 2011. The use of plastic film instead of netting when tying round bales for wrapped baled silage. *Biosystems Engineering*, 108(1), 1–8.

Borreani, G. and Tabacco, E. 2007. *Il Silomais: GuidaPratica*. QuadernidellaRegione Piemonte – Agricoltura, Torino, Italy.

Borreani, G. and Tabacco, E. 2014. Bio-based biodegradable film to replace the standard polyethylene cover for silage conservation. *Journal of Dairy Science*, 98, 386–394.

Borreani, G. and Tabacco, E. 2017. Plastics in animal production. In *A Guide to the Manufacture, Performance, and Potential of Plastics in Agriculture* (eds Michael Orzolek). 145–185. Elsevier, US.

Borreani, G., Tabacco, E. and Cavallarin, L. 2007. A new oxygen barrier film reduces aerobic deterioration in farm-scale corn silage. *Journal of Dairy Sciences*, 90, 4701–4706.

Borreani, G., Tabacco, E. and Deangelis, D. 2011. Special EVOH based films improve quality and sanity of farm corn silage. In *Proceedings of Agricultural Film 2011: International Conference on Agricultural and Horticultural Film Industry.* Applied Market Information Ltd., Barcelona, Spain, 1–15.

FAO (Food and Agriculture Organization). 1994. *Definition and Classification of Commodities*. Rome, Italy.

Griswold, K. E., McDonell, E. E., Kung, Jr., L. and Craig, P. H. 2010. Effect of bunker silo sidewall plastic on fermentation, nutrient content and digestibility of corn silage. *Journal of Animal Science*, 88(E-Suppl.2), 622.

Honig, H. 1991. *Reducing Losses during Storage and Unloading of Silage. Silage Conservation towards 2000*. G. Pahlow and H. Honig (eds). Institute of Grassland and Forage Research and Federal Research Center of Agriculture Braunschweig-Völkenrode, Braunschweig, Germany.

Levitan, L. C., Cox, D. G. and Clarvoe, M. B. 2005. *Agricultural plastic film recycling: Feasibility and options in the Central Leatherstocking-Upper Catskill region of New York State*. Accessed Feb. 17, 2014. Envir. Risk Anal. Prog. (ERAP), Cornell University, Ithaca, NY. http://ecommons.cornell.edu/bitstream/1813/33176/2/AgFilmRecyFeasibility-05Red.pdf

Lima, L. M., Dos Santos, J. P., De Oliveira, I. L., Gusmão, J. O., Bastos, M. S., Da Silva, S. M. and Bernardes, T. F. 2016. Lining bunker wall with oxygen barrier film reduces nutrient losses of corn silages. *Journal of Animal Science*, 94, 312.

Muck, R. E. 1998. *Influencing of Air on Ensiling*. ASAE Paper 981054. ASAE, St. Joseph, MI.

Norther, N. 2017. Poly tunnels. *Livestock Housing*, Issue 4. https://www.northernpolytunnels.co.uk/ (Accessed 25.06.2021).

Ruppel, K. A. 1993. Bunker soil management and its factors on haycrop quality. In *Silage Production from Seed Animal*. NRAES-67, Syracuse, NY. 23–25 February 1993. Northeast Reg. Agric. Eng. Ser., Ithaca, NY, 67–84.

Savoie, P. 1988. Optimization of plastic covers for stack silos. *Journal of Agricultural Engineering Research*, 41, 65–73.

Scarascia-Mugnozza, G., Sica, C. and Russo, G. 2011. Plastic materials in European agriculture: Actual use and perspectives. *Journal of Agricultural Engineering*, 42(3), 15–28.

Varda, M., Nishith, D. and Darshan, M. 2014. Production and evaluation of microbial plastic for its degradation capabilities. *Journal of Environmental Research and Development*, 8(4), 934.

12 Plastics as Cladding Material

12.1 INTRODUCTION

Plastic cladding materials play an important role to increase productivity. If India is to emerge as the world's economic force, our agricultural productivity should be at par with that of other countries currently listed as the world's economic forces. India is in desperate need of modern and efficient technologies capable of continually improving the production, viability, and sustainability of agricultural systems (Chandra et al., 1988).

The cropping trend has changed markedly toward diversification into the country's high-value horticultural and commercial crops. In order to make horticulture viable and sustainable, emphasis should be on improving the quality of resources to be used in order to improve relative profitability (Dalrymple, 1973). India is endowed with diverse agroclimatic conditions that contribute to the cultivation of all types of horticultural crops, almost throughout the year in one part of the country or another but the quality of produce under open field conditions, particularly in the case of high-quality flowers and vegetables, is largely not compatible with domestic and international market standards. Often because the crop is exposed to a variety of environmental factors that change regularly, there is no guarantee of stable production from open cultivation. Therefore, to satisfy the demand of quality-conscious customers, it is imperative to improve the productivity and quality of produce. In addition to ensuring vertical productivity growth, breakthroughs in production technology that incorporates market-driven quality parameters with the production system are required.

The most significant technology in this context is the Controlled Environment Agriculture (CEA), i.e., safe techniques for cultivation protected cultivation techniques (PCTs), such as greenhouse, nethouse, poly house, polytunnel, and glasshouse (Figure 12.1 (a–d)). Safe cultivation, which allows some control of wind speed, humidity, temperature, light intensity, and atmospheric composition, has contributed and will continue to make a major contribution to a better understanding of the requirements for growth factors and inputs for improving crop productivity in open fields (Jensen et al., 1995). With or without heat, safe technologies (windbreaks, irrigation, soil mulches) or structures (greenhouses, tunnels, and row covers) can be used. The primary focus is on growing horticultural crops of high value (vegetables, fruits, flowers, woody, ornamental, and bedding plants).

12.2 BENEFITS OF PROTECTED CULTIVATION

The advantages that can be obtained from a covered crop are as follows:

- Environmental management enables plants to be raised anywhere in the world at any time of the year, i.e., crops may be grown under poor climatic conditions if crops could not otherwise be grown under open field conditions
- The crop yields are per unit area, per unit volume, and per unit input base at the maximum level
- Microcosm regulation enables the production of products of higher quality that are free from insect attacks, contaminants, and chemical residues
- For export markets, high-value and high-quality crops can be produced
- Income from small and marginal holdings of land owned by the farmer can be increased through the cultivation of crops intended for export markets
- Protected cultivation can be used to create self-employment for the agricultural sector's skilled rural youth

FIGURE 12.1 Various structures for protected cultivation: (a) poly house; (b) nethouse; (c) glasshouse; and (d) poly tunnel

12.3 GREENHOUSE

Greenhouse is the most realistic way to achieve the goals of safeguarded agriculture, where the natural climate is altered to achieve maximum plant growth and yields by using sound engineering principles. A greenhouse is a framed or inflated structure covered by a transparent or translucent material in which crops can be grown under conditions that are at least partially managed by the atmosphere. The structure is sufficiently large to enable people to function within it in order to carry out cultural activities.

The earliest known covered farming is the cultivation of off-season cucumbers under a transparent stone for Emperor Tiberius in the 1st century. During the next 1,500 years, this technology was scarcely used. Glass lanterns, bell jars, and hotbeds covered with glass were used in the 16th century to shield horticultural crops from the cold. Low, portable wooden frames covered with oiled translucent paper were used to warm the plant atmosphere in the 17th century.

Primitive methods using oilpaper and straw mats to shield crops from the harsh natural environment were used in Japan as early as the early 1960s. Greenhouses in France and England were heated by manure and filled with glass panes during the same century. In the 1700s, the first greenhouse only used glass as a sloping roof on one side. Glass on both sides was used later in the century. Glasshouses were used for fruit crops such as melons, grapes, strawberries, and peaches and were seldom used for the cultivation of vegetables.

Protected agriculture was fully developed after the Second World War with the advent of polyethylene. The first use of polyethylene as a greenhouse cover was in 1948, when, instead of more costly glass, Professor Emery Myers Emmert at the University of Kentucky used the less expensive material. It was estimated that the total area of glasshouses in the world (1987) was 30,000 ha and most of these were located in Northwestern Europe. More than half of the world's plastic greenhouses are located in Asia, where China has the largest area, in contrast to glasshouses. According to 1999-figures, plastic greenhouses cover an area of 6,82,050 ha. Greenhouses are made of plastic and glass in most countries; the rest are plastic.

Cultivation in the plastic greenhouses in India is of recent origin. In India, 100 ha area is under greenhouse cultivation as per 1994–1995 estimates. The greenhouse has grown into more than a guardian of plants since 1960. It is now best understood as a Regulated Environmental Agriculture (CEA) method, with specific air and root temperature, water, humidity, plant nutrition, carbon dioxide, and light controls. Today's greenhouses may be regarded as factories for plants or vegetables. With the artificial environment and growing system under almost complete computer control, almost every part of the production system is automated.

12.3.1 Advantages of Greenhouses

The various benefits of using greenhouses for growing crops in a regulated setting are as follows:

1. Four to five crops can be grown in a greenhouse during the year due to the availability of necessary environmental conditions for plants
2. The productivity of the crop greatly improves
3. Superior quality goods can be obtained as they are grown in an atmosphere that is appropriately regulated
4. In a greenhouse, gadgets for the efficient use of different inputs, such as water, fertilizers, seeds, and plant protection chemicals, can be well managed
5. As the growing area is enclosed, efficient management of pests and diseases is feasible
6. The germination percentage of seeds in greenhouses is high
7. In a greenhouse, the acclimatization of tissue culture plants may be carried out
8. To take advantage of consumer requirements, agricultural and horticultural crop production schedules may be scheduled
9. Different types of growing medium can be used effectively in the greenhouse, such as peat mass, vermiculate, rice hulls, and compost used in intensive agriculture
10. In a greenhouse, export quality produce of international standards can be made
11. Using the trapped heat, the drying and related activities of the harvested commodity can be taken up when crops are not produced
12. Greenhouses are ideal for the use of computers and artificial intelligence techniques for irrigation automation, implementation of other inputs, and environmental controls
13. They generate self-employment for trained young individuals

12.4 TYPES OF GREENHOUSES

For crop production, greenhouse structures of different kinds are used. Although there are advantages for a specific application in each type, there is generally no single type of greenhouse that can be considered the best. Different greenhouse types are constructed to meet particular requirements. The various types of greenhouses are briefly listed below, based on form, utility, material, and construction.

12.4.1 Greenhouse Types Based on Shape

The uniqueness of the cross section of greenhouses may be considered a consideration for the purpose of classification. The styles of greenhouses based on the form/shape that are generally followed are given below.

12.4.1.1 Lean-to Greenhouse Type

When a greenhouse is situated against the side of an existing building, a lean-to style is used (Figure 12.2). It is constructed against a house, with one or more of its sides utilizing the existing framework. It is normally attached to a home, but other buildings can also be attached. With the required greenhouse covering material, the roof of the building is extended and the area is properly enclosed. Usually, it faces the south side. Single- or double-row plant benches with a total width of 7 to 12 feet are limited to the lean-to-style greenhouse. For sufficient sun exposure, it should be in the best direction. The benefit of the greenhouse lean-to type is that it is typically close to available energy, water, and heat. It is a structure that is less costly. This architecture allows the best use of sunlight and minimizes roof support requirements. It has the same drawbacks: limited space, limited light, restricted ventilation, and control of temperature. The height of the supporting wall restricts the design perspective's scale. Temperature regulation is more complicated because the wall on which the greenhouse is constructed will absorb the heat from the sun, while the greenhouse's transparent cover can easily lose heat. It's half a greenhouse, separated by the top of the roof.

12.4.1.2 Even-Span-Type Greenhouse

The even-span is the typical style and full-size structure, with equal pitch and width of the two roof slopes (Figure 12.3). This model is used for small-sized greenhouses, and it is built on level land. On one gable end, it is attached to a tower, and two or three rows of plant benches will accommodate it. The cost of an even-span greenhouse is more than the cost of a lean-to type, but it allows for more plants and has greater design flexibility. The even-span would cost more to heat, due to its size and greater amount of exposed glass space. The design has a better shape to maintain uniform temperatures during the winter heating season than a lean-to type for air circulation. If the structure is very similar to a heated house, a distinct heating system is required. Two side benches, two walks, and

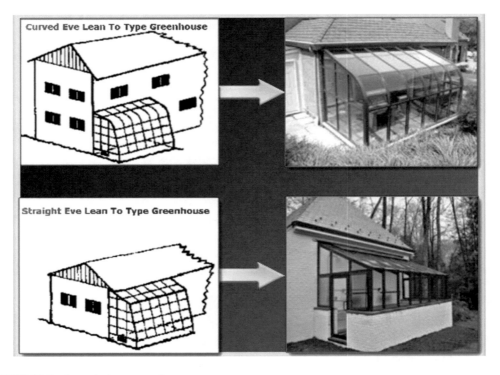

FIGURE 12.2 Lean-to-type greenhouse

Plastics as Cladding Material   247

FIGURE 12.3 Even-span-type greenhouse

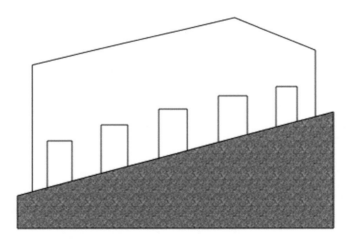

FIGURE 12.4 Uneven-span-type greenhouse

a large center bench will be housed there. For use in various regions of India, several single- and multiple-span types are available. The span usually ranges from 5 m to 9 m for the single-span form, while the length is approximately 24 m. It ranges in height from 2.5 m to 4.3 m.

12.4.1.3 Uneven-Span-Type Greenhouse

On hilly terrain, this sort of greenhouse is built. The roofs have an uneven width, rendering the framework adaptable to the hill's side slopes. This type of greenhouse is seldom used for automation nowadays, as it is not adaptable (Figure 12.4).

12.4.1.4 Ridge and Furrow-Type Greenhouse

Two or more A-frame greenhouses connected to one another along the length of the eave are used for designs of this kind. The eave acts as a gutter or furrow to take away rain and melting snow. Between greenhouses, the sidewall is removed, resulting in a structure with a single wide interior, and the interior space consolidation eliminates labor, lowers automation costs, increases personal management, and decreases fuel consumption, as there is less-exposed wall area from which heat escapes. The snow loads must be included in the frame requirements of these greenhouses, because, as in the case of individual freestanding greenhouses, the snow does not fall off the roofs but melts away. In spite of the snow loads, in northern countries of Europe and in Canada, ridge and furrow greenhouses are used successfully and are well suited to Indian conditions. The ridge- and furrow-type greenhouse is shown in Figure 12.5.

12.4.1.5 Sawtooth-Type Greenhouses

Except that there is a provision for natural ventilation in this form, these are almost comparable to the ridge- and furrow-style greenhouses (Figure 12.6). In a sawtooth-style greenhouse, a particular natural ventilation flow path emerges.

FIGURE 12.5 Ridge and furrow-type greenhouse

FIGURE 12.6 Sawtooth-type greenhouse

Plastics as Cladding Material

12.4.1.6 Quonset Greenhouse

It is a greenhouse that is supported by pipe purling along the length of the greenhouse to protect the pipe arches or trusses (Figure 12.7). Generally, polyethylene is the covering material used for this form of greenhouses. These greenhouses are usually less costly than the greenhouses connected to the gutter and are useful if a small isolated cultural area is needed. In freestanding-style or arranged in an interlocking ridge and furrow, these houses are linked. Truss members overlap enough in the interlocking form to allow a bed of plants to grow between the overlapping portions of adjoining houses. Thus, for a group of houses of this kind, a single broad cultural space exists, a structure that is best suited to the automation and movement of labor.

12.4.1.7 Greenhouse-Type Based on Utility

Based on the purposes or services, the grouping of greenhouses may be accomplished. Of the various utilities, the greenhouse artificial cooling and heating are more costly and elaborate. Thus, greenhouses are known as greenhouses for active heating and active cooling systems based on artificial cooling and heating.

12.4.1.8 Greenhouses Built for Active Heating

The temperature of the air within the greenhouse reduces throughout the night. Any volume of heat needs to be provided to prevent the cold bite of plants prior to freezing. Greenhouse heating specifications depend on the rate at which the outside atmosphere loses heat. Various types, such as double-layer polyethylene, thermopane glass (two layers of factory sealed glass with dead air space), or heating systems, such as unit heaters, central heat, radiant heat, and solar heating systems, are used to decrease heat losses.

12.4.1.9 Greenhouses for Active Refrigeration

It is beneficial to reduce greenhouse temperatures during the summer season relative to ambient temperatures for successful crop growth. This type of greenhouse consists of either an evaporative cooling pad with a fan or fog cooling, and hence appropriate changes are made in the greenhouse such that large amounts of cooled air are drawn into the greenhouse. This greenhouse is designed to cause 40% of the roof to open and almost 100% in some instances.

12.5 GREENHOUSE TYPE BASED ON CONSTRUCTION

The structural material is primarily influenced by the style of building, although the covering material also influences the form. The stronger the period, the stronger the substance can be, and more

FIGURE 12.7 Quonset greenhouse

structural members are required to make solid tissues. Easy structures like hoops can be adopted for smaller lengths. Thus, on the basis of the building, greenhouses can be categorized as follows.

12.5.1 Framed Timber Buildings

In general, only framed wooden structures are used for greenhouses with a span of less than 6 m. Without the use of a truss, side posts and columns are made of timber (Figure 12.8). Protected Cultivation & Post Harvest Technology 10 Pine Wood www.AgriMoon.Com 4 is widely used, as it is economical and has the needed strength. Locally accessible timber with good strength, toughness, and machinability can also be used for buildings.

12.5.2 Structures Framed by Pipes

Pipes are used to create greenhouses because there is a clear span of around 12 m. Using pipes, the side posts, columns, cross-ties, and purlins are usually constructed (Figure 12.9). The trusses are not used in this style.

12.5.3 Structures Framed by Truss

Truss frames are used where the greenhouse height is greater than or equal to 15 m. To shape a truss that involves rafters, chords and struts, flat steel, tubular steel, or angular iron are welded together. Under compression, struts are support members and chords are support members under strain. Each truss is bolted with angle iron purlins that run along the greenhouse length. Only in very large truss frame houses of 21.3 m or more columns are used (Figure 12.10). As these frames are ideally suited for prefabrication, most of the glasshouses are of the truss frame form.

FIGURE 12.8 A greenhouse with timber frame

Plastics as Cladding Material

FIGURE 12.9 A pipe-frame poly house

FIGURE 12.10 A greenhouse having truss roof

12.6 GREENHOUSE TYPE BASED ON COVERING MATERIALS

The main and significant part of the greenhouse system is the covering material. The greenhouse effect inside the building is directly affected by the covering materials and they affect the air temperature within the house. The styles of frames and the fixing process often differ with the material

FIGURE 12.11 Glass greenhouses

of covering. The greenhouses are categorized as glass, plastic film, and rigid panel greenhouses on the basis of the form of the covering material.

12.6.1 Glass Greenhouses

Prior to 1950, only glass greenhouses and glass as the covering material existed. Glass has the advantage of greater internal light intensity as a shielding medium (Figure 12.11). These greenhouses have a higher rate of air penetration, leading to lower interior humidity and improved control of diseases. For glass greenhouse building, lean-to-style, even-span-, ridge-, and furrow-type designs are used.

12.6.2 Plastic Film Greenhouses

In this type of greenhouse, lightweight plastic films, including polyethylene polyester and polyvinyl chloride, are used as a cover material. Plastics have become popular as a covering medium for greenhouses, as they are inexpensive and the heating cost is lower than for glass greenhouses. The key downside of plastic films is their short life. The highest quality ultraviolet (UV) stabilized film, for instance, will only last for four years. For the use of this covering material, the Quonset design and gutter-connected design are suitable as given in Figure 12.12.

12.6.3 Rigid Panel Greenhouses

Polyvinyl chloride rigid plates, fiberglass-reinforced plastic, rigid acrylic, and polycarbonate panels are used in the Quonset-type frames or ridge- and furrow-type frames as the shielding material (Figure 12.13). This composite is more breakage resistant as opposed to glass or acrylic, the light strength is constant in the greenhouse. Up to 20 years, high-grade panels have a long life.

12.7 TYPES OF GREENHOUSES BASED ON COST OF INSTALLATION

12.7.1 Low-Cost Poly House/Greenhouse

Polythene sheet of 700-gauge thickness is supported on bamboo ropes and nails (Figure 12.14). The temperature inside the greenhouse may be increased over that of outside.

Plastics as Cladding Material

FIGURE 12.12 Plastic film greenhouses

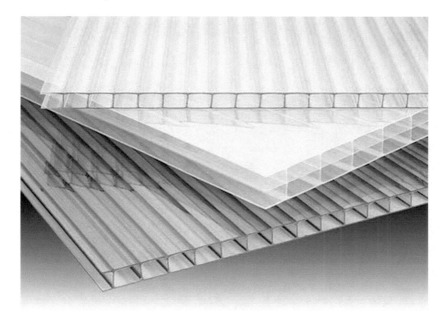

FIGURE 12.13 Rigid panel greenhouses

12.7.2 Medium-Cost Greenhouse

It costs more than just a low-tech greenhouse. Galvanized iron (GI) pipes are used in the Quonset-shape poly house frame. The thickness is 200 micron of a single-coated ultraviolet (UV) stabilized polythene (Figure 12.15). The exhaust fan is operated thermostatically. The life cycle of frames and glazing products is 20 years and 2 years, respectively.

12.7.3 Hi-Tech Greenhouse

The dome-shaped or cone-shaped designs consists of iron or aluminum frame. Peat, perlite, solarite, vermiculite, and rockwool soil media can be used which are extremely durable. The growing media

FIGURE 12.14 Low-cost poly house

FIGURE 12.15 Medium-cost greenhouse

can be used in these types of greenhouses. In India, as these materials are cheaper, coco fibers and rice husks are used as growing newspapers. Fertigation and pesticide sprays are also carried out using a fogging system (Figure 12.16).

12.7.4 MINIATURE FORMS OF GREENHOUSES

12.7.4.1 Plastic Low Tunnels

Low plastic tunnels are small greenhouse types that shield plants from fog, wind, low temperature, frost, and other environmental vagaries. Low tunnels are very basic structures that need very little

Plastics as Cladding Material

FIGURE 12.16 Hi-tech greenhouse

FIGURE 12.17 Plastic low tunnels

maintenance expertise and are easy to build and offer various advantages. For the construction of low tunnels, 100-micron films will be suitable (Figure 12.17).

12.7.4.2 Nethouses

In high-rainfall areas, nethouses are used for growing vegetable crops. The roof of the building is coated with the necessary material for cladding (Figure 12.18). The sides are constructed from assorted gauges of wire mesh. These systems are suitable for hilly areas in the northeast. Sweet pepper, an economically potential vegetable, is normally grown at high altitudes (>1,000 ft.) but more recently its cultivation is gaining popularity in the Northern Indian plains where under open

FIGURE 12.18 Nethouses

FIGURE 12.19 Walk-in tunnels

field conditions, its fruit size and productivity is very low due to temperature variations and insect pest attacks (fruit borer, aphid, mite, and whitefly). There is also a wide scope for the production of protected sweet pepper. Similarly, in the Indian plains, off-season capsicum production under protected cultivation is also becoming widespread.

12.7.4.3 Walk-in Tunnels
- Walk-in tunnels' structure is covered with UV film, suitable for all types of crops: flowers and vegetables (Figure 12.19).

FIGURE 12.20 Shading nets

- Designed to withstand wind up to 120 km/h and trellising loads up to 25 kg/m².
- Structure gable configuration can be 8 or 10 m wide.
- Height reaches 4.10 m (2″ pipe) for 8 m and 4.50 m (3″ pipe) for 10 m.
- 2 or 3 m height.

12.7.4.4 Shading Nets

There is a great range of plant types and varieties that thrive naturally in the most varied climatic conditions that have been moved from their natural environments to manage crop conditions by modern agriculture. Therefore, for the form and variety of plants, conditions identical to natural ones must be established. The precise form of shade needed for the different phases of its growth must be provided to each type of cultivated plant. The shading nets perform the purpose of supplying plants with acceptable microclimate conditions.

Shade nets are designed to shield crops and plants from UV radiation, but they also offer protection from environmental effects, such as fluctuations in temperature, heavy rain, and wind. Due to the regulated microclimate conditions created in the covered field, with shade netting, which results in higher crop yields, better growth conditions can be achieved for the crop. Both nets are UV-stabilized to suit the exposure of the area's predicted lifespan. They are characterized by high tear resistance, low weight, and a 30–90% shade value range for simple and fast installation. There is a wide variety of shading nets on the market that are specified on the basis of the percentage of shade they provide to the plant under them (Figure 12.20).

QUESTIONS

1. What is cladding material?
2. What are the benefits of protected cultivation?
3. What is a poly house? Describe different types of poly houses.
4. How do we design a poly house for a temperate region?
5. Describe different components of a poly house.
6. What is a hi-tech poly house?
7. Discuss a naturally ventilated poly house.

MULTIPLE-CHOICE QUESTIONS

1. Cultivation of greenhouse technologies has been promoted under which of the following?
    a) Mission for Integrated Development of Horticulture
    b) Rashtriya Krishi Vikas Yojana
    c) Integrated Scheme on Agriculture Cooperation
    d) Paramparagat Krishi Vikas Yojana
2. The greenhouse style recommended in hilly terrains is
    a) Even-span
    b) Uneven-span
    c) Lean-to
    d) Ridge and farrow
3. These are the features of higher energy levels and shorter wavelengths
    a) Infrared radiation
    b) Alpha radiation
    c) Beta radiation
    d) Ultraviolet radiation
4. The normal greenhouse effect is essential for the sustenance of life on earth as it helps in maintaining the average temperature of the earth to
    (a) 15°C
    (b) 33°C
    (c) –18°C
    (d) 50°C
5. The greenhouse gas present in a very high quantity is
    (a) Ethane
    (b) Carbon dioxide
    (c) Propane
    (d) Methane
6. The one which is not considered a naturally occurring greenhouse gas is
    (a) CFCs
    (b) Methane
    (c) Carbon dioxide
    (d) Nitrous oxide
7. A greenhouse that does not come under the classification based on the shape
    (a) Quonset-type (b) Curved-roof-type (c) Gable-roof-type (d) Forced ventilated
8. Which of the following does not come under environmental parameters?
    (a) Humidity
    (b) pH
    (c) Light
    (d) Temperature
9. Which one is not used as soilless media?
    (a) Vermiculite
    (b) Perlite
    (c) Coco peat
    (d) Vermiwash
10. Sand as growing medium improves the _____.
    (a) Aeration and drainage
    (b) Water-holding capacity
    (c) Nutritive status
    (d) pH

11. The direction of single-span greenhouse should be
    (a) East–West
    (b) North–South
    (c) North–East
    (d) South–West

ANSWERS

1	2	3	4	5	6	7	8	9	10
a	d	d	a	b	a	b	d	d	b

REFERENCES

Chandra, P. and Singh, J. K. 1988. *Instruction Manual for Establishment of a 4 x 24 m Green House*. Division of Agricultural Engineering, Indian Agricultural Research Institute, New Delhi.

Dalrymple, D. G. 1973. *A Global Review of Green House Food Production*. USDA. U.S. Department of Agriculture, Economic, Research Service, Washinton, DC. Report no 89.

Jensen, M. H. and Malter, A. J. 1995. *Protected Agriculture: A Global Review*. World Bank Technical Paper no 253. The World Bank, Washington, DC.

13 Plastics in Postharvest Management

13.1 POSTHARVEST MANAGEMENT

Postharvest management involves a number of unit operations, such as precooling, cleaning, sorting, grading, storage, and packaging. The moment a crop is harvested from the ground or picked from its parent plant, it starts to deteriorate. Hence, appropriate postharvest management strategies are indispensable to ensure better quality of crops, whether they are sold for fresh consumption or used to prepare processed products. Both quantitative and qualitative losses occur at all stages in the postharvest handling system due to various extrinsic and intrinsic factors. Extrinsic factors include surrounding temperature, relative humidity, gas composition, mechanical injury, etc., whereas crop variety, respiration rate, moisture content, microbial load, etc. are included in intrinsic factors. By and large, factors causing postharvest losses vary from place to place and crop to crop. Deterioration of fresh crops mainly results from weight loss, bruising injury, physiological breakdown due to ripening processes, temperature injury, chilling injury, or attack by microorganisms. All the crops counting fruits, vegetables, cereals, oilseeds, pulses, root crops, flowers, etc. are living organisms having a respiratory system. They respire even after harvest. During respiration, they take oxygen in and release carbon dioxide out. Respiration causes significant weight loss. It also accumulates heat in the container and changes the gas composition of the surroundings, thereby causing the deterioration of produce. Therefore, it is essential that the crops must be kept in good condition to have excellent quality with maximum shelf life. This constraint underlines the importance of postharvest management of crops.

The packaging industry in India has seen a strong diffusion of plastics as compared to global standards (Anonymous, 2014). However, the agriculture sector has still not explored the benefits of plastics to a considerable extent (Figure 13.1). The major applications of plastics in agriculture include mulching, irrigation, and protected cultivation.

In postharvest management, plastics are mainly used in packaging systems, transportation containers, grading and sorting machines, cold storage units, pack-houses, etc. Moreover, plastics have proved their potential to play a noteworthy role in the preservation of longevity and quality of harvested crops. Plastics are mainly favored for their extraordinary characteristics and versatility of applications. They are lightweight, corrosion resistant, moisture proof, highly versatile, adaptable, and can be molded into attractive shapes. In many cases, food-grade plastics are found as comparable as food-grade stainless steel used in the food processing industry. By considering the importance of plastics, an attempt has been made to discuss some of the important applications of plastics in the postharvest management of various crops. These applications are discussed one by one in the subsequent text.

Within the food system, plastics play an important beneficial role in food transportation, preservation, hygiene, and safety, increasing the lifespan of foods, the length of value chains, and contributing to food and nutrition security (Raghavi *et al.*, 2018). Therefore, it is important that these beneficial functions are not overlooked in the public and policy debates concerning this material, its uses, and impacts. However, recent decades have seen a correlation between substantial increases in plastic food packaging and upward trends in food waste (Seveda, 2012; Sangamithra *et al.*, 2014), suggesting that while plastic packaging can preserve food, in itself this might not be sufficient to reduce wastage. Recent calls to action on plastics are driven in part by observations that the widespread utilization of single-use or disposable plastics, coupled with poor recycling rates and waste

DOI: 10.1201/9781003273974-13

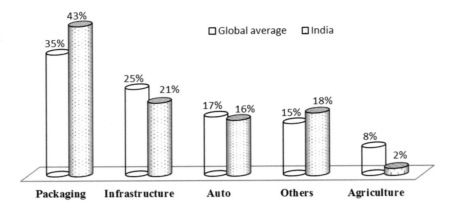

FIGURE 13.1 Use of plastics in different fields (*Source*: Anonymous, 2014)

management is contributing to visible buildups of plastic across natural environments and oceans around the world (Shahi *et al.*, 2011). To illustrate this flow, of the total 6,300 metric tons of plastic waste produced by 2015, only 9% had been recycled or repurposed, with the remaining 91% either incinerated, placed into landfills, or leaking into the natural environment (Yadav *et al.*, 2017).

The extent of the impacts of plastic pollution is still largely unknown and remains to be adequately explored. Among the evidence now beginning to emerge is that which reveals an increasing presence of microplastics, nanoplastics, and synthetic polymers in marine food chains, food products, and the air we breathe. Not surprisingly, negative consequences for human and planetary health are also now being hypothesized and investigated (Barboza *et al.*, 2018; Smith *et al.*, 2018). These interlinking concerns around sustainability come in addition to toxicology research pointing toward potentially harmful effects that chemicals or additives used in plastics may pose for humans as well as the suggestion that plastic packaging could be encouraging unhealthy diets. On a broader scale, it can be said that plastics are also linked to global warming and climate change, as around 99% of plastic monomers are derived from fossil fuels, the supply and demand for which contributes to greenhouse gas emissions (GHGs). Yet in agricultural production, evidence also suggests that plastic sheeting can deliver environmental benefits such as reduced GHG emissions. Taking such trade-offs into account, there are growing calls for improved data and evidence to better understand and address the various effects of plastics, while developing alternatives – where necessary – for the functions they serve.

13.2 FIELD HANDLING OF CROPS

It is the first postharvest unit operation carried out immediately after harvesting. It involves three major operations, namely collection, packaging, and transportation of crops. Crops are collected and subsequently transported, with or without packaging, to the desired destination.

Harvested crops are collected in field containers, such as bags, plastic buckets, plastic crates, woven baskets, gunny bags, etc. Harvesting bags with shoulder or waist slings are used for fruits with firmer skins like citrus. They are easy to carry and leave both hands free. Plastic buckets or other containers are suitable for fruits that are more easily crushed, such as tomatoes. However, presently, in India, plastic crates have replaced all other containers due to their strength, inertness, portability, and corrosion resistance. They are smooth with no sharp edges or projections to damage the produce. They are quite sturdy and hence do not bend out of shape when lifted or tipped. Packaging of harvested crops directly into packages in the field immediately after harvest reduces the damage caused by multiple handling. However, it is commonly used by commercial growers in developed countries but is not so common practice in rural areas. For example, tomatoes are packed

FIGURE 13.2 Plastic crates being used for tomatoes

in plastic crates (Figure 13.2) in order to avoid bruising owing to their softer texture. Cabbages are also being used to transport cabbages (Figure 13.3). For cauliflower curds, the best container has been found to be a plastic crate owing to its smooth inside finish, ease of cleaning, and the fact that it can be used over a long period of time, i.e., a maximum of 10 years. For best results, the layers of curds inside the plastic crate should be lined with plastic sheet liners so as to minimize abrasion damage to the curds during transit. Further, cauliflower curds are best contained in clean and undamaged plastic crates, thus damaged plastic crates (Figure 13.4) should not be used especially during transport. Plastic crates are also considered the best containers for transporting commodities.

Plastic crates can be either stackable (Figure 13.5) or nestable. Both stackable and nestable plastic containers are appropriate in use for the transportation of fruits. Stackable containers can be placed on top of each other, while nestable containers fit into each other, thus reducing storage space requirements during the transportation of empty crates.

13.3 MINIMIZING FIELD HEAT

Temperature management immediately after harvesting is the most essential step in extending the shelf life of crops. Freshly harvested fruits and vegetables contain significant amounts of field heat. This field heat needs to be removed prior to packaging, transportation, or storage. Conventional cost-effective practice is to harvest these crops in the evening so that they are subjected to lower atmospheric temperatures during night hours. Sometimes, these crops are immersed in cold water which cleans and cools the crop. Recently, farmers have started to use shade nets having 30–50% shading intensity to provide a cooler environment compared to the ambient air temperature. These

FIGURE 13.3 Plastic crates being used for cabbage

FIGURE 13.4 (a) Cauliflowers in clean and undamaged crates and (b) cauliflowers in damaged plastic crates

shade nets are mainly made of UV-stabilized plastics. In most cases, a shade net house used for crop production is also used for removing field heat from the crops up to a certain extent. This operation also helps achieve thermal equilibrium in the harvested crops.

13.4 PACKAGING OF FRESH AND PROCESSED CROPS

Packaging is one of the most important steps in the long and complicated journey of fresh and processed crops. Different types of packaging systems like flexible bags, crates, baskets, cartons, bulk bins, palletized containers, etc. are used for handling, transportation, and marketing of fresh and

FIGURE 13.5 (a) Stackable plastic containers for the handling and transportation of fruits and (b) Nestable plastic containers for the handling and transportation of fruits

processed crops. The principal roles of food packaging are to protect food products from outside influences and damages, to contain the food, and to provide consumers with ingredients and nutritional information (Coles, 2003). The goal of food packaging is to contain food in a cost-effective way that satisfies industry requirements and consumer desires, maintains food safety, and minimizes environmental safety (Marsh and Bugusu, 2007).

There are as many types of packages available as there are products to put in them. Packaging systems are available in a variety of materials such as plastic, corrugated fiberboard, wood, and even sustainable materials such as bioplastics and fibers that decompose. One of the most common plastic packaging containers is the clear clamshell, manufactured from polyethylene terephthalate (PET) and other plastics using mechanical or vacuum thermoforming. Although plastic containers are necessary for certain commodities, corrugated and non-corrugated fiberboard is the dominant material used in fresh produce packaging. Wooden containers, usually wire-bound, are a traditional form of produce packaging. They are an option for growers, although their use has gradually diminished over time because they are relatively heavy, expensive, and abrasive to the fruits and vegetables, and because they can present disposal issues. Sustainable packaging options are becoming increasingly more common and offer many advantages over traditional packaging containers. While beneficial to some, they are not appropriate for every operation. There are a variety of functional packaging options available to growers of fresh fruits and vegetables, it is important to select the appropriate format for each specific commodity. Regardless of the material used, for a given commodity, it is important to use standard packaging sizes during the postharvest process so that growers can readily calculate total harvest by weight, count, and volume and, thus, more easily communicate production volumes to their buyers. Also, some buyers require that packaging footprints conform to the dimensions of the standard grocery pallet, which measures 40″ × 48″ (101.60 cm × 121.92 cm).

13.4.1 Classification of Packaging Systems

Packaging systems can be classified as below.

1) Flexible sacks: made of plastics and nets
2) Wooden crates
3) Cartons (fiberboard boxes)
4) Plastic crates
5) Pallet boxes and shipping containers
6) Baskets made of woven strips of leaves, bamboo, plastic, etc.

Plastic nets are suitable only for hard crops, such as coconuts and root crops (potatoes, onions), whereas plastic films find application in a number of places. They can be easily observed that almost all the packaging systems involve plastics. In present times, plastics have become an integral part of the packaging systems of fresh and processed crops. Almost all the foods are being packaged in different types of plastics. The food industry is found to be heavily dependent on plastics in the form of packaging. Moreover, packaging standards have become stricter with the introduction of newer Indian norms closer to global standards which are also driving the use of plastics in packaging.

Different types of plastics are available in the market for packaging purposes. However, each material has its own appropriateness. The report indicates that polyethylene is the most used material, followed by polypropylene and polyethylene terephthalate (PET) (Anonymous, 2014). Plastics used in food packaging are commonly grouped as flexible plastics and rigid plastics. Flexible packaging consists of monolayer or multilayer films of plastics whereas multilayered laminated sheets of plastics include PE, PP, PET, and PVC. Polyethylene and polypropylene account for almost 62% of plastic usage in the packaging industry (Figure 13.6). With the increase in income, consumer preferences for packaged foods are expected to increase further.

In food packaging systems, plastics are used as liners, flexible bags, wraps, boxes, etc. In most of the paper boxes as in the case of tetra packs, plastic films are often used to line packing boxes in order to reduce water losses or to prevent friction damage.

13.4.2 Plastic Bags

Bags of polythene film are the chief containers for fresh fruits and vegetables packaging. Plastic bags are cheaper and involve very low costs in bagging and sealing. They are clear thereby allowing easy inspection of the stuffing and readily accept high-quality graphics. Plastic films are available in a wide range of thicknesses and grades and may be engineered to control the environmental gases inside the bag. The film material is permeable enough to maintain the correct mix of oxygen, carbon dioxide, and water vapors inside the bag. In addition to engineered plastic films, various patches and valves have been developed that are affixed to low-cost ordinary plastic film bags. These devices respond to temperature and control the mix of environmental gases.

Plastic bags may be subjected to rodent damage, which would compromise their ability to prevent moisture uptake, but hermetic packaging apparently reduces the attraction of rodents by preventing odors from escaping (FAO, 1989).

13.4.3 Shrink-Wrap

It is one of the most recent developments in the packaging of freshly harvested fruits and vegetables. In this type of packaging, the individual fruit or vegetable item is wrapped with polythene film. Shrink-wrapping has been used successfully to package kinnow, capsicum, cabbage, cauliflower,

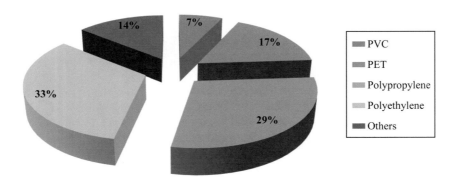

FIGURE 13.6 Share of different flexible packaging materials

potatoes, apples, cucumbers, bitter gourd, and various other fruits and vegetables. Shrink-wrapping with plastic wrap can reduce weight loss and shrinkage, protect the produce from disease, reduce mechanical damage and provide a good surface for stick-on labels (Dhall *et al.*, 2012). Various studies have been conducted to determine the efficacy of shrink-wrap packaging in extending the shelf life of perishables. In a study, green cucumbers were shrink-wrapped with Cryovac D955 (60 gauge) plastic film and stored at 12 ± 1°C, 90–95% RH as well as ambient conditions (29–33°C, 65–70% RH) (Dhall *et al.*, 2012). It was noted that at 12°C and 90–95% RH, shrink-wrapped cucumber showed a minimum physiological loss in weight (0.66%) as compared to unwrapped fruits (11.11%) at the end of refrigerated storage of 15 days.

13.4.4 RIGID PLASTIC PACKAGES

Rigid plastic packages include clamshells. Packages having top and bottom that are heat formed from one or two pieces of plastic are known as clamshells. Clamshells are inexpensive, versatile, provide excellent protection to the produce, and present a very pleasing consumer package. They are often used with smaller packs of high-value crops, such as small fruits, berries, mushrooms, etc., or food items that are easily damaged by crushing. They are also used extensively in the packaging of minimally processed, precut, and prepared salads. Molded polystyrene containers have been found as a substitute for corrugated fiberboard. At present these containers are not cost-effective, but as environmental pressures grow, they may be more common. Similarly, heavy-molded polystyrene pallet bins have been used as a substitute for wooden pallet bins. Although, at present, their cost is almost double that of wooden bins, they have a longer service life, are easier to clean, are recyclable, do not decay when wet, do not harbor disease, and may be nested and made collapsible.

13.4.5 BULK BINS

Bulk bins allow producers to store, handle, and transport a large amount of product with relative ease. Bulk bins are manufactured from plastic polymers, wood, or corrugated fiberboard. Plastic and wood bulk bins are often only used on the farm because they are quite heavy and bulky (Figure 13.7). In terms of transportation, this restricts the amount of product available for shipping because each state imposes weight limits on transported goods. All types of bulk bins require

FIGURE 13.7 Reusable plastic bulk bins (left) and wooden bulk bins (right) used mostly in farms and at packing houses

heavy lifting equipment, such as forklifts, to move the product from field to packinghouse or storage because the contents often weigh in excess of 1,000 pounds.

Despite their weight, bulk bins have many advantages. They allow for easy movement of large quantities of product and are usually stackable. The ability to stack products is oftentimes essential for operations that are limited by the size of their packinghouse or floor space. Bulk bins also eliminate the waste that results from smaller, compartmentalized packaging containers that are single-use and require more material to contain the same amount of product. Bulk bins produced from corrugated fiberboard (Figure 13.8) have been quite often used by wholesalers, retailers, or other repacking operations because they are lightweight, may be collapsible, and even sometimes reusable. Corrugated fiberboard bulk bins can also have graphics on the sidewalls that can be customized for a specific commodity or with producer information. However, for safety and sanitation, plastic bulk bins are preferred to wooden bulk bins and containers. The plastic surfaces are easier to clean, which should be done after every use. Because of their porous surface, wooden bulk bins and containers are very difficult, if not impossible, to clean and sanitize. If using a wooden container to transport product from the field to the packing house, you should wash the product after it has been removed from the wooden bulk bin. Alternatively, plastic liners may be used with wooden bins – both to reduce potential contamination and to minimize abrasion damage to the product from rough wood surfaces. Fasteners such as screws, nails, and bolts can also result in punctured and damaged products, requiring regular inspection of wooden bins. Corrugated fiberboard bins can be used more than once but are more susceptible to damage than plastic and wooden bins. If reused, these bins should be inspected for cleanliness and covered in a plastic liner after the initial use in order to reduce the risk of cross-contamination. Bulk bins, specifically those produced from plastic may be too expensive for some smaller producers. For commodities that are highly susceptible to bruising and mechanical damage during transportation and handling, the use of bulk bins would be inappropriate. However, other commodities, such as potatoes and watermelon, would be considered ideal candidates for bulk bins because they are able to resist most of the damage that occurs throughout

FIGURE 13.8 Zucchini summer squash in reusable plastic containers

the distribution chain. Considering these choices will help producers determine the best course of action for their specific operational needs.

13.4.6 REUSABLE PLASTIC CONTAINERS

Reusable plastic containers (RPCs; Figure 13.8) are becoming more prevalent in the fresh produce distribution system because producers and companies are trying to minimize their carbon footprint while simultaneously reducing costs. For many, RPCs have replaced corrugated fiberboard cartons as the preferred container for shipping fresh produce. Reusable plastic containers come in a wide variety of shapes and sizes, chosen primarily for their application to a specific commodity. Products that are hydrohandled or hydrocooled are especially suitable for packing in RPCs because RPCs are unaffected by moisture exposure.

Reusable plastic containers are more durable and more rigid than their corrugated fiberboard counterparts. They are also weather resistant. As the name implies, RPCs can be utilized over and over again with very little wear, compared to most corrugated fiberboard cartons, which are used only once and then discarded. Most RPCs fold flat to minimize required space for storage and transport. Reusable plastic containers are produced from recyclable polymers, such as polypropylene (PP) or high-density polyethylene (HDPE), and, therefore, can be melted and reformed into new products.

RPCs can be either purchased or leased. Depending on the frequency of use, purchased RPCs can provide significant cost savings over single-use corrugated fiberboard or wooden containers. While the initial cost of purchase and maintenance may seem too expensive, frequent use and economies of scale can allow for lower, long-term savings. Decide whether to use RPCS by considering the size and scale of the farming, packing, or shipping operation and the types of commodities with which the operation deals. Produce handlers must also have either closed distribution systems or return agreements with their receivers in order to utilize purchased RPCs. Otherwise, it will not be possible to recover the RPCs after they have been used (i.e., shipped). RPCs must be cleaned between each use because residues from previous crops can lead to contamination of new harvests. Adhering to strict sanitization and cleaning procedures is necessary in order to reduce the risk of contamination when RPCs are reused. Chemicals, such as detergents, acids, or alkalis, can be used to remove dust, dirt, product residues, and various other debris items from the surface. Scrubbing while also using chemicals increases the probability of eliminating any contaminants that may lead to food-borne illnesses. Note that cleaning procedures alone cannot guarantee the complete removal of all microorganisms. In order to reduce or eliminate microorganisms, the RPCs must also be treated with sanitizers or disinfectants. Sanitizing agents, such as chlorine-based compounds, iodine compounds, or ammonium compounds, are most often applied to food contact surfaces to destroy or reduce the number of microorganisms. While RPCs can provide significant savings to some producers, not all operations justify their utilization. The choice to purchase RPCs is dependent on the ability of the handler to maintain and clean those containers. Some producers may have the equipment and capacity to maintain and clean RPCs, while others may not. If a distribution system does not allow a handler's RPCs to be easily recovered, or if cleaning, sanitizing, and maintaining purchased RPCs is not feasible for an operation, then leasing is the better option. Leased RPCs are returned to the supplier by the receiver; the supplier is then responsible for cleaning and sanitizing the containers before reintroducing them into the distribution system. Producers should conduct a thorough cost analysis that considers customer (receiver) buy-in, material requirements, labor, transportation, inspections, maintenance, and cleaning before purchasing RPCs

13.4.7 INSERT TRAYS

Insert trays are used for a variety of reasons when packing fresh produce. Insert trays are typically formed from molded plastic or pulp to create an insert for a master container that has multiple

FIGURE 13.9 Thermoformed insert tray

cavities to contain individual fruit (Figure 13.9). The cavity is designed to accommodate a specific size and/or variety of commodities as well as to maximize the number of fruits per tray. Insert trays are relatively inexpensive, costing only pennies per tray and providing additional protection to the product to minimize damage. Plastic insert trays are usually thermoformed by placing large, heated, pliable sheets of plastic into molds where they conform to a customer's specified dimensions. This process is very similar to clamshell thermoforming. However, the sheets of plastic thermoformed for insert trays are much thinner than those used for clamshells. Plastic insert trays are typically formed from polypropylene (PP).

Foam trays manufactured from expanded polyethylene (EPE) and other polymers are common as well. Molded paper pulp insert trays are produced by combining water and recycled or reclaimed newsprint or corrugated paper to form a slurry. The slurry is then formed in a mold to produce a tray with cavities for the intended commodity. Insert trays provide additional physical protection to the product, facilitate hand sizing of produce items in field packing operations, and prevent the fruit-to-fruit spread of decay. Plastic insert trays help minimize abrasion damage, localize decay to individual fruit, and reduce moisture loss. Conversely, paper pulp insert trays do not perform well when wet. On the other hand, moisture retention in plastic trays, as a result of condensation or improper drying before or during the time of packing, can create areas of free moisture that encourage decay. This is a problem when the product rests in water trapped in the bottom of the plastic tray. Whether plastic or paper pulp, each tray is often used only once and then discarded. Therefore, the risk of food contamination is relatively low. While plastic and pulp trays both serve to separate the products and keep them from coming into contact with one another, paper pulp and plastic foam insert trays have an additional benefit. Molded pulp and foam insert trays provide more cushioning and support than plastic sheet insert trays. By absorbing shocks, the molded pulp and foam insert trays protect the product from bruising during transportation and handling, therefore further minimizing postharvest losses. Both plastic sheet and foam insert trays exhibit disposal issues. Paper pulp insert trays are oftentimes produced from recycled materials, making them a more eco-friendly choice. However, if a producer's distribution system allows retrieval, plastic insert trays can be cleaned, sanitized, and reused. Overall, insert trays are an attractive option for packers and producers because they are low cost, are hygienic, can be environmentally friendly, and result in fewer postharvest losses. Insert trays made of foamed polystyrene or PVC or PP are overwrapped with heat shrinkable or stretch films (Figure 13.10). A tight wrap

FIGURE 13.10 Tray with (a) stretch wrap and (b) shrink-wrap

FIGURE 13.11 Polyethylene terephthalate thermoformed (PET) clamp shell

immobilizes the fruits and keeps them apart. Trays of molded pulp, cardboard, thermoformed plastic, or expanded polystyrene are also used.

13.4.8 Clamp Shells

Clamp shell (Figure 13.11) containers are used extensively used as consumer packages for a wide variety of commodities like tomatoes, cherry, blueberries, kumquats, and strawberries. A clamp shell is a single-piece vessel consisting of two halves, sculpting the bottom and top, connected by a hinge with interlocking fasteners or snaps drafted for closure. Polyvinyl chloride (PVC) and polystyrene (PS) are a variety of polymers from which clamp shell containers are manufactured. However, polyethylene terephthalate (PET) is one of the most usual polymers used in clamp shell containers because it is easily recycled and is highly transparent. Clamp shells which are adequately drafted have an ample number of vent pores to facilitate the process of cooling and allow the produce to breathe and allow the bulk of moisture to evade. The process of thermoforming and injection molding is used to form the body/structure which can be attributed to various desired shapes and sizes.

These containers provide an area for attaching labels that include required information such as nutrition, weight, and manufacture identification in addition to marketing and consumer information. Clamp shell containers include complete product visibility, product security, and increased perceived value. The container also prevents the contents from being handled by multiple individuals throughout the distribution chain, providing an extra degree of food safety. Although they are recyclable, clamshell containers are produced from nonrenewable fossil fuels. Therefore, the polymers used to manufacture the containers are not the most environmentally friendly. However, clamshells made from biodegradable biopolymers such as polylactic acid (PLA), which is derived from corn starch or sugar cane dextrose, are available.

13.4.9 Sleeve Packs

These combine the low cost of bags and protective qualities and sales appeal of tray packs. Wraps of plastic film such as polyethylene or PVC, in the form of shrink-wrap, stretch film or cling film and regular net stocking or expanded plastic netting can also be used. The traditional fruits and vegetables retail trader packs the produce in the presence of consumers in the qualities and quantities required by them. The package normally used is a simple wrap of paper or a paper or polyethylene bag. Sleeve packs can be fabricated to contain from one to as many as ten fruits. The main advantage of sleeve packs is that they immobilize the produce at a fraction of the cost of tray packs and the produce can be seen from all sides without damage to the fruit.

13.4.10 Plastic Corrugated Boxes

The most commonly used material for a plastic corrugated box is polypropylene and HDPE. Its advantage over corrugated fiber board (CFB) is its low weight-to-strength ratio and its reusability. The printability is also excellent when compared to CFB boxes. But CFB box has an edge over and rigid plastic fiberboard boxes when cushioning properties are taken into consideration. The disadvantages are ultraviolet degradation and temperature resistance, which can be taken care of by the use of additives.

13.4.11 Plastic Sacks

These are flexible shipping containers which are generally used in food industries to transport raw materials viz. fruits and vegetables from the field. If the weight of the content is more than 10 kg then it is called sack otherwise bag. The commonly used materials for sacks are cotton, jute, and woven plastics (HDPE, polypropylene). These sacks are advantageous to use as they cost less, have high strength, are reusable, and require little space for the empties. The disadvantage of plastic-woven sacks is poor stackability due to the low coefficient of friction, which can be overcome by making anti-slip bags. However, plastic-woven sacks have the potential to fulfill the need in a cost-effective manner. These are made from either HDPE or PP (Figure 13.12). Plastic-woven sacks are mostly used for packaging cereals and have an added advantage over jute bags in that they are 5–6 times lighter in weight than jute bags. These plastic sacks have advantages over jute bags as given below.

i) Elongation at break of HDPE tapes is about 15–20% in comparison to jute bags, which is about 30%. Owing to this property, HDPE-woven sacks have better resistance to dropping
ii) HDPE-/PP-woven sacks do not impart any odor to the food product packed in them
iii) HDPE-/PP-woven sacks are not attacked by insects
iv) HDPE-/PP-woven sacks of strength equivalent to that of jute bags can be made using almost 70 times lower weight of the resin and hence are almost 60–65% cheaper than the jute bags
v) HDPE-/PP-woven sacks are the most hygienic material for packing cereals and pulses and one need not reuse the same owing to their low cost

FIGURE 13.12 HDPE/PP plastic-woven sacks

vi) The fabric allows diffusion of air/gases easily through the gaps between the filament thus facilitating ventilation of grains during earlier stages of harvest and penetration of fumigants

vii) Although HDPE/PP undergo degradation under UV light, it is possible to arrest the same by using appropriate UV stabilizers

viii) It is possible to laminate HDPE-woven sacks with LDPE. The laminated bags protect the product packed in the bag from moisture and also prevent the loss of products like flours due to spillage, which usually occurs through plain jute bags, which are commonly used for packing flour

Lamination of PP-woven sacks with polypropylene is also possible, as the grade of PP suitable for lamination is now available in the market.

13.4.12 Plastic Punnets

Lychees are graded, weighed, and packed in plastic punnets of 250 g capacity. About 8 numbers of punnets are placed in a CFB box (one-piece-%tray type) with ventilation holes. The capacity of the box is about 4 kgs. Lychees are also packed in 3-ply CFB boxes of regular slotted container (RSC) type of capacity 2 kg. These boxes are then put in cold storage till dispatch. From the cold store to the airport, the lychee boxes are transported in a refrigerated van. For export, the transportation is by air.

13.4.13 Plastic Tension Netting

Wide-mesh plastic tension netting is used to ensure the stability of pallet loads (Figure 13.13). This is considered to be the common alternative to stretch films and has been found to be better for stabilizing some pallet loads, such as those that require forced air cooling. Used plastic netting may be difficult to be properly handled and recycled.

13.4.14 Plastic Pouch

The flexible heat sealable pouch is a three/four-side-sealed pillow pouch with ventilation holes. The pouch material recommended is low-density polyethylene (LDPE) or polypropylene (PP) as it offers good printability, strength, permeability, and visibility.

FIGURE 13.13 Pallet stabilization using mesh plastic tension netting

FIGURE 13.14 Plastic pouches for *atta*

Milled products like *atta*, *maida*, *suji*, and *besan* are prone to deteriorate with changes such as rancidity, microbial spoilage, and insect attack. The deterioration is rapid when the moisture exceeds 13%. Prevention of Food Adulteration Act specifies maximum moisture content of 14% for wheat flour (*atta*) and *maida* and 14.5% for *suji*. The studies conducted on moisture absorption of milled products have revealed that moisture levels above 12% cause rapid deterioration due to hydrolysis of fat present in these products, hence moisture content of about 10% is recommended at the time of packing to ensure a longer shelf life of these products. Laminated plastic pouches are being used for the packaging of these products – *atta*, *maida*, *suji*, and *besan* (Figure 13.14).

Plastic pouches are also being used for packaging breakfast cereals (Figure 13.15). Breakfast cereals are low-moisture products, crispy in nature, and fortified with essential nutrients. Hence, the packaging material requirement includes the high moisture barrier properties and retention of nutrients throughout the storage period. Hot breakfast cereals are made from whole grains and must be cooked before eating. Cold breakfast cereals are products like "shredded wheat," corn flakes, granulated cereals, and breakfast cereals coated with sugar and are eaten by adding cold milk. These ready-to-eat cereals are processed by the addition of flavoring agents, precooking, and subsequent aeration and drying to create puffed crisp products. Because the grain is cooked

FIGURE 13.15 Plastic pouches for breakfast cereals

prior to expansion, flour can be mixed to obtain various flavors and texture effects. By extruding and expanding through different dies and with varying toasting and heating temperatures, a variety of different products like puffed rice, shredded and flaked wheat, puffed and toasted oats are obtained. Flavors are also added using synthetic sweeteners. The shelf life of the breakfast cereal also depends on the quality of oil contained in them. Breakfast cereals having low oil content such as wheat, barley, rice, and maize where the oil content is 1.5–2% have a comparatively long shelf life to cereals made from oats where the oil content in the product is about 4–11%. The shelf life of the breakfast cereal effect the moisture gain. Hence, while selecting a suitable packing medium for the packaging of breakfast cereals, the following factors are taken into account.

   i) **Loss of crispness**: The crispness is lost due to moisture absorbed by the product. Hence, the packing material should have good barrier properties to keep away the moisture from penetrating inside
  ii) **Lipid oxidation**: In dry breakfast cereals, lipid oxidation is one of the primary means of chemical deterioration. The grains used in breakfast cereals have a high ratio of unsaturated and saturated fat, which gives rise to lipid oxidation. To minimize oxidative rancidity, it is necessary that the package excludes light. Excluding oxygen may be of limited assistance in extending the shelf life. When a case study for storage stability of flaked oat cereal was conducted, it was found that PVC/PVDC copolymer coated with PP-LDPE performed to offer a good oxygen barrier. The use of antioxidants in packing materials can increase the shelf life of a product but is not permitted in most countries
 iii) **Mechanical damage**: The rigidity of the packing material could save the packed product from handling damages including transport
  iv) **Loss of vitamins**: This can be a problem when certain cereals are flavored with fruit. In such cases, loss of flavors would indicate the end of the shelf life of the cereal. Micronutrients present in cereals are not the major factor in determining the shelf life of cereal
   v) **Packaging materials**: Corn flakes are packed in polyester/foil/LDPE laminate packs, which are inserted in a duplex board printed carton. Other packaging materials that are used include 15-micron-BOPP/200-gauge LDPE laminate and 12-micron-metalized polyester/200-gauge LDPE laminate. The above laminates are less expensive as compared to the carton pack

13.4.15 EPS Tray/Stretch Wrapped

The vegetables are placed on a molded expanded polystyrene tray and the tray is wrapped with a cling film.

13.4.16 Leno Bag – 5 Kg

Leno bags are made from polypropylene on a circular weaving loom (Figure 13.16). Leno is a kind of weaving in which the adjacent warp tapes are twisted around consecutive weft tapes to form a spiral pair effectively locking each weft in place.

13.4.17 Consumer Packs for Whole Food Grains

Although cereals and pulses are primarily packed in bulk packs owing to the development of supermarkets and due to an increase in consumer awareness on the quality of food products, branded commodities are now being sold in the market on a large scale. The packaging material used for consumer packs of whole cereals and pulses are as follows: (a) packs made of printed LDPE/LLDPE film, (b) packs made of polypropylene (PP) or biaxially oriented polypropylene (BOPP) film, (c) high-molecular high-density polyethylene film packs, and (d) laminates made from BOPP/LDPE, cast polypropylene (CPP)/LDPE, and polyester/LDPE are used in a few branded commodities. Although laminates are essential for expensive products like basmati rice (Figure 13.17) where the flavor-retention of the product is very important, it may prove to be very expensive for low-value products. The three types of films mentioned above (a, b, and c) offer adequate protection to the consumer packs and they are the most cost-effective packaging materials for consumer packs of cereals and pulses. Laminate mentioned in (d) helps in protecting the product from insect attacks, as polyester is a tough material due to which the insects find it difficult to puncture the laminate.

13.4.18 Biodegradable Films

Environmental pressure is growing continuously due to human interruptions. Under such circumstances, disposal and recyclability of the packaging material of all kinds is becoming a very

FIGURE 13.16 Leno bag

FIGURE 13.17 Plastic pouches for 1 kg branded rice

important issue. It has been studied that common polyethylene takes about 200 to 400 years to break down in a landfill. However, an addition of 6% starch reduces this time to about 20 years or less. Hence, nowadays, packaging material companies are developing starch-based polyethylene substitutes that break down in a landfill as fast as ordinary paper. This move toward biodegradable or recyclable plastic packaging materials may be driven by cost in the long term but lowered by legislation in the near term. Some authorities have proposed a total ban on plastics.

13.4.19 Modified Atmospheric Packaging

Modified atmospheric packaging (MAP) is almost impracticable without using suitable plastics. MAP of fresh fruits and vegetables is based on modifying the concentrations of O_2 and CO_2 in the atmosphere generated inside the package (Mangaraj et al., 2009). It is desirable that the natural interaction occurring between respiration of the product and packaging generates an atmosphere with low levels of O_2 and/or a high concentration of CO_2. The growth of harmful microorganisms is thereby reduced and the shelf life of the product is extended. In a modified atmosphere packaging, gases of the inside atmosphere and the external ambient atmosphere try to equilibrate by permeation through the package walls at a rate dependent on the differential pressures between the gases of the headspace and those of the ambient atmosphere. In this context, the barrier to gases and water vapor provided by packaging material must be considered. Thus, it can be stated that the success of the MAP largely depends upon the barrier (packaging) material used. These packages are made of plastic films with relatively high permeability for gas.

The applications of polymeric films for MAP are most often found in flexible package structures; they may also be used as a component in rigid or semirigid package structures, for example, as a liner inside a carton or as lidding on a cup or tray. The plastic film used in MAP is low-density polyethylene (LDPE), linear low-density polyethylene (LLDPE), high-density polyethylene (HDPE), polypropylene (PP), polyvinyl chloride (PVC), polyester, i.e., polyethylene terephthalate (PET), polyvinylidene chloride (PVDC), polyamide (nylon), and other suitable films (Mangaraj et al., 2009).

Packaging films with a wide range of physical properties are used in MAP. There are several groupings in MAP films such as in the plural, vinyl polymers, styrene polymers, polyamides, polyesters, and other polymers. Polypropylene is part of the polyolefin group and is used largely in MAP, in both forms: continuous and perforated. Although various types of plastic films for packaging are available, relatively few have been used to pack fresh fruits and vegetables and even fewer have a permeability for gas that makes them suitable for MAP. It has been recommended that the permeability for CO_2 should be 3 to 5 times the permeability for O_2. Many polymers used to formulate packaging films are within this criterion.

13.4.19.1 Major Requirements for Plastics Films for MAP

a. These should be flexible, be semirigid, or have the semirigid lidded tray
b. Their permeability must be more for carbon dioxide than oxygen
c. These should have sealing reliability
d. These should have optimum water vapor transmission rate

Much of the plastic material used in the bagging of fresh produce is unsuitable owing to poor moisture and gas permeability. This often leads to condensation, leads to high CO_2 and low O_2 levels in bagged produce, and results in flavor deterioration and fermentation or failure of the fruit to ripen. Thus, the use of plastic packaging designed for the marketing of fresh produce should incorporate consideration for factors such as O_2 uptake, CO_2 production, and the production of heat and ethylene by the produce. Low-density polyethylene (LDPE) film is generally used for the packaging of fresh fruits and vegetables, owing to its high permeability and softness when compared to high-density film. Polyethylene can be easily sealed, has good O_2 and CO_2 permeabilities, has low-temperature durability, has good tear resistance, and has a good appearance. This film is therefore used for the production of modified atmosphere packaging (MAP) which can be manipulated to match the characteristic respiration of produce, by reducing O_2 levels in order to slow down the respiration rate, metabolic rate, and senescence of the produce. Effective O_2 levels must be maintained between 2% and 10% in MAP systems if fermentation of the produce is to be prevented. Elevated CO_2 levels reduce the sensitivity of fresh produce to ethylene and slow down the loss of chlorophyll. At CO_2 levels ranging between 1% and 5%, however, fruits fail to ripen, internal breakdown occurs, and off-flavor development ensues. Oxygen and carbon dioxide transmission rates for MAP films should, therefore, match the respiration rate of the produce to be stored.

13.5 STORAGE

Significant volumes of fresh and processed food materials are stored before ultimate consumption. From very simple to the most modern storage structures are utilized for the storage of agricultural crops. Clamps, root cellars, evaporative cooling structures, bunkers, cover and plinth, godowns, ventilated structures, and such types of storage structures are used conventionally in India (Kale *et al.*, 2016; Kale and Nath, 2018). However, in recent times, with introduction of mechanical refrigeration systems and automatic conveyors and elevators, the storage structures have seen a paradigm shift. Presently, cold storage, modified and controlled atmospheric storage, hypobaric storage, vertical metal silos, hermetic storage, etc. are used for bulk storage of various crops. It can be seen that plastic is extensively used in almost all types of storage structures as construction material. One thing can be noticed that out of all these structures, cover and plinth (CAP) storage is impossible without the use of suitable plastics. In India, CAP is still used to store very large volumes of wheat and paddy. In fact, CAP storage is a necessity in India as Indian grain production has increased faster than storage capacity (Bhardwaj, 2015).

The storage of food grains under large polythene has been practiced in India as well as in other countries for a long time. In the CAP storage, outdoor stacks of bagged grains are covered with a waterproof polythene sheet (Figure 1.3). The advantage of CAP storage is its low cost. It is considered that the cost of CAP storage is only one-fourth of the cost of godown storage. However, CAP storage is vulnerable to wind damage and needs to be inspected frequently to detect damage. The system requires careful management if severe losses are to be avoided. Careful quality control is achieved with regular sampling. Security is also a problem and extra fencing is to be included in the cost calculation.

For CAP construction a plinth with hooks for the ropes lashing the stack is constructed on a suitable site. Dunnage is provided and the covers are made of black polythene sheet of 250-micron-thickness shaped to suit the stack. The covers are held down by nets and nylon lashing. Condensation is prevented by placing a layer of paddy husk-filled sacks on the top of the stack under the polythene. For typical 150-ton CAP storage, the commonly constructed size is 8.55 m × 6.30 m for 3,000 bags each of 50 kg capacity. It is generally provided on a raised platform where grains are protected from rats and the dampness of the ground. The grain bags are stacked in a standard size wooden dunnage. The stacks are covered with 250–350 micron low-density polyethylene (LDPE) sheets from the top and all four sides. Wheat grains are generally stored in such CAP storage for 6–12 months. It is the most economical storage structure and is being widely used by the Food Corporation of India for bagged grains (Jain and Patil, 2012).

13.6 TRANSPORTATION

Transportation is an inevitable postharvest operation in the journey of crops from farm to fork. Crops are transported many times after harvesting till consumption. Transportation occurs within the field, field to the packhouse, field to cold storage, packhouse to cold storage, field to the warehouse, cold storage to distant markets, markets to consumers, etc. Freshly harvested grains are generally transported in bulk/gunny bags using trolleys, whereas fresh fruits and vegetables are transported using rigid crates, sacks, wooden boxes, etc. One of the most important containers used for transporting fruits and vegetables is the returnable/reusable molded plastic crates (Figure 13.18). These reusable boxes are molded from high-density polythene (HDPE) and are widely used for transporting crops in many countries.

Returnable plastic crates can be made to almost any specification. They are strong, rigid, smooth, easy to clean, and made to stack and nest when empty in order to conserve space. In spite of the cost, their capacity for reuse can make them an economical investment. However, they are found economical only in large quantities. They are attractive, have many alternative uses, and are subject to high pilferage. Returnable plastic crates are easy to clean due to their smooth surface and are hard in strength, giving protection to products. They can be used many times, reducing the cost of transport. They are available in different sizes and colors and are resistant to adverse weather conditions. However, plastic crates can damage some soft produce due to their hard surfaces, thus liners are

FIGURE 13.18 Returnable plastic crates (*Source*: Kitinoja, 2013)

recommended when using such crates. One of the major drawbacks is that these crates deteriorate rapidly when exposed to the sunlight unless treated with an ultraviolet inhibitor.

The use of polyethylene film bags for wrapping whole bunches of bananas for transport has been found to be most suitable for reducing wastage. The use of wooden crates having internal dimensions of 42 cm × 32 cm × 29 cm has also been recommended for the long-distance transportation of bananas. Mandarins either can be individually wrapped in cling films or are packed in consumer packages such as plastic bags, plastic mesh, trays of molded pulp, paper board, plastic, or foamed plastic. Losses in first-grade tomatoes can be reduced from 15% to 3% by using upright cone baskets together with dry grass as a packaging material between the layers of fruits. Packing of tomatoes in sealed unventilated polyethylene provides a modified atmosphere which extends storage life. Printed plastic bags are used to reduce light transmission to potato tubers. Plastic oven ventilated bags of 25 kg and 50 kg of capacity are used for onions and potatoes. Palletization and containerization will go a long way in establishing both internal and international trade on a firm footing.

QUESTIONS

1. Discuss plastics application in postharvest management.
2. What is a plastic bag?
3. What is a biodegradable film?
4. Describe an application of plastic in packaging.
5. Write short notes on (i) insert trays and (ii) reusable plastic containers.
6. Discuss the advantages of plastic packaging material over wooden packaging material when used as bulk packaging material.
7. Write short notes on (a) sleeve packs and (b) clamp shells.
8. What is the precooling of fresh produce? Write down its advantages.

MULTIPLE-CHOICE QUESTIONS

1. Postharvest management involves
   a) Precooling and cleaning   b) Sorting and grading
   c) Storage and packaging   d) All of these
2. Both quantitative and qualitative losses occur at all stages in postharvest handling system due to
   a) Extrinsic factors   b) Intrinsic factors
   c) Both a and b   d) None of these
3. _____ management immediately after harvesting is the most essential step in extending shelf life of crops.
   a) Temperature   b) Relative humidity
   c) Dewpoint   d) None of these
4. Polyethylene and polypropylene account almost ___ of plastics usage in the packaging industry.
   a) 52%   b) 92%
   c) 62%   d) None of these
5. Storage structure conventionally used in India involves
   a) Clamps   b) Bunkers
   c) Root cellars   d) All of these
6. MAP stands for
   a) Modified atmospheric packaging   b) Modified aerobic packaging
   c) Multi-aerobic packaging   d) None of these

7. Drying_____
   a) Reduces weight of product   b) Reduces volume of product
   c) Extend shelf life   d) All of these
8. Tarpaulins made of _____ are used for drying purposes.
   a) High-density polythene   b) Low-density polythene
   c) Polypropylene   d) All of these
9. In refractive window drying system _____ air is circulated over the product layer to achieve drying
   a) Hot   b) Cold
   c) Both a and b   d) None of these
10. Poly house dryers consist of
    a) Drying chamber   b) Exhaust fan
    c) Chimney   d) All of these
11. Fresh horticultural produce that is hydrocooled are especially packed in:
    a) Insert trays   b) Reusable plastic containers
    c) Wooden crates   d) All of the above
12. Which of the following polymers is most commonly used for making clamp shells?
    a) Polystyrene   b) LDPE
    c) HDPE   d) None of the above

ANSWERS

1	2	3	4	5	6	7	8	9	10	11
d	c	a	c	d	a	d	a	b	d	b

REFERENCES

Anonymous. 2014. *Potential of Plastics Industry in Northern India with Special Focus on Plasticulture and Food Processing. A Report on Plastics Industry.* FICCI, New Delhi.

Bhardwaj, S. 2015. Recent advances in cover and plinth (CAP) and on-farm storage. *International Journal of Farm Sciences*, 5(2), 259–264.

Coles, R. 2003. Plastic in food packaging. In Coles, R., Mcdowell, D., and Kirwan, M. I. (eds.), *Food Packaging Technology.* Blackwell Publishing, CRC Press, London, UK, 1–31.

Dhall, R. K., Sharma, S. R. and Mahajan, B. V. C. 2012. Effect of shrink wrap packaging for maintaining quality of cucumber during storage. *Journal of Food Science and Technology*, 49(4), 495–499.

FAO Training Series: No. 17/2. (1989). Food and Agriculture Organization of the United Nations, Rome.

Jain, D. and Patil, R. T. 2012. Modelling of thermal environment in covered and plinth storage of wheat as effect of colour of plastic sheet. *Journal of Agricultural Engineering*, 49(1), 36–42.

Kale, S. J. and Nath, P. 2018. Kinetics of quality changes in tomatoes stored in evaporative cooled room in hot region. *International Journal of Current Microbiology and Applied Sciences*, 7(6), 1104–1112.

Kale, S. J., Nath, P., Jalgaonkar, K. R. and Mahawar, M. K. 2016. Low cost storage structures for fruits and vegetables handling in Indian conditions. *Indian Horticulture Journal*, 6(3), 376–379.

Kitinoja, L. 2013. *Returnable Plastic Crate (RPC) Systems can Reduce Postharvest Losses and Improve Earnings for Fresh Produce Operations.* PEF White Paper, (13–01).

Mangaraj, S., Goswami, T. K. and Mahajan, P. V. 2009. Applications of plastic films for modified atmosphere packaging of fruits and vegetables: A review. *Food Engineering Reviews*, 1(2), 133.

Marsh, K. and Bugusu, B. 2007. Food packaging-roles, materials, and environmental issues. *Journal of Food Science*, 72(3), 39–55.

Raghavi, L. M., Moses, J. A. and Anandharamakrishnan, C. 2018. Refractance window drying of foods: A review. *Journal of Food Engineering*, 222, 267–275.

Sangamithra, A., Swamy, G. J., Prema, R. S., Priyavarshini, R., Chandrasekar, V. and Sasikala, S. 2014. An overview of a polyhouse dryer. *Renewable and Sustainable Energy Reviews*, 40, 902–910.

Seveda, M. S. 2012. Design and development of walk-in type hemicylindrical solar tunnel dryer for industrial use. *International Scholarly Research Network ISRN Renewable Energy*, 2012, 1–9. doi:10.5402/2012/890820.

Shahi, N. C., Khan, J. N., Lohani, U. C., Singh, A. and Kumar, A. 2011. Development of polyhouse type solar dryer for Kashmir valley. *Journal of Food Science and Technology*, 48(3), 290–295.

Yadav, Y. K. 2017. Performance evaluation of solar tunnel dryer for drying of garlic. *Current Agriculture Research Journal*, 5(2), 220–226.

14 Plastics in Horticulture

14.1 INTRODUCTION

Plasticulture is not the cultivation of plastics. Plasticulture deals with the use of plastic materials in agriculture without including nonagricultural applications (such as insulation of buildings, packaging of fresh produce, etc.). It can be an old technique (mulching or drainage) where the plastics have expanded on their known technique or an entirely new one (local irrigation or floating row coating), inexpensively using only plastic materials. In the 1980s, the world's plastic culture comprised over 2 million tons of annual plastic consumption. A variety of polymers were used, from expanded polystyrene in nursery trays to polypropylene (PP) cables for knitting plants.

With the current plastic revolution, there have been significant advances in horticultural production worldwide in terms of new technologies and the most efficient use of resources. Plastics offer intensive expansion in vertical and horizontal sizes for high-quality crops (vegetables, cut flowers, fruit, beds, potted plants, nurseries, houseplants). This technology is of particular interest in China, Egypt, Japan, Israel, and the Netherlands, where population densities are high, soil and water restrictions are very strict, and climatic conditions are most favorable for plastics, such as in southern and northwestern Europe, Mediterranean countries and the surrounding East. The potential is equally good for industrialized and developing countries. Plastic greenhouses, high tunnels, low tunnels, row covers, and plastic soil mulch can now be seen from the equator to the Arctic and Antarctic regions (Figure 14.1).

Plastics are utilized extensively in nearly all aspects of horticulture, from crop raising to harvesting, packaging, marketing, and shipping. Plastics serve different purposes as covering structures, mulches, energy curtains, air handling systems, shade cloths, irrigation system components, and growing/harvesting containers. This chapter mainly focuses on the use of plastics in nurseries and/or greenhouse containers, which are also referred to as trays/pots.

14.2 HISTORY

Containers are used far and wide in horticulture, particularly when plants are raised under protected conditions, i.e., in nurseries and greenhouses. This ubiquitous use of containers in horticulture dates back to the time when the practice of horticulture was pioneered. There exist many historical examples vis-à-vis container application in horticulture from the Middle and the Far East, and Western Europe for growing, harvesting, and transporting plants. The existing status quo of the plant container is the consequence of extensive trials of different materials being employed for the production of containers (Currey, 2016).

The commencement of application of containers in horticulture is very difficult to determine precisely. The use of plant containers can be tracked down as far as the Neolithic time period (almost 10000 BC) as in ancient Egyptian horticulture (Janick, 2002). The plant containers were employed by Egyptians for storing harvested grapes, figs, and pomegranates. Nevertheless, harvesting is not the sole purpose which was rendered by plant containers, but they were also used for live plants, both for transporting plants and for ornamental reasons.

Another historical example of using a container is the hanging gardens of Babylon and Nineveh. The convention of bonsai is another early illustration of the utilization of containers for growing plants. Bonsai is mostly related to the production of dwarf trees relative to their counterparts grown on the ground outdoors. Bonsai is a word of Japanese origin meaning "plant in pot." Even though the term is of Japanese origin and is closely related to Japanese culture, bonsai is believed to have

FIGURE 14.1 Plastic greenhouse with plastic mulching for growing crops

originated from China. The bonsai prototypes have been spotted in burial places going back to somewhere in the range of AD 25 and AD 220. However, the practice developed into its current form around AD 700 and was referred to as "potted landscape" or "penjing." The practice of bonsai was basically initiated by the elite social class and these dwarfed tree structures were mostly used as gifts.

A recent historical paradigm of the use of containers dates back to 17 to 19 centuries when the plant containers were used in protected structures, i.e., greenhouses, also referred to as "orangeries" or "solariums" in European culture. Orangeries were protected structures, usually connected to buildings, which protected orange and other sensitive fruits during temperate winter months and allowed the production of fruits out of season or year round. In orangeries, the plants were raised in containers so that their transfer to land could be facilitated in favorable conditions/seasons. In Europe, the plants were raised in containers under protected structures to produce fruits out of season (Nelson, 2012). The capacity to produce fruits out of season under temperate conditions of Europe was an indicator of wealth.

Different materials have been used and tested down the line under different conditions. Before 1970, different materials for containers were tried in greenhouse trials, which included a huge spectrum comprising glass "Mason" jars, wax milk cartons, drinking glasses (Chapman, 1941), traditional clay pots (Kozlowski, 1943), buckets (Kozlowski 1943), foam cups (Kaufmann 1968), or tin cans. In several instances, tar paper was used for making plant containers by hand. The commercial manufacturing of these containers was followed by attempting different materials/compositions for neoteric container types. Traditionally, three container materials have dictated the horticulture industry: clay (terracotta), wood, and petroleum-based plastics (Bahr, 1937; Post, 1950; Laurie and Ries, 1950; Ball, 1977).

The primitive material used to manufacture containers is clay, commonly called terracotta with regard to horticulture containers. Terracotta-based containers are fired at moderately low temperatures, the resultant of which is porous hardened materials (Currey, 2016). Other materials were wood, metal, and sometimes even stone vessels. These options lasted until the 1950s. These were heavy, difficult to disinfect, and rot or break easily. Shipping was also difficult and expensive.

Efforts to reduce cost and weight have led to experiments with pots, recycled paper bins, peat bins, and canned food. Their shortage paved the way for plastic pots in the 1960s. Petroleum-based plastics have been the industry standard for the production of plant containers for the past 75 years (Schrader, 2016). The plastic container can be designed in any color and molded in almost any shape, allowing for the optimization of design possibilities taking different parameters (function, shape, and aesthetics) into consideration. The petroleum-based plastic containers are durable and impermeable to water leading to greater water use efficiency (Koeser *et al.*, 2013).

14.3 NURSERY MANAGEMENT

14.3.1 Advanced Plastic-Growing Pots

The primary goal of protected cultivation and nursery management is to produce healthy crops and achieve economy in production (Figure 14.2). The final product should have a prolonged shelf life which is critical to achieving success in gardening and landscaping.

Plastic containers present an affordable and suitable alternative to conventional soil production and provide the following benefits to the horticulture industry.

- Contain sufficient media to grow crops without the replacement during the growing season
- Standardization of weight and size can be easily achieved from simple experiments
- Easily adaptable to automation with the installation of smart sensors (Pokorny, 2013)
- Compatible with equipment like conveying trays, carts, and benches essential for automation
- Heavy-duty use is possible and long-lasting which makes them suitable for use in poly houses for growing perennial and nursery crops
- Aesthetics in the retail production greenhouses and gardens and durable for shipping
- Can be used in the manufacture of attractive hanging baskets, aesthetically pleasing urns, and attractive merchandising pots
- Reuse and recyclability are possible
- Adaptability to latest production systems and new growing media
- Makes retail gardening a successful venture

Recycled plastic is being used chiefly for manufacturing plastic pots by the majority of plastic pot manufacturers. Nursery Supply Inc., Chambersburg, Pennsylvania, US, reported that their finished plastic containers were produced from 100% recycled plastic, with the exclusion of additives and colors used in the production. A closed-loop technique of manufacturing plastic was used by Jordan

FIGURE 14.2 Plastic nursery growing trays and pots. (Image from indiamart.com)

Plastics, Inc., in South Haven, Michigan, US, which helped archive sustainability in plastic production. Plastic recycling technology in combination with the use of recycled plastic containers from nurseries and retail garden centers has helped achieve economy in plastic pot production (Schrader, 2013b).

Another category of pots, commonly called injection-molded pots (Figure 14.3), are rigid and strong. These pots also offer the advantages of affordability and ease of handling when they are filled. On the other hand, vacuum and thermoformed containers (Figure 14.4) provide exceptional strength and have very large capacities. These containers are mostly used in aquatic landscaping as landscape ponds. Plastic pots, containers, and trays are adaptable to automation and work well for shipping due to the fact that these are accurately formed. An added advantage is that they can be manufactured with different colors on the inside and an entirely different scheme of colors on the outside. Due to this property of rendering them attractive colors and printability, they can be useful in point of purchase (POP) and marketing campaigns.

14.3.2 Growing Media

A growing medium serves four functions for a plant (Figure 14.5). It offers physical support to the growing plant, provides aeration and water supply, and also maintains a source of nutrients to the plant. Young plants are fragile and must remain upright to photosynthesize and grow. With larger nurseries in separate containers, the growing medium must be heavy enough to withstand the wind. Plant roots need a constant supply of oxygen to convert photosynthesis from leaves into energy so roots can grow and absorb water and minerals. A by-product of this respiration is carbon dioxide, which must be released into the atmosphere to prevent toxic concentrations from building up in the root area. This gas exchange takes place in large pores (macropores) or air spaces in the nutrient medium. Because nurseries grow quickly, they need a medium with good porosity. Young plants require large amounts of water for their growth and development and this water supply must come

FIGURE 14.3 Injection-molded pots used in nursery production and the mold used for producing them

FIGURE 14.4 Thermoformed pots for nursery

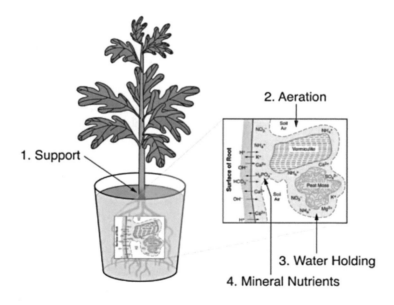

FIGURE 14.5 Primary functions of a growing medium (Dumroese *et al.*, 2008)

from a nutrient medium. The growing media is made in such a way that it can hold water in the small pores (micropores) between the particles. Many culture media contain high levels of organic matter, such as peat and compost, because these materials have interiors that hold water like a sponge. Therefore, the nutrient medium must have sufficient porosity to absorb and store the large amounts of water required by plants. Most of the essential minerals needed for seedlings to grow quickly must be taken through the roots of the growing medium.

Incidences of rots in the plant roots increase with the widespread usage of plastic pots as growth containers. The main reason behind the increased rot incidence is ascribed to the wetness of growing media for prolonged amounts of time as the plastic pots could not dry down or breathe as is for pots made of clay. Thus, there are two major challenges for healthy root growth in containerized plants. First, a very shallow layer of the growing medium was available in the growth containers which could become saturated very swiftly (Barrett *et al.*, 2016.). Second, the limited capacity of the growing containers resulted in inadequate water storage in the root zone between irrigation events (Bunt, 2012). The field soil used to be a main component of the growing media during those days. A standard mix comprises sand, peat moss, and field soil in the ratio of 1:1:3. The resulting mix is poorly aerated and heavily results in the need for the preparation of a special soil mix. A good growing media must have physical properties that help maintain a favorable balance between air and moisture storage between and during irrigation events in order to prevent root asphyxia and drought stress (Caronand Nkongolo, 1997). The incapability of soil to achieve this favorable balance at limited volumes becomes the key driver in the innovation and development of growing media for containerized plant production.

The topsoil that is suitable for plastic pots is difficult to find since most topsoils are contaminated with herbicides and agrochemicals and are heavily infested with pathogens. This creates the need for the development of special growing media that is free from pathogens and diseases, lightweight, reliable, and able to support plant growth. New growing media techniques were developed by the horticulturists in both Europe and the United States. These growing media have been a crucial innovation facilitating careful control over air, water, and nutrient supply to the root zone, while excluding soil-borne pathogens. The meaning of a "good growing media" is context specific, but there are general considerations that are applicable to all growing media. In addition to an appropriate physical structure, a growing medium must provide a suitable biological and chemical environment in which plant roots can effectively take up nutrients. Economic and practical considerations must also be met, in addition to easy availability and manageability (Carlile *et al.*, 2019).

14.3.3 Soilless Peat

Plants take in many elements through their roots: Around 50 elements are present in different plants. However, not all are considered important elements. An essential element is defined as an element that is necessary for the normal life cycle of a plant and whose role cannot be taken over by other elements. It is believed that 20 elements are essential for most plants to grow and they are divided into macronutrients and micronutrients. Most of the basic trace elements were recognized in the 20th century, from the 1930s onward, thanks to the development of pure chemicals and more sensitive methods for analyzing trace concentration levels. The ranges of concentrations of the main elements in plant tissues and the annual amounts required for maximum yield are given in Table 14.1.

Lite was developed as early as 1960 at Cornell University by Professors Ray Sheldrake and James Boodley. Peat serves as the ideal constituent of growing medium taking into account its affordability and high performance (Bragg, 1990). The existence of the vast reserves of peat in the northern hemisphere makes it a readily available and relatively cheap resource (Robertson, 1993). Consequently, it has become the material of choice throughout plant production systems from propagation to saleable "finished plant" material. Media mixtures called "composts" were developed by European researchers at places like John Innes Center. For the consideration of new materials for growing media, economic parameters and performance have become very important. Recently emphasis is being given to organic substrates derived from industrial, agricultural, and even kitchen and municipal wastes. A number of flexible materials are available which can be used across a wide range of horticultural sectors. A combination media of rock-wool and sawdust presents another affordable alternative as a growing medium due to its ready availability, lightweight, moisture retention properties, and economic cost. Renewable and environmentally sustainable substrates obtained from organic sources should be considered as possible options for the growing media.

Improved plant performance and reduced labor cost could be achieved with modern automated transplanting machines, machine learning algorithms, and robotics. Long-distance shipping and

TABLE 14.1
Ranges of the Essential Element Concentrations in Nutrient Solutions and Plant Tissues, and the Required Annual Amounts for Maximum Yields (Silber and Bar, 2008)

Element	Chemical Symbol	Form Available to Plants	Nutrient Solution	Plant Tissues	Annual Consumption
Macronutrients			mg/L	g/kg	kg/ha/year
Calcium	Ca	Ca^{2+}	40–200	2.0–9.4	10–200
Magnesium	Mg	Mg^{2+}	10–50	1.0–2.1	4–50
Nitrogen	N	NO_3^-, NH_4^+	50–200	10–56	50–300
Phosphorus	P	HPO_4^{2-}, $H_2PO_4^-$	5–50	1.2–5.0	5–50
Potassium	K	K^+	50–200	14–64	40–250
Sulfur	S	SO_4^{2-}	5–50	2.8–9.3	6–50
Micronutrients			mg/L	µg/g	g/ha/year
Boron	B	$H_3BO_3^-$, HBO_3^-	0.1–0.3	1.0–35	50–250
Copper	Cu	Cu^+, Cu^{2+}	0.001–0.01	2.3–7.0	33–230
Iron	Fe	Fe^{3+}, Fe^{2+}	0.5–3	53–550	100–4,000
Manganese	Mn	Mn^{2+}	0.1–1.0	50–250	100–2,000
Molybdenum	Mo	MoO_4^{2-}	0.01–0.1	1.0–2.0	15–30
Zinc	Zn	Zn^{2+}	0.01–0.1	10–100	50–500

transportation of liners, plugs, pots, seedlings, and finished plant products are possible now. There is increased interest from retailers to sell new products such as plastic hanging baskets, flower bowls, decorative pots, and garden baskets. Attractive, uniform, versatile, well-designed, and uniform-sized plastic pots and containers allow for the inventiveness of growers, plant breeders, design engineers, retailers, and marketers to reach the market and flourish. The innovation of reliable, versatile, molded plastic-growing pots rendered the horticulture sector the ability to become the economic powerhouse.

14.3.4 Nursery Containers

A nursery container could be anything that holds growing media, drains, doesn't break down prior to outplanting, and allows for healthy root development which can be easily removed with the least unsettling influence on the plant. Plants were most commonly sold and transported bare-root prior to the invention of containers, generally with roots coated in a slurry and bundled in damp burlap or some other cloth to retain moisture. After the Second World War, the commercialization of nursery crops in containers started in earnest. Nurseries adapted to this need by increasing container plants, shipping plants from the nursery to the point of purchase and installation for longer distances, and streamlining production, especially with regard to container size. Containers have been seen as a way of prolonging plant life and health and reducing transport costs (because container substrates are lighter than field soil). In addition, the development of container nurseries permitted land not suitable for field production to be used and planting activities could take place independently of land or environmental conditions. In areas far from markets but where land was inexpensive and the climate was ideally adapted to rapid plant growth through long growth stages and predictable rains, growers gradually built containers that allowed plant development (Sun Belt states).

Container manufacturing nurseries developed specimen plants in big clay containers long before nurseries started using plastics. Weight, size constraints, and breakage potential, however, contributed to the substitution of clay containers with tar paper or metal containers. Commercial food cans used in restaurants have been reused as containers because they are bigger, more durable, cheaper, and the plant scale of the landscape industry is more acceptable. Used metal food cans have progressively come to dominate the production of container nurseries. Today some farmers still use the term "can" interchangeably with "container." Before, for use in nurseries, metal cans were processed in many other ways to dull any sharp corners (from which the top of the bottom could be removed by the restaurants). As the manufacturing of container nurseries became more common, metal containers were custom built to include rust-resistant paint and tapered sides for nesting and stacking containers more easily.

14.3.5 Petroleum-Based Plastic Nursery Containers

Plastic containers were eventually produced primarily for the development of container nurseries, and these metal cans were gradually replaced. Over the years, various types of container goods, such as injection molded, blow molded, pressure formed, vacuum formed, and thermoformed, have emerged from plastic container manufacturers, each with differing characteristics and advantages for different plant development, shipping, or marketing needs. For much of its history, plastic containers have been produced using virgin petroleum-based resins. Recycled material is now used in many traditional plastic nursery containers, and some containers are entirely made of recycled plastics. Although not biodegradable, they are a recycled commodity that represents a step toward sustainability. The sustainability of using petroleum-based plastic nursery containers could be significantly improved by high rates of plastic nursery container recycling or increased use of recycled plastics in nursery containers. Plastic nursery vessels are lightweight, durable, manufacturers are familiar, are well automated, and can be reused or recycled as an industry standard. Plastic nursery containers are, however, rarely recycled in use, despite attempts by nurseries and manufacturers. Used containers must be collected from the customer, sorted by resin form, thoroughly cleaned to remove the substrate and other pollutants, and sent to recycling facilities in order to recycle a plastic nursery container. Together these measures have built major recycling barriers.

14.3.6 Alternatives to Petroleum-Based Plastic Containers

Containers made of petroleum-based alternatives are very attractive to consumers and farmers. Studies have shown that consumers will prefer and be paying premium rates for "carbon-neutral" or majority-recycling or waste goods (Figure 14.6). The rapidly changing disponible capacity and

FIGURE 14.6 Examples of well-known fiber-based alternatives to petroleum-plastic containers. Shown from left to right: paper fiber, coir fiber, wood fiber, and peat fiber (Schrader,2016)

prices of petroleum are subject to petroleum-based synthesis plastics that lead to substantial changes in plastic container costs. Alternative waste or recycled containers minimize waste to sites of waste and American dependence on foreign oil. Furthermore, containers made from nonpetroleum materials can sidetrack oil issues and naturally decompose, thus avoiding or reducing waste from sites. Finally, several other containers provide work efficiency, because before planting the container must not be removed, instead, it is planted alongside the plant. These savings in labor in nurseries can also be increased (when planting into larger containers) and landscape contractors and customers can benefit as well.

The alternative lifetime of the container is based on the components and additives and should equate to the crop production period. For some plants, some alternative containers can biodegrade too rapidly in the sunshine, irrigation, and fertilizer nursery conditions. The mold stabilizes or reinforces containers and prolongs the lifetime of the container which is complemented by adhesives, binders, and other compounds. These additives can be synthetic, natural, or mixed and can or cannot be biodegradable or reusable. To accommodate different crop production times, alternative containers with different life spans are needed. Different formulations for long-term use are required, for example, with woody ornamentals and for shorter growing cycles, such as perennials or yearly ones. These alternative components are typically compostable containers that cannot easily biodegrade when planted.

14.3.7 Physical Properties of Nursery Containers

For plantable, compostable, biodegradable, natural, or organic containers, no official requirements or guidelines exist. In addition, particular conditions of the site and atmosphere influence the decomposition rate. Manufacturers and advertisers may also boast widely about the biodegradability of containers. Alternative container considerations include physical properties of the strength of the container, water loss through the surface of the container (affecting the volume and frequency of crop irrigation), algal and fungal growth on the outer container (affecting appearance) and if planted with the plant, the rate of decomposition of the container in the soil (Evans, *et al.* 2010). The components of alternative containers continue to develop and more recent alternative containers may not be protected by the findings reported here.

14.3.8 Pot-In-Pot

The pot-in-pot is mainly used in pot sizes from #7 to #45 for large trees and shrubs. The permanent socket pot is inserted into the ground and a removable pot with media and plant is placed inside it (Figure 14.7). Plants do not blow away, the roots remain cooler, and harvesting only includes removing the lipped pot.

14.3.8.1 Air-Pruning Pots

There are several benefits to growing woody plants in containers; however, the growth of "circling roots" or "girdling roots," often referred to as "pot-bound" plants, is a persistent issue. Circling roots can inevitably hurt or destroy plants once they grow in the landscape. Horticulturists suggest slashing three-to-five vertical cuts in the root ball to avoid the growth of girdling roots and then spreading the roots in the final planting hole. This calls for extra labor and is sometimes not done, so plants suffer.

Various techniques have been developed by manufacturers and researchers to minimize or remove circular roots that grow in traditional nursery pots. One of the best approaches is to coat a root-regulating copper hydroxide compound, such as spin-out, within the regular pot. Research shows that circular roots with no decrease in plant growth or total root mass are significantly reduced. In order to facilitate root branching, other approaches involve redesigning the expanding

FIGURE 14.7 Components in a pot-in-pot production system. A) Containers on a high-density planting using in-row fabric floor management. Each tree is staked and has an irrigation spray stake from main lines covered by the fabric. B) Floor management is with an in-between row grass cover (Geneve 2016)

container using open slots, grooves, or ledges. Smart Pot, Fabric Pot, Root Builder, Compressor, Air Pot, and Root Trainer are some common brands for air-pruning pots (Crawford, 1997).

14.3.8.2 Reusable and Recycling

When it comes to reusing and recycling the vast amount of plastic generated, our green industry has not always been so green. Landfills are not an appropriate choice, and burning will release potentially hazardous substances into the atmosphere unless it is closely regulated. It is possible to reuse containers, but there are problems related to sorting, cleaning, and controlling control. It is not possible to reuse greenhouse coverings and mulch films because they deteriorate under UV radiation. In addition, horticultural plastics are getting dirty.

In the horticultural industry, many different plastic types are used, including low-density polyethylene, high-density polyethylene, polypropylene, and high-impact polystyrene. These need to be carefully sorted and treated separately to ensure the highest quality of the recycled product. And even when properly sorted, the consistency of recycled plastic relative to virgin material is reduced by different pollutants, including soil and organic matter. Any time plastic goes through the recycling process, the consistency of plastic is diminished by pollutants and structural deterioration, so that it can become more difficult to remanufacture the same product.

Except for vegetable and ornamental crops where propagators and growers are in near proximity, reusable trays are rarely used. Otherwise, the cost of transportation is prohibitive. Even though it is possible to recycle plastic pots, most are not, and end up as garbage. As the preferred way of disposing of plastic pots and trays, there is a great interest in recycling or repurposing. The issue is that recycled items are internationally sold and there is no demand for recycled flowerpots. For nonplastic or recyclable containers, some customers are also willing to pay a premium. Meeting this market interest provides an incentive for farmers and retailers to sell more eco-friendly goods. It is not always as easy to recycle plastic containers, packs, and flats as to recycle soda and water bottles. One of the most easily recycled plastics is plastic drink bottles, with polyethylene terephthalate plastic recycling code 1. Other plastic types, including high-density polyethylene, polypropylene, and polystyrene, are made from nurseries and greenhouse containers. Although all municipal recycling typically accepts plastic code 1, fewer recognize the plastic products from which plant pots are made.

From the use of virgin plastic to the use of most recycled plastic for the manufacturing of containers, plastic manufacturers have made great strides. As it recovers plastic from the waste stream, this is a big step toward sustainability. At the end of their useful life, however, typical plastic pots too often end up in landfills. Many greenhouses, nurseries, and garden centers are operating at the local level to collect pots, flats, and recycling packs. In both cases, pots are sorted and reused where possible in the greenhouse industry and are shredded and recycled into landscape timbers and other outdoor products if discovered to be unusable.

14.3.8.3 Eco-Friendly Growing Containers

Eco-friendly material is classified as "recyclable," "biodegradable" or "compostable." Recyclable means items such as aluminum, glass, or plastic can be recycled and processed into new goods. Biodegradable simply implies that in time the substance would break down into carbon dioxide, water, and biomass; but it could take a very long time! Compostable means materials which more significantly are not only biodegradable but which could also disintegrate at normal temperatures within months. One of the oldest compostable containers has been developed since 1953 from naturally occurring sustainable peat moss produced by the Jiffy Company. Farm fibers, manures, processed chicken feathers, nuts, and hulls are made from other compostable pots. In traditional plastic pot production, some plastic pot manufacturers substitute 40% chicken feather bioresins; however, chicken feather products are very slow to break down, even though they are biodegradable. Cow pots are made of composted odorless cow manure and are planted in soil for months but decay takes place in approximately four weeks within the soil. Fiber pots can be made out of recycled newspapers, wood fibers, or coir in around a year's time. All natural materials have major problems because of their moisture fragility and their complexity in automated planting and transportation processes.

14.3.8.4 Bioplastic – An Alternative to Petroleum-Based Plastics

Bioplastics are similar to conventional plastics and are made from a combination of biopolymers and petrochemical polymers. Plastics based on biopolymers are manufactured from renewable sources. Protein from poultry waste feathers is extracted from soybeans (Glycine max). Lipids come from animal fats and plant oil sand. These raw materials are usually mixed with petrochemical processing fossil-based polymers to minimize cost, increase quality and both (Riggi *et al.*, 2011). The nondegradable and nonrenewable nature of petroleum-based plastic triggered researchers to develop materials that are environment-friendly. Bioplastics and bioplastic composites are a category of emerging sustainable materials that exhibit beneficial characteristics for use in containers for horticulture production. Bioplastic horticulture, produced from bio-based materials, releases nutrients as plastic degrades and provides a variety of environmental benefits as well. Bioplastics are better described as plastics in which all carbon is extracted out. Biocomposites are composed of bioplastic resin mixed with natural fibers or fillers (Currey, 2016). With the advancement of bioplastics, the demand is closer to sustainability through the reduction of the carbon footprint and dependence on fossil fuels. A common bioplastic derived from biomass, such as vegetable fats and oils, cornstarch, tapioca, microbiota, and other sustainable resources, is polylactic acid (PLA). PLA is a translucent plastic that can be transformed into molded containers, agricultural films, sheets, and three-dimensional printing and spinning by extrusion and injection. In the soil or sunlight, PLA plastics can degrade. One of the issues with PLA is that it is only stable up to 110°F, whereas higher temperatures are tolerated by other compostable resins and can be used for food containers. The anaerobic fermentation of feedstock produces polylactic acid and is primarily employed by starch mixtures due to its slow soil biodegradation. Bioplastics can be processed on petrochemical plastic equipment that removes the need for the production of new industrial machines (Koeser *et al.*, 2013a). The benefits of bioplastics are their physical properties, including low weight, structural durability, rigidity, and decay resistance, and mostly comparable to conventional plastics, allowing their incorporation easily into a range of short- and long-term plant production systems. The majority of bioplastic containers are to be removed at the end of plant development and either composted or anaerobically digested. If the container was not removed until transplanting, the slow degradability inherent to bioplastics would impact the root system. Some containers, such as a bioplastic-based sleeve style, the soil wrap, will degrade in the soil and are known as plantable pots.

QUESTIONS

1. Describe the application of plastics in horticulture.
2. Explain the use of plastics in nursery production.
3. What is a poly house production system?
4. What is a hi-tech poly house?
5. Discuss naturally ventilated poly houses.

MULTIPLE-CHOICE QUESTIONS

1. Mulches are mainly used to reduce
   a) Transpiration
   b) Evaporation
   c) Evapotranspiration
   d) All of the above
2. What type of mulches reduces soil temperature?
   a) Organic mulch
   b) Gray polythene
   c) Black polythene
   d) Transparent polythene
3. Low-cost greenhouse tunnels are suitable for?
   a) Cucumber
   b) Tomato
   c) Cabbage
   d) Pumpkin
4. Mist chamber is mostly used for
   a) Hardening of seedlings
   b) Hardening of cuttings
   c) Rooting of leaf cuttings
   d) All of the above
5. Nursery container is used for
   a) It is holds growing media
   b) It doesn't break down prior to outplanting
   c) It allows for healthy root development
   d) All of the above
6. Plastic containers are used for
   a) They are affordable and a suitable alternative to conventional soil production
   b) They contain sufficient media to grow crop without replacement during the growing season
   c) Standardization of weight and size can be easily achieved from simple experiments
   d) All of the above

ANSWERS

1	2	3	4	5	6
d	a	a	c	d	d

REFERENCES

Bahr, F. 1937. *Fritz Bahr's Commercial Floriculture: A Practical Manual for the Retail Grower.* A.T. De La Mare Company, New York.

Ball, V. 1977. *Ball Bedding Book: A Guide for Growing Bedding Plants.* Geo. J. Ball, Batavia, IL.

Barrett, G. E., Alexander, P. D., Robinson, J. S. and Bragg, N. C. 2016. Achieving environmentally sustainable growing media for soilless plant cultivation systems–A review. *Scientia Horticulturae*, 212, 220–234.

Bragg, N. 1990. *A Review of Peat Reserves and Peat Usage in Horticulture and Alternative Materials.* Report for the Horticultural Development Company (HDC), Petersfield, UK, 59.

Bunt, B. R. 2012. *Media and mixes for container-grown plants: a manual on the preparation and use of growing media for pot plants.* Springer, Dordrecht.

Carlile, W. R., Raviv, M. and Prasad, M. 2019. Organic soilless media components. In Soilless culture. doi:10.1016/B978-0-444-63696-6.00008-6.

Caron, J. and Nkongolo, V. K. N. 1997. Aeration in growing media: Recent developments. In *International Symposium on Growing Media and Hydroponics 481*, May, 545–552.

Chapman, A. G. 1941. Tolerance of shortleaf pine seedlings for some variations in soluble calcium and H-ion concentration. *Plant Physiology*, 16(2), 313–326.

Crawford, M. A. 1997. *Update on Copper Root Control.* Gen. Tech. Rep. PNWGYR-419. Portland.

Currey, C. J. 2016. A brief history of containers in horticulture. In J. A. Schrader, H. A. Kratsch, and W. R. Graves (eds.), *Bioplastic Container Cropping Systems: Green Technology for the Green Industry*, 27–35. Sustainable Hort. Res. Consortium, Ames, IA.

Dumroese, R. K., Luna, T. and Landis, T. D. 2008. *Nursery Manual for Native Plants: Volume 1: A Guide for Tribal Nurseries. Agriculture Handbook 730.* U.S. Department of Agriculture, Forest Service, Washington, DC, 302 p.

Geneve, R. L. 2016. An introduction to pot-in-pot nursery production©. In *Proceedings of the 2016 Annual Meeting of the International Plant Propagators' Society 1174*, 19–22.

Janick, J. (2002). Ancient Egyptian Agriculture and the Origins of Horticulture. *Acta Horticulturae*, 582(582), DOI:10.17660/ActaHortic.2002.582.1

Kaufmann, M. R. 1968. Water relations of pine seedlings in relation to root and shoot growth. *Plant Physiology*, 43(2), 281–288.

Koeser, A., Lovell, S. T., Evans, M. and Stewart, J. R. 2013. Biocontainer water use in short term greenhouse crop production. *HortTechnology*, 23, 215–219.

Kozlowski, T. T. 1943. Transpiration rates of some forest tree species during the dormant season. *Plant Physiology*, 18(2), 252–260.

Laurie, A. and Ries, V. H. 1950. *Floriculture: Fundamentals and Practices.* McGraw Hill, New York.

Nelson, P.V. 2012.*Greenhouse operation and Management.* Seventh edition. Prentice Hall, Boston, MA, 607p.

Pokorny, K. 2013. Building the case for biopots. *Greenhouse Management Magazine*, December. Available from: http://www.greenhousemag.com/article/gm1213-biocontainers-research.

Post, K. 1950. *Florist Crop Production and Marketing.* Orange Judd Company, New York.

Riggi, E., Santagata, G. and Malinconico, M. 2011. Bio-based and biodegradable plastics for use in crop production. *Recent Patents on Food, Nutrition & Agriculture*, 3(1), 49–63.

Robertson, R. A. 1993. Peat, horticulture and environment. *Biodiversity & Conservation*, 2(5), 541–547.

Schrader, J. A. 2016. Bioplastics for horticulture: An introduction. In J. A. Schrader, H. A. Kratsch, and W. R. Graves (eds.), *Bioplastic Container Cropping Systems: Green Technology for the Green Industry.* Sustainable Hort. Res. Consortium, Ames, IA.

Silber, A. and Bar-Tal, A. 2008. Nutrition of substrate-grown plants. In M. Raviv and H. Lieth (eds.) *Soilless Culture: Theory and Practice.* Elsevier, San Diego, CA, 291–339.

15 Plastic Mulching

15.1 INTRODUCTION

The word *mulch* is derived from an old English word *molsh*, "soft and moist." A layer of material, either permanent or temporary, covering the soil surface is called mulch. That is, mulching is an agriculture cropping technique that provides a favorable environment to the soil for effective and efficient crop production. It may be applied to the uncovered soil or around existing crops. The material used for mulch is mostly natural/organic (rice straw, wheat straw, a dried maize plant, sugarcane leaves, wool, animal manure, etc.) derived either from animals or from plants, and mulching is carried out by most farmers traditionally. Natural mulches help in maintaining soil organic matter and tilth (Tindall *et al.*, 1991) and provide food and shelter for earthworms and other desirable soil biotas (Doran 1980). Natural mulches have many advantages but are also associated with a few disadvantages.

The growing civilization, floods, droughts, erosion, degradation of soil, faulty management practices, lack of information, etc. lead to low productivity of vegetable crops and, thus, become a major concern for the world. The production required in terms of both quality and quantity and its availability throughout the season for the rapidly growing population are urging the scientists to evolve new techniques and advanced ways. To overcome all the problems, a favorable soil–water–plant relation is created by placing mulch over the soil surface. So, a trend started using synthetic mulches, which includes the combinations of paper and polyethylene, and foils and waxes. Introduction of resin mulches, thin sheets of plastic, paper, and petroleum materials have proved to have increased benefits over natural mulches. Waggoner *et al.* (1960) described microclimatic changes caused by various mulches such as polyethylene film, straw, paper, and aluminum films. The age-old technique of mulching has been revolutionized by the incorporation of plastic film into it. The use of polyethylene film as mulch in plant production saw its beginnings in the mid-1950s. Professor Emery M. Emmert of the University of Kentucky was one of the first to recognize the benefits of using LDPE (low-density polyethylene) and HDPE (high-density polyethylene) films as mulch in vegetable production.

The use of soil cover and mulching is also known to be beneficial through their influence on soil moisture conservation, solarization, and control of weeds. This also results in moisture conservation, less soil compaction, and higher CO_2 levels around plants. Plastic mulch maintains higher soil temperature at night which favors the root activity (Kader *et al.*, 2019). It also reduces the weed population and improves the microbial activities of soil by improving the environment around the root zone. Continuous use of mulches is helpful in improving the organic matter content of the soil which, in turn, improves the water-holding capacity of the soil. Hence, drip irrigation in combination with plastic mulch offers a sound scientific basis for increasing crop yields (Sharma and Meshram, 2015). Mulching with drip irrigation system is an effective method of manipulating the crop-growing atmosphere to increase yield and improve product quality by ameliorating soil temperature, conserving soil moisture, reducing soil erosion, improving soil structure, and enhancing organic matter content. The benefit of mulching is particularly buffering the properties of soil temperature and conservation of soil moisture in different agro-climatic areas. The practice of mulching in vegetable production has been advised so as to cut down the cost incurred in cultivation and obtain quality produce and maximum return with increased profits. For high-value crops, such as tomatoes and cucumbers, it is cost-effective to cover the ground with heavy paper or plastic sheets to protect the soil, save water, and prevent weed growth. The application of mulch under micro-irrigation is illustrated in Figures 15.1 and 15.2.

FIGURE 15.1 Using different color mulches

FIGURE 15.2 Use of black-, white-, and silver-colored mulches

15.2 PLASTIC MULCHING

The technology of using plastics in the agricultural division is known as plasticulture. The science of plasticulture had its beginning as early as 1924 when Warp (1971) developed the first glass substitute for widespread agricultural use. Plastic mulches are primarily used to protect seedlings and shoots through insulation and evaporation prevention, thus maintaining or slightly increasing soil temperature and humidity (Tarara, 2000). In the late 1800s, tar-coated paper mulches were used, long before polyethylene was available (Rivise, 1929). The plastics in use are made from inorganic and organic raw materials (carbon, silicon, hydrogen, nitrogen, oxygen, and chloride). The basic materials used for making plastics are extracted from oil, coal, and natural gas (Seymour 1989). Plastic mulching has become a widely used technique for its instant economic benefits such as higher yields and improved crop quality (Lamont, 1999). Plastic mulch films were first used in the late 1950s in university research and have been used commercially for vegetable production since the early 1960s (Hussain and Hamid, 2003; Lamont 2004 a, b). Primarily it was used in cold regions and then eventually was used for all types of weather and soil conditions to enhance soil temperature. The use of plastics in agriculture started in developed countries and then emerged in developing countries. The techniques got enhanced with time, starting with a simple method of mulching, then row covers and small tunnels, and finally to plastic houses. The numbers published by Plastic Europe (2018) about the global plastic production demonstrate an increase from 335 million tons of plastics in 2016 to 348 million tons in 2017. Sintim and Flury (2017) describe the rapid growth of the market for agricultural films: the global use of agricultural films accounted for 4.4 million tons in 2012 and is expected to grow to 7.4 million tons in 2019. Plastic mulch has witnessed a tremendous increase in peanut (*Arachis hypogaea*) production, which is called a white revolution in China (Hu *et al.*, 1995). The world consumption of low-density polyethylene mulching films in horticulture is at present around 700,000 tons/year (Espi *et al.*, 2006). Plastic film mulching can increase soil temperature (Liu *et al.*, 2003; Peng *et al.*, 1999), and higher temperatures can favor not only N mineralization (Wilson and Jefferies, 1996) but also plant N uptake (Liu *et al.*, 2003). Plastic mulch was first noted for its ability to increase soil temperature in the 1950s (Emmert, 1957). Plastic mulches alter the crop microclimate by changing the soil energy balance (Liakatas *et al.*, 1986; Tarara, 2000). Heating properties of plastic, such as reflectivity, absorptivity, and transmittance, and their interaction with solar radiation have a direct effect on soil temperature under the plastic mulch (Schales and Sheldrake 1963). Plastic mulch protects the soil from water and wind erosion and hail damage (Garnaud, 1974).

Higher soil temperatures increase nutrient availability, enhance nutrient uptake by roots, increase the number and activity of soil microorganisms, and speed up plant germination and growth (Tindall *et al.*, 1991). However, after six decades of research for an extensive historical review, the knowledge of the sustainability of plastic mulches remains vague in terms of both environmental and agronomic perspectives.

15.3 CLASSIFICATION AND COLOR OF MULCHES

Throughout the 1960s, a variety of plastic films made of various types of polymers were tested for mulching. Although there were some technological variations between LDPE, HDPE, and flexible PVC, they were minimal. Polyethylene is favored because it has a higher permeability to long-wave radiation, which can raise the temperature surrounding plants at night. Because LLDPE is more cost-effective to utilize, it now makes up the great majority of plastic mulch.

Advancement in plastic chemistry has resulted in the development of films with optical properties that are ideal for a specific crop in a given location. Horticulturists need to understand the optimum above and below ground environment of a particular crop before the use of plastic mulch. Thus, these are of two types:

a) Photodegradable plastic mulch: This type of plastic mulch film gets destroyed by sunlight in a shorter period
b) Biodegradable plastic mulch: This type of plastic mulch film is easily degraded in the soil over a period of time

15.3.1 Color of Film

The soil environment can be managed precisely by a proper selection of plastic mulch composition, color, and thickness. Films are available in a variety of colors including black, transparent, white, silver, blue, red, etc. But the selection of the color of plastic mulch film depends on specific targets. Generally, the following types of plastic mulch films are used in horticultural crops.

a) Black plastic film: It helps in conserving moisture, controlling weed, and reducing outgoing radiation

b) Reflective silver film: It generally maintains the root zone temperature cooler
c) Transparent film: It increases the soil temperature and is preferably used for solarization

15.4 ADVANTAGES OF PLASTIC MULCH

Plastic mulches have the following advantages

1. It improves soil structure by preventing the formation of clusters and clods
2. It prevents humans and animals from walking over the soil, thus, helping in maintaining the soil stability
3. By providing a microclimate it causes the early germination of seeds
4. Early crop production results in higher yield and more economy
5. Most crops are sensitive to winter, vegetable crops in general. Thus, plastic mulching helps retain the heat and regulates the soil temperature by warming it up as the cold season approaches
6. Opaque plastic mulch obstructs the penetration of sunlight thus suppressing the weed growth effectively and saves from the hassle of pulling it individually and hence less investment in weedicides
7. As the soil remains undisturbed under plastic mulch, roots grow and spread evenly and efficiently and it helps in reducing root damage
8. In some cases, reflective silver and white plastic mulch help in controlling insect infestation and the spread of viral diseases
9. It helps in maintaining soil nutrient contents by preventing from leaching or by flooding
10. As the soil is completely covered it reduces the impact of direct rainfall resulting in less chance of soil erosion
11. Plastic mulch has a longer-lasting effect as compared to organic mulch
12. A combination of mulch films with fertilizers delivers the maximum output
13. The moisture barrier qualities of plastic film prevent soil moisture from escaping. Water vaporizes from the soil surface beneath the mulch film, condenses on the film's bottom surface, and falls back as droplets. As a result, moisture is kept for several days, extending the time between irrigations. Irrigation or rainwater enters the soil either through gaps in the mulch surrounding the plant area or through the unmulched area

15.5 LIMITATIONS OF PLASTIC MULCH

1. When compared to organic mulches, they are more expensive to employ in commercial production
2. Due to the high temperature of the black film, there is a risk of the young pants "burning" or "scorching"
3. Runoff is more
4. There is a chance of environmental pollution
5. Sometimes, there is difficulty in moving the machinery
6. It can be detrimental to the livestock
7. In the case of thin films, weed penetration can occur

15.6 AREAS OF APPLICATION

Mulching is mainly employed in the following:

1. Conservation of moisture in rain-fed areas
2. Irrigation frequency reduction and water conservation in irrigated areas

3. Stabilization of soil temperature under greenhouse cultivation
4. For the control of soil-borne infections using soil solarization
5. Reduce the amount of rain that falls, minimize soil erosion, and keep the soil structure intact
6. In areas where only high-value crops are grown

15.7 EFFECT OF DIFFERENT COLOR MULCHING

The color of plastic films selected for soil mulching determines the performance of the radiant energy, impacting the microclimate around the cultivated plants. The response of plants to the colored film is the result of interaction among the quality of light reflected by the surface of plastic film, the capacity for transmission of solar radiation, and the increase in soil temperature. The impact of plastic films on soil temperature and crop canopy microclimate depends on their properties: reflection, transmission, and absorption of light (Ham et al., 1993). The root zone temperature influences physiological processes in roots like the uptake of water and nutrients (Ibarra-Jimenez et al., 2008).

Plastic mulches, particularly black, white, and clear, are nothing new in the commercial vegetable production arena. However, it is fairly recently that these and other colored mulches have made their way into the realm of home gardening. Plastic mulches offer a variety of benefits: they can extend the growing season by warming garden soil, they can improve weed and insect management, they can help retain soil moisture, and they can increase crop yield and quality. Black, white, brown, red, silver, green, or blue – which color is the best? That depends on the crop(s) you're growing and the effect(s) you desire from the mulch. Because research on colored mulches is not entirely conclusive, most of these plastics are sold with a "for trial use only."

15.7.1 White Mulch

White plastic may be of less interest in cool weather since it tends to keep soil temperatures cool rather than warming the soil as done by black plastic. The benefits of white plastic in keeping weeds at bay, retaining soil moisture, and keeping the soil cool around the roots of crops such as peas, broccoli, cabbage, and cauliflower may be more easily and inexpensively achieved with a biodegradable mulch, e.g., straw. Brown infrared transmitting (IRT) plastic mulch is a fairly recent innovation. It warms garden soil better than black plastic early in the growing season and also controls weeds. IRT is a technology that combines the weed-suppressing properties of black plastic with the heat-absorbing qualities of clear plastic.

15.7.2 Black Mulch

The most widely used, available, and inexpensive of the colored mulches is black plastic mulch which has excellent weed suppression ability because of its opacity. It is also useful for warming soil during the growing season, particularly if as much of the plastic as possible is in contact with the soil below. Research at Pennsylvania State University has shown that soil underneath black plastic can be up to 5°F warmer at a 2-inch depth and up to 3°F warmer at a 4-inch depth than uncovered soil at the same depths. This means that plants can be set out earlier than on bare soil and may result in earlier maturing fruit. The use of black-, white-, and silver-colored mulch is illustrated in Figure 15.3.

15.7.3 Red Mulch

Certain crops performed better when grown in red mulch as opposed to black mulch: tomatoes, which yielded 20% more fruit; basil, the leaves of which had greater area, succulence, and fresh

FIGURE 15.3 Use of different-colored mulches

TABLE 15.1
Specification of Different Types of Mulch

Thickness	Life	Width Available	Colors Available
15–20 micron	Single-season crop	600 mm to 1,800 mm	Transparent, Black, White, Silver + Black, Red or as per customers' requirement
20–30 micron	2nd or 3rd season crops	600 mm to 1,800 mm	Transparent, Black, White, Silver + Black, Red or as per customers' requirement
50 micron onwards	Orchids for longer life	600 mm to 1,800 mm	Transparent, black, white, silver + black, red or as per customers' requirement

Source: https://tilakpolypack.com/mulch-film/

weight; and strawberries, which smelled better, tasted sweeter, and yielded a larger harvest. Penn State researchers found that the yield increases for tomatoes and eggplants on the red mulch as compared to black mulch.

15.7.4 GREEN MULCH

Green IRT mulch has been shown to encourage earlier ripening and greater yields. As with brown plastic, one can look for the IRT designation on green mulch for sale by garden suppliers.

15.7.5 BLUE MULCH

The cucumbers all produced significantly more fruit when plants were grown in blue mulch than in black mulch.

15.8 SPECIFICATIONS

The specifications of different types of mulches regarding their thickness, life, width, and colors are given in Table 15.1.

Plastic Mulching

15.9 PARAMETERS OF PLASTIC MULCH

The parameters that are of importance for plastic mulching are listed as follows:

a) **Thickness**: Except when employed for solarization, the thickness of the film has little effect on the mulching. However, several recent references suggest that film thickness has an effect on crop production. Because it is sold by weight, it is desirable to utilize as thin a film as feasible, but proper consideration should also be paid to the film's lifetime. Mulch film used should be 60–75 micron thick (240–300 gauge); however, with the development of film extrusion technology, it is now possible to produce a 15-micron-thick film. The easy shredding of these films when pulled demonstrates their mechanical weakness.
b) **Width**: This is determined by the interrow spacing. Normally, a film with a width of one to one and a half meters can be easily adapted to various situations.
c) **Perforations**: Perforations may be desirable in some instances while being harmful in others. Under unperforated conditions, water capillary circulation and fertilizer distribution will be better and more uniform. Perforation, on the other hand, is better for preventing water stagnation around the plants. However, it has the drawback of increasing weed growth.
d) **Mulch Color**: The color of the mulch also plays an important role because it directly affects the temperature of the soil, soil salinity, and the air temperature around the plants.

15.10 SELECTION OF MULCH

Mulch selection is influenced by environmental conditions as well as main and secondary factors of mulching. Table 15.2 shows the selection of mulches under different conditions.

15.11 TECHNIQUES OF MULCH LAYING

i. Mulch should be put in a wind-free environment
ii. The mulch material must be held tight and spread on the bed without creases (Figure 15.4)
iii. The borders (10 cm) must be embedded in the soil in little furrows at a 45° angle, around 7-10 cm deep

TABLE 15.2
Selection of Mulch

Rainy season	Perforated mulch
Orchard and plantation	Thicker mulch
Soil solarization	Thin transparent film
Weed control through solarization	Transparent film
Weed control in cropped land	Black film
Sandy soil	Black film
Saline water use	Black film
Summer cropped land	White film
Insect repellent	Silver color film
Early germination	Thinner film

Source: http://www.agritech.tnau.ac.in/agricultural_engineering/plastic_mulching.pdf

FIGURE 15.4 Laying of plastic mulch

15.11.1 Mulching Techniques for Vegetables or Close-Spacing Crops

For short-term crops like vegetables, a thin film is utilized. One meter of the film is taken and folded in "than" form every one meter along the length of the film for one row of crop. Following are some technique mulching can be used as.

- A punch or a larger diameter pipe with a hammer or a heated pipe end can be used to make round holes at the center of the film
- The mulch film is wrapped along the length of the planting row, with one end (along the width) anchored in the soil
- Before mulching, thoroughly till the soil and add the needed amount of FYM and fertilizer
- Mulch film is then put on all sides (4–6″) into the soil to keep it intact
- Seeds are sown directly through the mulch film's openings
- Seedlings might be put immediately into the hole in the case of transplanted crops
- The procedure of piercing the hole is the same for mulching grown seedlings. The mulch film is then unrolled over the saplings, with one end of the film buried in the soil along the width. The saplings are gripped in the hand and fitted into the perforations on the mulch film from the lower surface during the unrolling procedure so that it can spread on the upper side

15.12 IRRIGATION TECHNIQUES FOR MULCHING

- The lateral conduits for drip irrigation are installed beneath the mulch film
- If inter-cultivation is required, it is preferable to keep the laterals and drippers on top of the mulch film and control the water flow through a short pipe or holes drilled in the mulch film
- Irrigation water passes through semi-circular pores in the mulch sheet during flooding

15.13 PREVENTIONS IN MULCH LAYING

- Don't stretch the film too much. It should be elastic enough to accommodate temperature-related expansion and contraction as well as the effects of cultural operations
- For the black film, the slackness should be higher due to expansion and shrinkage. In this color, the effect is at its peak
- The film should not be shown during the hottest part of the day.

QUESTIONS

1. What is mulching?
2. Enlist different types of plastic mulch.
3. Discuss the parameters of plastic mulch.
4. Discuss the effect of different color mulching on crop production.

MULTIPLE-CHOICE QUESTIONS

1. Retaining soil moisture is one of the major factors affecting dry-land farming. Water is lost through evaporation from soil surfaces and many methods are applied to prevent the rate of evaporation. One such method is done by applying some material on the soil surface and these materials are called _____.
   a) Coolants
   b) Evaporation inhibitors
   c) Mulches
   d) Surface reactants
2. The major types of mulches are
   a) Artificial mulches: paper and plastic mulches
   b) Inorganic; soil mulch and stone mulch
   c) Natural mulches
   d) All of the above
3. Thermal insulation, leading to greater deposition of dew. In addition, organic mulches also contribute to:
   a) Enhanced humification and microbial activity leading to higher nutrient availability
   b) Improvement of soil structure, by incorporation of plant residues and by encouraging soil fauna
   c) Mulching is of profound importance in dry regions or/and
   d) All of the above
4. The major effects of mulches, irrespective of their composition, are attributed to
   a) Suppression of weeds, by restricting the amount of light on the ground surface and thus hindering their germination and growth
   b) Conservation of soil by preventing erosion
   c) Maintenance or improvement of soil structure, by eliminating or mitigating the severity of rain and wind action or/and
   d) All of the above
5. Mulching helps in _____.
   a) Soil fertility
   b) Moisture conservation
   c) Improvements soil structure
   d) Soil sterility
6. What is the use of biochar in farming?
   a) Biochar can be used as part of the growing medium in vertical farming
   b) When biochar is part of the growing medium, it promotes the growth of nitrogen-fixing microorganisms
   c) When biochar is part of the growing medium, it enables the growing medium to retain water for a longer time
   d) All of the above
7. The distance of trees near to the greenhouse should be about ____ times the height of the greenhouse.
   (a) 1.5

- (b) 2.5
- (c) 3.5
- (d) 4.5

8. One kilogram weight of poly film can be accommodated in _____.
   - (a) 5.4 m^2
   - (b) 3.4 m^2
   - (c) 2.4 m^2
   - (d) 1.4 m^2

9. For soil solarization use, UV-stabilized transparent sheet has a size of _____ micron
   - (a) 25
   - (b) 35
   - (c) 45
   - (d) 55

10. Physical method of soil disinfection is by _____.
    - (a) Weedicide
    - (b) Fungicide
    - (c) Solarization
    - (d) Formaldehyde

ANSWERS

1	2	3	4	5	6	7	8	9	10
c	d	d	d	b	d	b	a	a	c

REFERENCES

Doran, J. W. 1980. Microbial changes associated with residue management with reduced tillage. *Soil Science Society of America Journal*, 44, 518–524.

Emmert, E. M. 1957. Black polyethylene for mulching vegetables. *Proceedings of the American Society for Horticultural Science*, 69, 464–469.

Espi, E., Salmeron, A., Fontecha, A., Garcia, Y. and Real, A. I. 2006. Plasticfilms for agricultural applications. *Journal of Plastic Film and Sheeting*, 22, 85–102.

Garnaud, J. C. 1974. *The Intensification of Horticultural Crop Production in the Mediterranean Basin by Protected Cultivation*. FAO of theUnited Nations, Rome.

Ham, J. M., Kluitenberg, G. J. and Lamont, W. J. 1993. Optical properties of plastic mulches affect the field temperature regime. *Journal of the American Society for Horticultural Science*, 118(2), 188–193.

Hu, W., Shufen, D. and Qingwei, S. 1995. High yield technology for groundnut. *International Arachis Newsletter*, 15, 20–30.

Hussain, I. and Hamid, H. 2003. Plastics in agriculture. In Andrady, A. L. (ed.), *Plastics and the Environment*. Wiley, Hoboken, 185–209.

Ibarra-Jimenez, L., Zermeno-Gonzalez, A., Lozano-Del, R. J., Cedeno-Rubalcava, B. and Ortega-Ortiz, H. 2008. Changes in soil temperature, yield and photosynthetic response of potato (*Solanum tuberosum* L.) under coloured plastic mulch. *Agrochimica*, 52, 263–272.

Kader, M. A., Singha, A., Begum, M. A., Jewel, A., Khan, F. H. and Khan, N. I. 2019. Mulching as water-saving technique in dryland agriculture: Review article. *Bulletin of the National Research Centre*, 43(147), 2–6. doi: 10.1186/s42269-019-0186-7

Lamont, W. 1999. Vegetable production using plasticulture. http://www.agnet.org/library/eb/476/.

Lamont, W. 2004a. Plastic mulches. In Lamont, W. (ed.), *Production of Vegetables, Strawberries, and Cut Flowers Using Plasticulture*. Natural Resource, Agriculture, and Engineering Service (NRAES), Ithaca, NY.

Lamont, W. 2004b. Plasticulture: An overview. In Lamont, W. (ed.), *Production of Vegetables, Strawberries, and Cut Flowers Using Plasticulture*. Natural Resource, Agriculture, and Engineering Service (NRAES), Ithaca, NY.

Liakatas, A., Clark, J. A. and Monteith, J. L. 1986. Measurements of the heat balance under plastic mulches part I. Radiation balance and soil heat flux. *Agricultural and Forest Meteorology*, 36, 227–239.

Liu, X. J., Wang, J. C., Lu, S. H., Zhang, F. S., Zeng, X. Z., Ai, Y. W., Peng, B. S. and Christie, P. 2003. Effects of non-flooded mulching cultivation on crop yield, nutrient uptake and nutrient balance in rice-wheat cropping systems. *Field Crops Research*, 83, 297–311.

Peng, S., Shen, K., Wang, X., Liu, J., Luo, X. and Wu, L. 1999. A new rice cultivation technology: Plastic film mulching. *International Rice Research Newsletter*, 24, 9–10.

Rivise, C. W. 1929. Mulch paper. *Paper Trade Journal*, 89, 55–57.

Schales, F. D. and Sheldrake, R. 1963. Mulch effects on soil conditions and tomato plant response. *Proceedings of the National Agricultural Plastics Congress*, 4, 78–90.

Seymour, R. B. 1989. Polymer science before & after 1989: Notable developments during the lifetime of Maurtis Dekker. *Journal of Macromolecular Science, Part A: Pure and Applied Chemistry*, 26, 1023–1032.

Sharma, U. K. and Meshram, K. S. 2015. Evaluate the effect of mulches on soil temperature, soil moisture level and yield of capsicum (*Capsicum annuum*) under drip irrigation system. *International Journal of agriculture Engineering*, 8, 54–59.

Sintim, H. and Flury, M. 2017. Is biodegradable plastic mulch the solution to agriculture's plastic problem? *Environmental Science and Technology*, 51(3), 1068–1069, DOI:10.1021/acs.est.6b06042

Tarara, J. M. 2000. Microclimate modification with plastic mulch. *Hortscience*, 35, 169–180.

Tindall, J. A., Beverly, R. B. and Radcliffe, D. E. 1991. Mulch effect on soil properties and tomato growth using micro-irrigation. *Agronomy Journal*, 83, 1028–1034.

Waggoner, P. E., Miller, P. M. and De Roo, H. C. 1960. Plastic mulching: Principles and benefits. *Bulletin. Connecticut Agricultural Experiment Station*, 634.

Warp, H. 1971. Historical development of plastics for agriculture. *Proceedings of the National Agricultural Plastics Congress*, 10, 1–7.

Wilson, D. J. and Jefferies, R. L. 1996. Nitrogen mineralization, plant growth and goose herbivory in an arctic coastal ecosystem. *Journal of Ecology*, 184, 841–851.

16 Hydroponics and Vertical Farming

16.1 INTRODUCTION

Sustainable agriculture is crucial for food security in India. The advancement in technology plays an important role in enhancing food production. Farming has moved from the traditional system of soil-based farming to hydroponic and soilless techniques. It is a method of growing plants using mineral nutrient solutions without soil. Soilless culture is also a type of technique of hydroponics. Growing plants without soil has also been achieved through water culture without the use of any solid substrates. Due to urban expansion, the demand for food increases which, in turn, boosts food production. Conventional crop-growing methods in soil involve more space, a large volume of water, and a lot of labor. Under such circumstances, soilless culture can be introduced as a feasible and successful alternative. It can be used in less space and without any need for agricultural soil.

Open-field soil-based agriculture is facing some major challenges, most importantly the decrease in per capita land availability. In the future, it's impossible to feed the growing population by cultivating in the open-field system of agricultural production. Naturally, soilless culture is becoming more relevant in the present scenario to cope up with these challenges.

With the global population set to exceed 10 billion people by 2050, the challenge of providing food for everyone in a sustainable, efficient, and cost-effective way is rising in significance. Shedding the restrictions of seasonal weather patterns, overcoming transportation challenges, and significantly enhancing yields, the growing trend of vertical farming could herald the future of food production. Human populations have been farming the earth for food for thousands of years. But with the sharp rise in the number of people on our planet over recent centuries as a result of the industrial revolution, increased living standards, and falling mortality rates, the pressure on traditional farming has continually increased (Barak *et al.*, 1996; Chang *et al.*, 2012). Despite the fact that modern technology has increased output rates, agricultural cultivation currently occupies more than 11% of the world's total land area, posing environmental difficulties such as habitat destruction and soil degradation, as well as putting enormous strain on our planet's resources (Ritter *et al.*, 2001; Resh, 2012; FAO, 2013; Kumar, 2019). Furthermore, as our cities expand, the distances between suitable farming land and large populations which consume its produce are growing, raising the impact of transportation. Adding to these challenges is a changing climate that is disrupting seasonal weather patterns and the lack of suitable soils in close proximity to rapidly expanding areas. One potential solution is the quite literally growing trend of vertical farming; a concept that sees the sprawling crop farms of old, condensed into much smaller factory-like sites where conditions can be optimized and yields significantly increased.

The vertical farm is essentially an indoor farm that uses soilless technology (hydroponics or aeroponics) to grow food. Rather than being restricted to two dimensions, they stack levels on top of each other and make use of artificial lighting. The combination of highly controllable growing conditions, optimal light levels at all times full year-round growing and harvesting give vertical farms an incredible yearly output for a given area of land. Facilities like AeroFarms in New Jersey (US) see crops produced in an enclosed environment where almost everything from the lighting and ambient temperature to soil conditions and nutrients is carefully controlled (Figures 16.1–16.2). The facility uses extensive vertical racking to optimize space, as compared to conventional crop farms enabling it to be located on a far smaller site and much closer to an established urban area. Such a location reduces the extent of haulage or "food miles" required to transport the produce to

FIGURE 16.1 Vertical farming at AeroFarms, New Jersey

FIGURE 16.2 Vertical farming of rice paddy at Tokyo

consumers, thereby cutting carbon dioxide emissions. Geography aside, the creation of controlled conditions delivers many benefits. For starters, agricultural production is not affected by seasonal weather patterns, which are extremely prone to disturbance as a result of climate change. Lighting, water, and temperature may all be controlled in a vertical farm to reduce climate risk and boost output rates.

As a result, facilities like MIRAI's facility in Tokyo, the world's most populous metropolis, may produce yields 50 to 100 times higher than a standard crop farm. The utilization of a controlled environment also reduces losses from birds and insects, which must be factored in on traditional farms, reducing the need for toxic pesticides and boosting the quality of products. Vertical farms also maximize the amount of nutrients available to crops, overcoming the difficulty of finding enough good farming land in close proximity to a major city. Many times, the soil is completely removed, and crops are grown atop membranes that are sprayed with nutrient-rich liquids. However, vertical farms do have their limitations and critics have pointed to the level of energy required to maintain such refined environments. While these concerns are valid, several vertical farms are powered by renewable technologies and recycle many of their resources. The use of energy-efficient LED lighting reduces power consumption, while the blue and red shades of lights are even more economical to run. The optimized crop production process also allows vertical farmers to reduce the amount of water used, and many vertical farms are served by rainwater-harvesting techniques. Even the water that condenses within the controlled atmosphere is collected and recycled by some. This closed-cycle approach has added the benefit of preventing nutrients and fertilizers from damaging the land or being washed away to rivers and streams. Though the cost and availability of lands for vertical farms in urban areas can prove challenging, many facilities are finding a home in repurposed shipping containers, former factories, and disused warehouses. More ambitious plans, such as the one proposed by Studio NAB, may see the vertical farming concept expanded to encompass fish and honey production, as well as reconnecting consumers with the food production process and creating long-term jobs in the surrounding community. While the vertical concept still represents a smaller part of the global food production industry, the benefits it offers to our ever-expanding population could come to tilt the farming landscape by 90 degrees.

16.2 DOES AGRICULTURE NEED TO CHANGE?

According to the FAO, we will need a 70% increase in our global food production by 2050. If we take technological improvements into consideration, will we have enough land to meet this increase? Global food yields increase every year on an average with the projected non-compounding yearly growth rate of 0.65%. Yet despite yield improvements, the demand will still lead to an expected 12% increase in cropland. A profitable market of biofuels exists and is expected to increase by 2050. The global biomaterial market is also expected to grow significantly. Taken together, these will increase the cropland required by an additional 1.2 km^2, taking our 2050 cropland increase up to 21%. This is an area around the size of India and we are losing our existing land because farmable land is set to be turned into an urban environment. With the global urban footprint expanding by 33%, the desertification turns previously farmable land sterile. Combining these effects, the loss of land relative to the current cropland is 18%. That's a considerable amount of land which raises the concern of loss of farmable land at an alarming rate. As of 2005, the world was 71% ocean and 29% land; 66% of that land is considered farmable and out of that 51% is already being used for agriculture, with 40% being forests and 6% shrubland. If we factor in our 2050 model and take cropland increase and land losses into consideration, we get the following; we still have enough land left to feed the world of 2050 but we essentially have no spare farmable land left globally. This model doesn't even account for the anticipated yield loss related to climate change. These factors make the likelihood of cutting down significantly more global forests in 2050 almost inevitable.

Thus, the land required to keep the global population sustained points us to our next global challenge, and vertical farming in this context has been called the future of agriculture as it claims to solve many of the problems related to traditional farming. Vertical farming utilizes hydro, aero, and aquaponics. These methods use much less water than typically used in soils. For plant growth, hydroponics uses a circulating water and fertilizer mix instead of soil. Aeroponics utilizes an open membrane and a water moist spray with a nutrient mix. Aquaponics uses hydroponics and an aquatic ecosystem to balance nutrients in both systems.

Thus, vertical farming has the most promise to positively impact our global issues.

16.3 ENVIRONMENTAL IMPACTS OF AGRICULTURE

16.3.1 Water Use Problem

Agriculture is the world's largest water user, accounting for roughly 71% of all water consumption. Irrigation is the technique of diverting water from its normal course to hydrate crops in locations where there isn't enough rain to do so. Based on their yearly rainfall, these locations are classified as arid or semiarid environments. The population growth tendency in arid and semiarid regions has stimulated agricultural development in previously undeveloped areas. As the human population expands, so does the demand for food production to be closer to the settlement. Farmers are being compelled to cultivate land that would not normally sustain agricultural development due to the rising demand for food. Irrigation boosts dry and semiarid lands' agricultural yield (Goddeck, 2015).

Evaporation of irrigation water before the crops have a chance to absorb it is a problem with contemporary farming practices. About 5% of the water in a conventional open-air sprinkler system that sprays water over the full area of crops evaporates or blows away before it reaches the ground. About 5% equates to hundreds of thousands of gallons of water on a massive scale. Plants retain less than 5% of the water taken by roots for cell expansion and plant growth when it reaches the roots; the rest is evaporated into the atmosphere (A. McElrone, 2013).

16.3.2 Water Use: Solution

The use of a closed building as the farming location solves the problem of evaporation and reduces overall water consumption of crop irrigation utilizing vertical farming. Evaporation of irrigation water can be controlled when irrigating crops in a closed/controlled system inside a building used for vertical farming by keeping the watering system contained from contact with the surrounding air with water systems that only encompass the roots of the plants, such as aquaponics, or by containing and reusing the evaporated and transpired water that does end up in the air.

Over the course of a year, a vertical farming facility uses up to 97% less water than a traditional outside farm (A. McElrone, 2013). Water use for crop cultivation may be drastically reduced by using more efficient water delivery systems and having a facility that is a more controlled environment than a field.

16.3.3 Land Use: Problem

Many human-populated areas have unsuitable growing conditions for crops. Arable land is defined as land that is suitable for crop growth. Approximately 40% of the Earth's surface is currently arable, with some areas performing better than others. Dr. Navin Ramankutty's data from 1999 revealed that around 30% of the world's total arable land was already in use. According to recent data, the number could be as high as 80%. To put this in context, agriculture had only covered around 11% of the world's surface in all of human history until 1999, but in the previous 20 years, it has grown to cover more than double the percentage of the Earth's total land surface. Humans will run out of space to grow the crops required to feed an ever-increasing population if the rate of land conversion continues. The possibility of running out of arable land is a problem not only for humans but also for the animals that dwell in those areas and the global ecology.

16.3.4 Land Use: Solution

Instead of clearing fresh land for more traditional horizontal farms, vertical farming involves stacking new farms on top of each other. The primary goal of vertical farming is to be able to grow crops in the same quantities as a horizontal farm while using a fraction of the land space. Construction of a vertical farm inside existing erected structures around or within a city or community, rather than developing fresh tracts of land from previously undisturbed ecosystems, is another land use saver.

Hydroponics and Vertical Farming

16.3.5 Chemical Use: Problem

In locations where natural conditions and animals do not support good harvests, humans utilize fertilizers and insecticides to make their crops more productive. These chemical insecticides and nutrient-rich fertilizers runoff or seep into many habitats where they either don't belong or are damaging to the health of those ecosystems and the humans who rely on them in large enough quantities. Sediment is swept up in enormous amounts and spread abnormally along with the nutrients and toxins poured into ecosystems by agricultural runoff.

16.3.6 Chemical Use: Solution

Vertical farming prevents runoff by using a nearly closed system of watering system and more direct fertilizing techniques. This stops the spread of environmental problems caused by agricultural runoff. Because vertical farms are indoors, producers can nearly entirely control the environment inside the farm, keeping any dangerous consequences of the growing process contained. Pests and, as a result, pesticides aren't found inside vertical farming facilities because pests can't get to the crops in large enough numbers to be a problem.

16.4 SOILLESS CULTIVATION

Protected cultivation as well as other modes of controlled environment cultivation has been evolved to create favorable microclimates. It is emerging as a specialized production technology to overcome biotic and abiotic stresses and to break the seasonal barrier to production. It also ensures round the year production of high-value vegetables, like capsicum and tomato, especially during the off-season. Adopting soilless culture in protected cultivation with technical practices like fertigation, drip irrigation, plant protection, and climate control ensures better yield and water use efficiency (Figures 16.3–16.5). The protected cultivation system can control the growing environment through the management of the amount and composition of nutrient solution and the growing medium. In the present time, manufactured media, such as perlite, rock wool, expanded clay, and other materials in plastics, grow bags, and containers, are used as growing media. Certain organic products, such as coconut coir, rice hulls, sawdust, composted plant material, wood chips, etc., are also used successfully for

FIGURE 16.3 Soilless cultivation inside the poly house

FIGURE 16.4 Soilless cultivation under protected conditions

FIGURE 16.5 Soilless tomato crop cultivation inside the poly house

poly house soilless culture of vegetables. The total area covered under protected cultivation in India is 30,000 ha. During the past decade, this area must have increased by 10%.

16.4.1 History of Soilless Cultivation

A small soilless cultivation farm named Safeway Farm was first established in Shanghai during 1935–1945, which occupied more than 2,000 m^2, and tomato was the main vegetable planted there. In the 1980s with imported sets of equipment on soilless cultivation, Shanghai started its research on soilless cultivation. In early 1996, five sets of greenhouse facilities were introduced from the Netherlands and Israel by the municipal government of Shanghai, some of which were used for soilless vegetable cultivation. Today in Shanghai, the area of vegetable soilless cultivation is about

50 ha. In recent years, soilless cultivation has developed dramatically in the suburbs of Shanghai, especially in some relaxation and sightseeing Agri-parks, which is pushing function diversification of soilless cultivation.

Market acceptance, price, and lack of skilled farmers are key factors which limit the development of soilless cultivation. We are happy to see that soilless cultivation is gradually accepted in the market, especially in relaxation and sightseeing agriculture and family cultivation. We should pay more attention to facilities improvement and popularization and enhancement of technology which should be practical, standard, and simple.

16.4.2 Advantages of Soilless Cultivation

- Soilless agriculture can be performed in controlled environments to address many of the concerns we now have. That is why it is expected to be the future method of farming in many parts of the world
- It helps in the intensive production of crops under full or partially controlled conditions
- It guarantees flexibility and intensification of crop production systems in areas with adverse growing conditions
- Precise control over the supply of water and nutrients is possible through it
- Soilless farming also helps in the elimination of soil-borne diseases, reduction of labor requirement, more crops per year, etc.
- The plant gets everything it needs in the right proportions and at the right time
- Everything is utilized most efficiently
- Apart from these advantages, it is a very suitable system of production in the context of limited resources, such as land, labor, and water

16.4.3 World Scenario of Soilless Cultivation

With the advent of civilization, open-field/soil-based agriculture is facing some major challenges, most importantly, a decrease in per capita land availability. In 1960, with 3 billion people in the world, per capita land was 0.5 ha, but presently, with 6 billion people, it is only 0.25 ha, and by 2050, it will reach 0.16 ha. Due to the rapid urbanization and industrialization as well as the melting of icebergs, arable land under cultivation is further going to decrease. Again, soil fertility status has attained a saturation level, and productivity is not increasing further with the increased level of fertilizer application. Besides, poor soil fertility in some of the cultivable areas, less chance of natural soil fertility buildup by microbes due to continuous cultivation, frequent drought conditions and unpredictability of climate and weather patterns, rise in temperature, river pollution, poor water management and wastage of huge amounts of water, the decline in groundwater level, etc. are threatening food production under conventional soil-based agriculture. Under such circumstances, in near future, it will become impossible to feed the entire population using an open-field system of agricultural production only. Naturally, soilless culture is becoming more relevant in the present scenario, to cope up with these challenges. In soilless culture, plants are raised without soil. Improved space and water-conserving methods of food production under soilless culture have shown some promising results all over the world.

16.5 HYDROPONICS

Plants grown hydroponically are not planted in conventional soil, unlike so many other agricultural crops. The roots are instead immersed in a nonorganic growth medium. The plant's roots are then sprayed with nutrient-rich water. This is the foundation of any hydroponic system.

To understand how hydroponically grown plants grow and mature, we need to understand the purpose that soil serves in supporting the life of a plant. One of the most important functions of soil is its ability to retain water and nutrients and supply these to the roots of the plant. In hydroponics, this need is overcome by placing the plant in an inorganic growing medium, like vermiculite, perlite, rock wool, or an expanded clay substrate. Alternatively, the plants can also be placed in a simple container without any substrate or float on the water itself. The nutrient demands of the plants are then met through regular applications of nutrient-enriched water to the root zone of the plant (Figure 16.6).

16.5.1 Types of Hydroponic Setups

With all of the possibilities of the hydroponic method, a newbie can easily become enthralled. Let's have a look at some of the various hydroponic uses.

1. **The Nutrient Film Technique (NFT):** Plants are placed in sloping hollow pipes or channels in this setup. Nutrient-rich water is continuously injected via the root system's channels. No additional growing medium, such as vermiculite or rock wool, is required. The NFT is a closed system that pumps water to the upper or higher ends of the channels, where it subsequently runs down the hill (Figure 16.7).
2. **Wick Systems:** This is one of the most basic hydroponic setups available, making it ideal for novices. A nutrient-rich water reservoir is installed under the plants. A wick extends downward from the growth media into the reservoir. The water travels from the reservoir into the growing media via capillary action, keeping the roots nourished and hydrated. This concept requires very little specialized equipment and may be easily used with buckets, rope, and a drill (Figure 16.8).
3. **Ebb and Flow:** Plants implanted in a growing medium are put atop a reservoir supplied with nutrient water, often known as the flood and drain system. The water gradually flows back into the reservoir after the pump is switched off. This is another basic procedure that requires little expert equipment other than the pump and pipe. The water gradually flows back into the reservoir after the pump is switched off (Figure 16.9).
4. **Deepwater Culture (DWC):** Plants are grown in pots with growing material or light coverings that float on the surface of the water. The roots of plants then grow into the water. The water is oxygenated by a pump to keep the roots healthy, but no further machinery is necessary. A pump oxygenates the water to keep the roots healthy, but apart from that, no other mechanization is required. Because of this, the deepwater culture system is ideal for novices or do-it-yourself projects (Figure 16.10).

FIGURE 16.6 Nutrient-enriched water being placed into the system

Hydroponics and Vertical Farming

FIGURE 16.7 Nutrient film technique

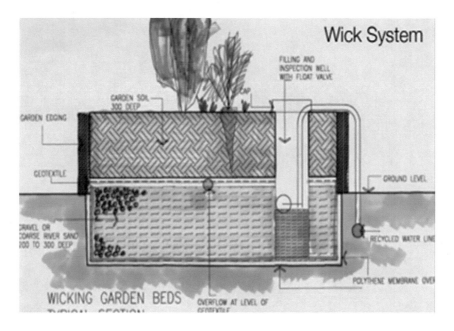

FIGURE 16.8 Wick system

5. **Drip Systems:** Drip lines and a pump are used in this procedure. Drip lines and a pump are used in this method to convey water from an underground reservoir to the growing material that surrounds the plants above. Each plant is usually given its own dripper, which allows for fine-tuning of fertilizer concentration and watering strength. The water can be either recycled to save time and money or drained out of the system to maintain the reservoir's appropriate nutrient contents (Figure 16.11).
6. **Aeroponics:** Plants are suspended vertically or horizontally in the air here. A pump then mists the nutrition medium over the roots of the plants at predetermined intervals or continuously, depending on the plant's needs. This is a more complex system that requires

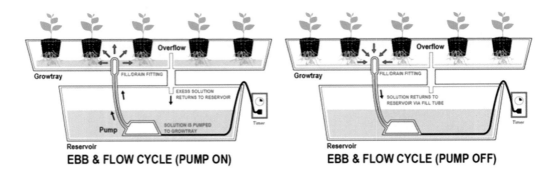

FIGURE 16.9 Ebb and flow

FIGURE 16.10 Deepwater culture

FIGURE 16.11 Drip system

pumps, misters, and timers, among other things, and should only be considered by experienced growers who can justify the expenditure.

16.6 ADVANTAGES OF HYDROPONICS

No matter the type of hydroponics system you decide on, you will always have to consider the advantages and disadvantages before you start your production. Let's start with the advantages of hydroponically grown plants.

First, you can set up a hydroponic operation, big or small, almost anywhere. Because no soil is needed, you can grow a considerable number of plants indoors even.

You can maximize your yield because the roots of hydroponically grown plants are kept compact. This is possible because when they are provided with the perfect amount of water and nutrients, roots do not need to grow in search of more.

You will also be free from the constraints of poor soil conditions and environmental pollutants, allowing you to grow almost any herb or leafy vegetable anywhere.

The money saved on water and nutrient inputs is another advantage offered to hydroponic growers. The majority of the water and nutrients provided to the plants in conventional agriculture are lost through drainage, runoff, and evaporation. This challenge is solved with hydroponics by just giving the plants the exact amount of water and nutrients they require.

The savings are further improved in circulatory hydroponics systems, as the same water source is pumped through the system over and over again.

Hydroponic systems also allow you to overcome the constraints imposed by a change of season. Climate-controlled systems allow the grower to fine-tune the temperature and lighting to suit the needs of their plants. This can open up the possibility of producing crops in their off-season, which can ensure you receive a premium for your harvest and limit the need for expensive and polluting international airfreight.

Plants grow faster as a result of climatic control. Because you can alter the environmental conditions as needed, the plants can always be provided with the optimal conditions for growth. The lack of soil will also eliminate any soil-borne diseases or weed infestations. If this problem is solved, then the money spent on the usual control methods will no longer be required. The labor and time spent mitigating pest and weed problems will also be removed. These savings either can be reinvested into expanding production or will result in improved profit margins. Furthermore, the plants will be less susceptible to soil-borne pathogens and will not need to compete with weeds for water or nutrients.

Lastly, one will be able to change crops or cultivars easily and quickly if market demand fluctuates away from or toward certain products. As a result, one will be able to stay on top of consumer trends and meet demand as it arises.

16.7 DISADVANTAGES OF HYDROPONICS

It is easy to see why hydroponics is a favorite among commercial growers and hobbyists alike. However, there are a few issues that every grower should think about before getting started.

While a small, backyard setup may not be too costly, the startup of a commercial hydroponics operation will be expensive. One will therefore need to make sure that one has a solid business plan and contingencies in case of unforeseen obstacles. This may be a barrier for small-scale farmers who cannot afford the initial investment or be approved for necessary loans. Related to the expenses of setting up a hydroponic production, one will need the expertise and experience required to make a commercial setup a success. One does not want to waste money on blunders and disasters that could have been avoided with the right knowledge and experience. This is less of an issue for hobbyists, as making mistakes and learning from them is all a part of the hydroponics experience.

Last, and this is especially so for large, electric-powered systems, the risk posed by power and water outages is a big one. Because the roots of plants are not anchored in water-retaining soil, prolonged periods without water or nutrient applications will be devastating and will result in large-scale losses. It is therefore wise to invest in generators and boreholes if one is a commercial grower.

16.8 AEROPONICS

Aeroponic systems rely solely on nutrient-rich mist to feed their plants. The idea is based on hydroponic systems, in which the roots are held in a soilless growing media, such as coco coir, and nutrient-rich water is pumped over them on a regular basis. Aeroponics eliminates the need for a growing medium, allowing the roots to dangle in the air and be puffed by specially designed misting devices on a regular basis (Figure 16.12). Seeds are "planted" in pieces of foam stuffed into tiny pots in aeroponics systems, which are exposed to light on one end and nutrition spray on the other. As the plants grow, the foam also holds the stem and root mass in place.

16.8.1 Equipment Considerations

An enclosure (usually a plastic bin with holes drilled for each plant) is required for all aeroponics systems, as well as a separate tank to hold the nutrient solution. There are a few other factors to consider when designing an aeroponic system to fit one's needs in addition to these fundamental components. Some aeroponics systems, like a regular planting bed, are meant to be utilized horizontally. However, towers and other vertical techniques are becoming more popular; because the roots must spread out; this is a smart way to save space. Vertical systems are especially popular because misting devices can be put at the top of the system, allowing gravity to spread the moisture.

Another distinction in aeroponics is the difference between high-pressure and low-pressure systems. Low-pressure systems, which use a basic fountain pump to spray water through the misters, are affordable and easy to build. Because low-pressure misters can only produce a light spray, similar to a tiny sprinkler, rather than real mist, this method is frequently referred to as "soakaponics." True mist requires higher water pressure than an ordinary pump can produce, as moisture floats in the air and more effectively supplies nutrients to the roots. Professional aeroponics systems, on the other hand, rely on a pressurized water tank capable of holding 60–90 psi, as well as high-quality misters capable of providing the finest possible puff of moisture.

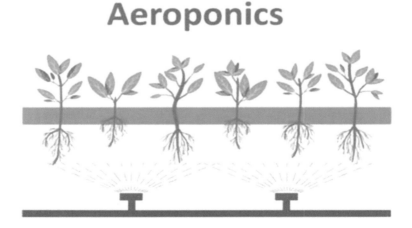

FIGURE 16.12 A simplified aeroponics system

Hydroponics and Vertical Farming

16.8.2 How Does Aeroponics System Work?

Plants are normally inserted into the platform top holes on top of a reservoir and placed into a sealed container in an aeroponic system (Figure 16.13) The aeroponics system is shown in Figure 16.14.

One will need to make a support collar to keep stems in place because there's no root zone medium for them to anchor in. These collars must be hard enough to hold plants upright and keep roots in place while yet being flexible enough to allow roots to grow.

The pump and sprinkler system extracts vapor from the nutrient-rich solution (which is a hydro-atomized spray mixture of water, nutrients, and growth hormones) and sprays the mist in the

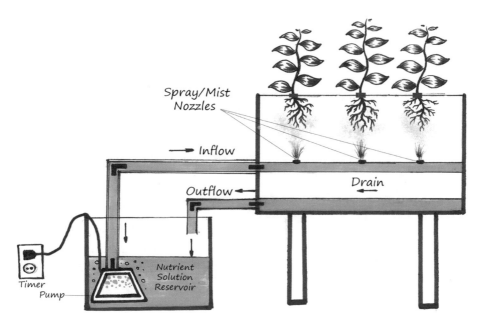

FIGURE 16.13 Aeroponic system using platform top holes and sealed container

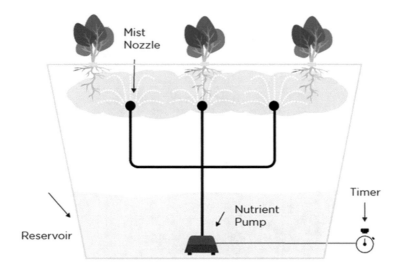

FIGURE 16.14 Working of aeroponics system

reservoir, engulfing and absorbing the dangling plant roots. This spray delivers just the right amount of moisture to encourage the plant's growth and development.

The timer provides the plants with preprogrammed spray intervals and durations. Some people consider that growing plants in an aeroponic system would be frailer compared to hydroponics. That, however, is not the case. The secret to an aeroponic system is the amount of oxygen available to the roots without the use of a root zone media.

16.8.3 Types of Aeroponic Systems

1. **Low-Pressure Aeroponics (LPA):** Due to the ease in setting it up, its availability at any hydroponic shop, and its low cost, this is the most often utilized aeroponic variety by most hydroponic hobbyists. The size of the droplets produced by a low-pressure aeroponic system differs significantly from that of a high-pressure aeroponic system. A pump powerful enough to transport water into the sprinkler heads to spray water about is required for this system, just as it is for any other hydroponic system (Figure 16.15).
2. **High-Pressure Aeroponics (HPA):** This sort of aeroponics is more complicated and more expensive to put up due to the specialized equipment required. As a result, rather than home growers, they are frequently used in commercial production. To atomize water into small water droplets of 50 microns or less, the HPA must operate at extremely high pressure. This system produces such little droplets that it provides more oxygen to the root zone than the LPA, making it more effective (Figure 16.16).
3. **Ultrasonic Fogger Aeroponics:** Ultrasonic fogger Aeroponics, or commonly called fogponics, is another interesting type of aeroponic system. Growers would utilize an ultrasonic fogger to atomize water into extremely fine droplets, as the name implies. These are quite small and appear in the guise of fog. Though plant roots find it easier to absorb water in little amounts, the fog produced has little moisture, and when run over time, it is more likely to produce salt, which can clog these foggers than other aeroponic varieties (Figure 16.17).

16.8.4 What Can Be Grown with Aeroponics?

In theory, anything is possible. In practice, leafy greens, culinary herbs, marijuana, strawberries, tomatoes, and cucumbers are grown in aeroponics systems, which are similar to hydroponics systems. Root crops, for example, are impracticable in a hydroponic system but ideal in aeroponics since the roots have lots of room to grow and are easily harvestable. Because of their size, fruiting shrubs and trees are impracticable in aeroponics systems.

Other vegetable crops are feasible, but their nutrition requirements are more complicated.

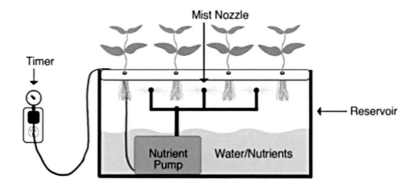

FIGURE 16.15 Low-pressure aeroponics

Hydroponics and Vertical Farming

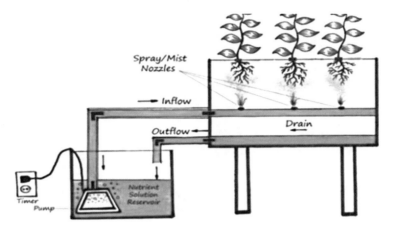

FIGURE 16.16 High-pressure aeroponics

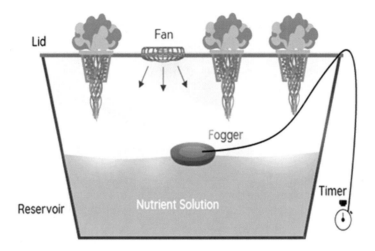

FIGURE 16.17 Ultrasonic fogger aeroponics

16.8.5 Advantages of Aeroponics

Who'd have guessed that bare roots could live, let alone thrive? It turns out that removing the growing medium is extremely liberating for a plant's roots: the increased oxygen exposure leads to faster growth. Aeroponic systems save a lot of water as well. Closed-loop systems use 95% less water than plants grown in the ground. The nutrients are also recycled because they are stored in the water.

Aeroponics' eco-friendly reputation is strengthened by the ability to grow huge quantities of food in small places, in addition to this efficiency. The method is mostly used in indoor vertical farms, which are becoming more widespread in cities, to reduce the environmental costs of transporting food from field to plate. Furthermore, because aeroponics systems are completely enclosed, no nutrient runoff pollutes the neighboring waterways. Rather than using harsh chemicals to combat pests and diseases, the growing equipment may simply be sterilized as needed.

16.8.6 Drawbacks of Aeroponics

Aeroponics systems necessitate a certain amount of dexterity in order to function properly. The nutrient concentration in the water must be kept within strict limits, and even a little fault of the equipment can result in crop loss. Those dangling roots will swiftly desiccate if the misters do not spray every few minutes-perhaps because the power goes out. The misters, on the other hand, require regular maintenance. Aeroponic systems rely on electrical power to pump water through tiny misting devices, which has one major disadvantage in terms of the environment. While they can be utilized in a greenhouse with natural light, they are more commonly utilized with energy-intensive glow lights. However, solar power or other alternative energy sources can be used to overcome this disadvantage.

16.9 FEASIBILITY AND SUITABILITY OF THESE TECHNOLOGIES IN INDIAN BACKGROUND

Vertical farming under soilless circumstances is practiced in the United States, Europe, Japan, and Singapore, but it has yet to be introduced and implemented on a significant scale in India. It will take a tremendous effort to merge vertical farming with traditional agriculture. The majority of agricultural activities are carried out solely for the purpose of earning a living; farmers are not technologically savvy, and it is not economically viable for everyone due to its capital-intensive nature.

Although the above conditions are not on our part, as time goes by, vertical farming would be a necessary need because of the decrease in the cultivable land tract. Seeking the scenario, scientists from the Indian Council for Agricultural Research (ICAR) are working at Kolkata on a module to grow vegetables and fruits in multistoried structures (Awadhesh, 2019). If the project of the ICAR gets success, people of urban India would be able to consume daily doses of fresh vegetables and fruits grown next to their residences.

LONG ANSWER-TYPE QUESTIONS

1. Discuss in detail why traditional agricultural practices need to be changed to vertical farming.
2. What is vertical farming? Why is it considered to be next-generation farming?
3. What are the different types of hydroponic setups? Discuss in detail.
4. Explain with the help of suitable diagrams the different types of aeroponic systems.
5. Differentiate between hydroponics and aeroponics.

SHORT ANSWER-TYPE QUESTIONS

1. How does the AeroFarms in New Jersey utilize vertical farming?
2. What is the purpose of soil in affecting the growth and life of plants?
3. How is the yield maximized in hydroponics?
4. How does the aeroponics system work?
5. What are the drawbacks of aeroponics?

MULTIPLE-CHOICE QUESTIONS

1. What is an advantage of vertical farming?
    a) It is fast
    b) It is year-round
    c) It increases jobs
    d) It makes food tasty
2. What will be the global population in 2050?
    a) 10 billion
    b) 15 billion

c) 20 billion
d) 25 billion
3. How much more will a vertical farm grow than a traditional farm?
   a) 10 times more
   b) 200 times more
   c) 50 times more
   d) 20 times more
4. Vertical farming utilizes
   a) Hydroponics
   b) Aeroponics
   c) Aquaponics
   d) All of the above
5. What is the foundation of the hydroponics system?
   a) Plants are immersed in soil
   b) Plant roots are immersed in nonorganic growth material
   c) Plant roots are immersed in organic growth material
   d) All of the above
6. In a wick hydroponic system, water travels down from the reservoir through
   a) Hygroscopic action
   b) Capillary action
   c) Gravitational action
   d) Adhesion action
7. The Ebb and flow system is also known as
   a) Sprinkler system
   b) Flood and drain system
   c) Down water culture system
   d) Wick system
8. Fogponics is the other name given to
   a) High-pressure aeroponics
   b) Low-pressure aeroponics
   c) Ultrasonic fogger aeroponics
   d) Drip system aeroponics
9. Which type of aeroponic system is used for commercial production?
   a) Low-pressure aeroponics
   b) High-pressure aeroponics
   c) Wick system aeroponics
   d) Ultrasonic fogger aeroponics
10. A closed loop system uses
    a) 95% more water than plants grown in the ground
    b) 95% less water than plants grown in the ground
    c) 80% less water than plants grown in the ground
    d) 80% more water than plants grown in the ground

ANSWERS

1	2	3	4	5	6	7	8	9	10
b	a	c	d	b	b	b	c	b	b

REFERENCES

Awadhesh, K. 2019. Integration of vertical farming and hydroponics: a recent agricultural trend to feed the indian urban population in 21st century. *Acta Scientific Agriculture*, 3(2), 54–59.

Barak, P., Smith, J., Krueger, A. and Peterson, L. 1996. Measurement of short-term nutrient uptake rates in cranberry by aeroponics. *Plant, Cell and Environment*, 19(2), 237–242.

Chang, D. C., Park, C. S., Kim, S. Y. and Lee, Y. B. 2012. Growth and tuberization of hydroponically grown potatoes. *Potato Research*, 55(1), 69–81.

FAO Statistical Yearbook. 2013. *World Food and Agriculture*.

Goddeck, S. 2015. Challenges of sustainable and commercial aquaponics. *Sustainability*, 7(4), 4199–4224.

Kumar, A. 2019. Integration of vertical farming and hydroponics: A recent agricultural trend to feed the Indian urban population in 21st century. *Acta Scientific Agriculture*, 3(2), 54–59.

McElrone, A. J. 2013. Water uptake and transport in vascular plants. *Nature Education Knowledge*, 4(5), 6.

Resh, H. M. 2012. Plant culture. In *Hydroponic Food Production*, Seventh revised edition, CRC Press, Boca Raton, FL, 1–513.

Ritter, E., Angulo, B., Riga, P., Herrán, C., Relloso, J. and San Jose, M. 2001. Comparison of hydroponic and aeroponic cultivation systems for the production of potato minitubers. *Potato Research*, 44, 127–135.

17 Design of Protected Structures

17.1 INTRODUCTION

Protected cultivation practice can be defined as a cropping technique wherein the microclimate surrounding the plant body (sunlight, temperature, humidity) can be manipulated to meet the requirements of growing a better crop is controlled partially/fully as per the requirement of the plant species grown during their period of growth. With the advancement in agriculture, various types of protected cultivation practices suitable for a specific type of agroclimatic zone have emerged. Among these protective cultivation practices, greenhouse/poly house cum rain shelter is useful to increase crop productivity.

Joseph Fourier, in 1824, first stated that a layer of specific gases, individually called greenhouse gas covers the earth. This gas mostly comprises carbon dioxide (CO_2), which traps heat energy emitted by the earth thus maintaining the temperature in such a way that the plants on earth are nurtured properly (Aldrich and John 1989). This phenomenon was termed the greenhouse effect and Svante Arrhenius (1896) first investigated it quantitatively, which described it as the process by which an atmosphere warms a planet. The greenhouse gas consists of carbon dioxide (CO_2), methane, chlorofluorocarbon (CFCs), nitrous oxides (NO_2), and others. Carbon dioxide (CO_2) is estimated to account for 50% of the greenhouse effect, methane for 20%, CFCs for 14%, and the remainder by other components including water vapor.

The sustainability of mountain agriculture in the north-western Himalayan states, which include Jammu and Kashmir, Himachal Pradesh, and Uttarakhand, is influenced by various mountain-specific parameters like physiography, ecosystem diversity, fragility, marginality, biodiversity, and cultural heterogeneity. More than two-thirds of people in this region draw their sustenance from agriculture and allied activities and it need not be over-emphasized that the population explosion and the laws of land inheritance are steadily declining the average size of landholding in the region. The Kashmir Valley of Jammu and Kashmir essentially has a typical temperate climate with severe protracted winter with often subzero temperatures and frost conditions which limit growing season whereas summer is warm enough suiting to the cultivation of rice and maize. Predominantly monocropping is practiced with very few Rabi crops (cropping intensity 123%) with extreme temperate and cold conditions in winter (−5 to −10°C), except for a few evergreen trees, the landscape turns gray and trees go under hibernation. Most of the agricultural activities are confined to the summer season. Due to this reason, human beings experience an acute shortage of greens and fresh vegetables, fruits, livestock due to acute shortage of feeds and fodders experience morbidity with a serious drop in milk, meat, egg, and fish production. Due to heavy snowfall and rains in the winters snow avalanches and landslides cause closure of the National Highway linking the Kashmir Valley to the rest of the country and which prevents new arrivals of supplies resulting in the steep rise in prices. Large imports of fruits, vegetables, milk, meat, egg, fish, etc. cause a severe drain on the state's economy. For attaining near-self-sufficiency in food and nutritional security measures to get over constraints imposed by severe protracted winter it is a must to apply modern agricultural structures and environment-control measures. Low-cost poly houses can go a long way in enabling year-round domestic fruit and vegetable supplies with due economic dividends (Raina *et al.*, 1999; Stauffer, 2004; Shahi *et al.*, 2005).

Varied types of climatic conditions in the north-western Himalayas region, ranging from subtropical to temperate, have endowed the region with different agro-ecological zones and livelihood production systems that offer vast scope for growing large varieties of crops like food grains, oilseeds, vegetables, fruits, and other high-value crops (Table 17.1). There is little room

TABLE 17.1
Total Area under Protected Cultivation in Major Greenhouse-Producing Countries

S.No	Country	Greenhouse Area (ha)
1.	China (2010)	2,760,000
2.	Korea (2009)	57,444
3.	Spain	52,170
4.	Japan	49,049
5.	Turkey	33,515
6.	Italy	26,500
7.	India	25,756
8.	Mexico	11,759
9.	Netherlands	10,370
10.	France	9,620
11.	United States	8,425
12.	Germany	3,430

for horizontal expansion and the only options available are vertical expansion and precision farming of high-value crops with high cropping intensity to get over the constraints to productivity, particularly the inputs and the environment. It is to be recognized that without excess to economically superior cropping systems that use improved technology and high pay-off inputs, the social and economic conditions of mountain populations and dependence on subsistence farming are generating greater problems of environment and natural resource management. In the context of climatic change being experienced in the region, the use of modern technologies like the application of precision farming techniques becomes all the more essential (Nelson, 2012; Prasad and Kumar, 2007).

Using this concept, as a basic idea, a microclimate can be created manually for the best possible growth of the plant in comparison to open-field situations. Off-season and more (both quality and quantity) production is the obvious goal for the creation of said microclimate. The structure required to create such a microclimate of crop is identified as the greenhouse, and the technology required to generate the desired crop microclimate is termed greenhouse technology. The greenhouse is a structure that allows a specific portion of sunlight inside and provides a favorable micro crop-climate in respect of temperature, humidity, rain, and wind for optimum growth of the crop/plant grown inside. However, it is not easy to create the above-stated condition (Fisher and Bill, 1976; Jenson and Malter, 1995). It requires high-precision technical knowledge of the construction of a greenhouse under different climatic conditions, climate-controlling systems, and the best method of crop husbandry. The technicality of these three aspects and application of the same can create a greenhouse for growing plants. The manual tends to address the shortcomings in hilly and mountainous regions for large-scale adoption of greenhouse technology in the state.

17.2 WORLD SCENARIO

Greenhouse technology is one of the most promising areas for agriculture in the context of climate change scenarios. Technology is very essential for cold climatic regions of the country. China started protected cultivation in the 1990s and today the area under protected cultivation in China is more than 2.5 Mha (Dalrymple, 1973; Chandra and Singh, 1988) and 90% area is under vegetables. In China, low-cost protected technology, viz. plastic mulches, plastic low tunnels, and walk-in

Design of Protected Structures

TABLE 17.2
Wind Classification Based on Speed

S. N	Descriptive Term	Particulars	Speed (km/h)
1	Calm	Smoke rises vertically	<1
2	Light air	Smoke bends from the vertical and drifts slowly with the wind; wind vane not affected	1–5
3	Light breeze	Wind felt on face; leaves rustle, ordinary vane moved by wind	6–11
4	Gentle breeze	Leaves and small twings in constant motion; wind extends light flag	12–19
5	Moderate breeze	Raises dust and lose paper; small branches moved	20–28
6	Fresh breeze	Small trees begin to sway; crested wavelets form on island waters	29–38
7	Strong breeze	Large branches in motion; whistling heard in telegraph wires; umbrellas used with difficulty	39–48
8	Moderate gale	Whole trees in motion; inconvenience felt when walking against the wind	50–61
9	Fresh gale	Breaks twigs off trees; generally, impedes progress due to difficulty experienced in walking against the wind	62–74
10	Strong gale	Slight structural damage occurs (chimney pots and slates on roof removed)	75–88
11	Whole gale	Trees uprooted and considerable structural damage occurs for instance kutcha houses blown down (seldom experienced inland)	89–102
12	Storm	Widespread damage (very rarely experienced)	103–117
13	Hurricane	-	>118

tunnels are being used on 80% of the total area under protected cultivation and perhaps this is the basic reason that today China is the largest producer of vegetables in the world. In recent years, Israel has taken big advantage of this technology by producing quality vegetables, flowers, fruits, etc. in water-deficit desert areas for meeting not only its small domestic demand but also the huge export demands. India, at present, is the second-largest producer of vegetables in the world. It is estimated that annual returns per unit area from protected cultivation could be 10–100 times than those of open-field cultivation. The area under all forms of protected cultivation in India is around 25,756 ha (Anonymous, 1999; Bartok and Susan, 1982). The total area in major greenhouse-producing countries is shown in Table 17.2.

17.3 PRINCIPLES OF PROTECTIVE CULTIVATION

The greenhouse is generally covered by transparent or translucent material such as glass or plastic. The greenhouse covered with a simple plastic sheet is termed a poly house. The greenhouse generally reflects back 43% of the net solar radiation incident upon it allowing the transmittance of the "photosynthetically active solar radiation" in the range of 400–700 nm wavelength. The sunlight admitted to the greenhouse is absorbed by the crops, floor, and other objects. These objects, in turn, emit longwave thermal radiation in the infra-red region for which the glazing material has lower transparency. As a result, the solar energy remains trapped in the greenhouse, thus raising its temperature. This phenomenon is called the "greenhouse effect." This condition of a natural rise in greenhouse air temperature is utilized in cold regions to grow crops successfully. However, in the summer season due to the above-stated phenomenon ventilation and cooling is required to maintain the temperature inside the structure well below 35°C. The ventilation system can be natural or a forced one. In the forced system, fans are used, which draw out 7–9 m^3 of air/s/unit of power consumed and are able to provide 2 air changes/minute (Boodley, 1996). The various types of cooling systems employed are as follows:

1. Roof shading
2. Water film covering
3. Evaporative cooling, which includes the following
   (i) Fan and pad cooling, (ii) high-pressure mist system, and (iii) low-pressure mist system

However, in cold regions, natural ventilation is sufficient to maintain the desired temperature in the poly house. This can be achieved using the agro-shade net or by providing doors (on opposite sides) in order to facilitate cross ventilation. Most importantly, to take maximum advantage of sunlight, the single-span poly house should be oriented east–west while the multi-span poly house should be oriented north–south as much as possible.

17.4 SELECTION OF FILM FOR POLY HOUSE

Generally, a 200-micron film is used for the poly house. While selecting the film for the poly house, the following points should be taken into consideration:

- Light transmission of the film should be good
- Light diffusion
- Anti-drip effect
- Anti-dust effect

17.5 BENEFITS OF GREENHOUSE TECHNOLOGY

The benefits which can be derived from greenhouse cultivation are as follows:

1. The environment control allows raising plants anywhere in the world at any time of the year, i.e., crops could be grown under inclement climatic conditions when it would not be otherwise possible to grow crops under the open-field conditions
2. The crop yields are at the maximum level per unit area, per unit volume, and per unit input basis
3. The control of the microcosm allows the production of higher-quality products that are free from insect attack, pathogens, and chemical residue
4. High-value and high-quality crops could be grown for export markets
5. Income from the small and the marginal landholdings maintained by the farmer can be increased by producing crops meant for the export markets
6. It can be used to generate self-employment for the educated rural youth in the farm sector

17.6 EFFECT OF WIND ON STRUCTURAL DESIGN OF POLY HOUSE

The poly houses need to be designed structurally as per the prevailing conditions. The proposed site of the poly house should not be located in the vicinity of industry on its downwind side because of possible pollution effects on poly house crops. The area should have sufficient windbreaks on all sides to minimize the wind damage to the poly house. In areas where snow is expected the trees for windbreaks should be at least 30 m away from the poly house for checking snowdrifts. The prevailing winds should not adversely affect either the structure or operation of the poly house facility. Wind direction if known at the location also influences poly house orientation. In a naturally ventilated poly house, the ventilators should open on the leeward side. A freestanding poly house should have its long axis perpendicular to the wind direction. The windbreaks should be at least 30 m on the north and west of the poly house. During periods of heavy winds, all windows and doors must be properly closed to minimize structural damage. The wind classification based on speed is illustrated in Table 17.2.

Design of Protected Structures

Since the prevailing wind in an agro-climatic zone varies in speed and intensity from location to location, the structural design of poly houses varies from place to place. Power available in wind is proportional to the wind speed cubed. Also, the power in the wind is a function of air density, so it declines with altitude as air thins. The structural design of a poly house must provide safety from wind, snow, or crop load damage. Therefore, frames should be minimum in size while providing adequate strength to resist expected loads over the planned life of the poly house. Design load includes the weight of the structure (dead load), load brought on because of the use of structure (live load), and load from snow and winds. The size of structural members should be selected keeping in view the total expected design load and a suitable safety factor. Air-inflated double-layer polyethylene glazing performs better structurally in intense wind conditions than would be predicted based on calculations utilizing standard wind loading coefficients which are determined for rigid structural components. The air space between the layers which allows substantial deformation in the event of severe wind loading serves to reduce the maximum stresses that occur in the film and structural components. Polyethylene is usually applied in a double layer with a small fan inflating the air space between the layers. This increases the strength and stability of the glazing film. The wind direction is relative to the poly house geometry and the placement of ventilation openings as well as wind speed influences air movement.

17.7 CONSTRAINTS OF CLIMATE IN HILLY AND MOUNTAINOUS REGION

Winters in mountains are severe and prolonged which reduces the growing season from 2.5 to 7 months. The subfreezing temperatures are experienced during winters and these go on declining as the altitude increases (Table 17.3).

Due to extremely low temperatures restricting growth, frosts, subfreezing temperatures, and snowfall especially in higher altitudes (1,500 amsl to 5,000 amsl), large tracts are mono-cropped and demand cold-tolerant varieties with timely maturity. Late-sown crops do not reach maturity before winter starts.

17.8 PROTECTED VEGETABLE CULTIVATION AT HIGH ALTITUDES

Generally, low-cost poly houses are used for vegetable production in high hills. A poly house is a framed structure covered with a transparent cover/shielding/cladding which is transparent to solar short-wave radiation and translucent to infra-red radiation creating a greenhouse effect. A poly house maintains inside temperature higher than ambient, higher RH than ambient as well as higher CO_2 levels which are favorable to crop growth. Shielding also provides protection from insects, predators, and vectors of diseases. In the summer months, natural and forced ventilation, evaporative cooling, shading, and misting techniques are used to lower temperature, depending upon the degree of sophistication adopted.

TABLE 17.3
Agricultural Season at Different Altitudes

Altitude (m above msl)	Duration	Month
2,670	April–October	7.0
3,000	May to mid-October	5.5
3,300	Mid-May–mid-September	4.0
4,000	Mid-June–August	2.5

17.9 LOW-COST POLY HOUSE TECHNOLOGY FOR VEGETABLE PRODUCTION

Poly house is a framed or inflated structure covered with a transparent material in which crops are grown under controlled or partially controlled conditions. The purpose of the poly house is to provide and maintain an environment suitable for crop production with increased productivity. Major controlled factors are light, temperature, humidity, and carbon dioxide concentration. There is always a large and sustained demand for fresh vegetables all-round the year. Therefore, to facilitate production in "off-season," poly house cultivation is one of the most viable alternatives.

Hi-tech poly house technology requires a huge initial investment, as a result, the production cost is also very high, making them less cost-effective. Hence, the small and medium farmers cannot afford the technology. For this, a low-cost poly house made of locally available materials, like galvanized iron (GI) pipe and ultra-violet (UV) transparent polyethylene, has been found quite effective for nursery raising and "off-season" vegetable production. The size of a poly house varies from a few square meters to a few hectares, but it is always better to start with a small poly house. There are some factors to determine the size of a poly house facility, like fund availability, market facility, labor requirement, etc. It is estimated that an area ranging from 100 m^2 to 250 m^2 under the poly house is sufficient for a single-person retail business.

17.10 SITE SELECTION

Poly houses are better utilized if well connected with the market, for the supply of inputs and sale of its produce. Good quality water availability and proper drainage facilities are other important factors while choosing a site. The site should be shadow-free and windbreaks on all sides to minimize wind damage. A low-cost poly house has two distinct segments, i.e., frame and glazing material. Frame life is 15–20 years and glazing material's life is 3–4 years. The frame made up of 25 mm and 20 mm GI and MS pipe (class B) is an important component of the poly house as it provides support to glazing material. Polyethylene films are used as the glazing material. UV-stabilized polyethylene film (200-micron thickness, 7 m width) could be used for the purpose to ensure efficiency and economy.

17.11 PROSPECTS FOR PROTECTED CULTIVATION

The crop productivity is greatly influenced by the growing environments and management practices. Under open-field conditions, it is not possible to control the growing environment of plants thereby affecting the productivity and the quality of the crops. In the case of mountainous regions where we are faced with the issue of livelihood to rural people due to declining landholdings and the rising cost of production, protected farming is a good option. Efforts have been made to promote protected farming to increase cropping intensity, productivity and improve product quality for more income and employment. Off-season produce can fetch greater net returns, achieving a greater degree of food and nutritional security and widening livelihood opportunities for the rural people. However, to ensure sustainable agricultural development of hill regions, together with generating additional income opportunities through on-farm and off-farm avenues, protected farming needs to be promoted and practiced on a large scale. The farmers should be taught the art and science of protected farming and more R&D be undertaken to make it even more remunerative. Central and state governments should launch massive extension schemes promoting it and networking the farmers for optimal returns. Options in protected farming are many but important ones include greenhouses, poly houses, tunnels, clothes, and mulching coupled with micro-irrigation better still fertigation. With the advancement of science and technology as well as institutional innovations, there is greater scope for promoting profitable protected cultivation in the hill and mountainous regions of the country on a larger scale.

Design of Protected Structures

17.12 PRINCIPLES OF GREENHOUSE DESIGN

A solar greenhouse is designed to trap enough solar radiation for the photosynthesis process and to provide the interior climatic conditions required for growing crops all year round. When the outside conditions are very cold, heat is stored during the day in the ground and walls of the greenhouse and released during the night to keep the greenhouse air warm. During winter, the greenhouse traps enough energy during the day to ensure that the vegetables do not freeze at night. The temperature variation between day and night should be minimized to reduce thermal stress to the plants. Overheating during the day can be prevented using natural ventilation for cooling, regulated by manually operated shutters. Ventilation also regulates the humidity and thus helps to limit diseases and pests. A solar greenhouse has the following characteristics.

- Picks up solar radiation
- Stores this radiation as heat in the mass of the walls and the ground during the day
- Releases this heat during the night to warm the interior space
- Is insulated to retain this heat
- Can be ventilated to avoid overheating

17.13 SITE CHARACTERISTICS THAT AFFECT THE DESIGN

17.13.1 Wind

If the door of the greenhouse is exposed to wind, the infiltration of cold ambient air will be lower. The door must always be located to the opposite side of the prevailing wind.

17.13.2 Climate (Altitude)

Temperature decreases with altitude; thus, a similar greenhouse will be more efficient at lower altitudes than on the high plateau. The design can be adapted to colder climates by increasing the thermal mass and ground insulation and reducing the width, among others. The greenhouse design should take into account the normal lowest winter temperatures at the site.

17.13.3 Snow

Heavy snowfall can damage the polythene laid over the greenhouse if a considerable weight of snow remains on it. In snowy areas, the polythene needs to slope more steeply so that the snow slips off. The polythene sheet must slope more steeply in areas with high snowfall.

17.14 DIFFERENT TYPES OF POLY HOUSES

Design for heavy snowfall areas has been standardized by the College of Agricultural Engineering and Technology, SKUAST-Kashmir. Ridge-type greenhouses that have been fabricated in polycarbonate sheets have been found to withstand heavy snowfall loads and have been recommended for construction in these areas. Three models have been developed and the first one is a naturally ventilated greenhouse and the second is a hi-tech greenhouse. The third is a walk-in tunnel-type greenhouse with a horizontal angle of greater than 40° is given to the arcs so that they can sustain heavy snowfall during winters (Figures 17.1–17.3). The designed models of the greenhouse can be installed depending upon the suitability of the site.

The design of a greenhouse for a specific location is influenced by the site characteristics, the climate, and the expected amount of snowfall. In all the designs the door must be constructed on the wall opposite to the prevailing wind to limit infiltration of cold air. Poly house crop production is

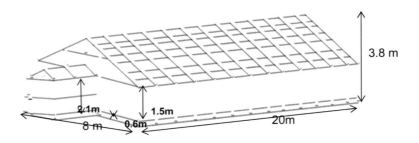

FIGURE 17.1 Naturally ventilated greenhouse. Naturally ventilated greenhouse (160 m²) ridge type

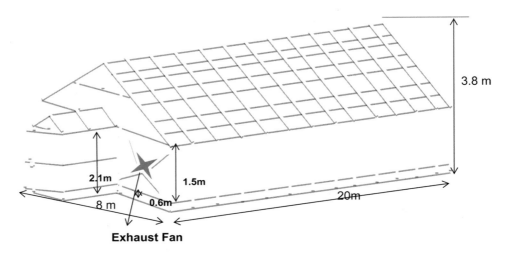

FIGURE 17.2 Hi-tech greenhouse. Hi-tech greenhouse (160 m²) ridge type. Estimate for the hi-tech greenhouse (20 m × 8 m × 3.8 m) 160 m² ridge type

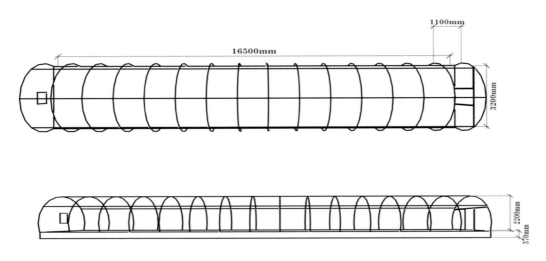

FIGURE 17.3 Quonset-shape poly house

becoming popular in Jammu and Kashmir state because of its wide applicability to most problems in the agriculture and horticulture sector. The crop productivity is greatly influenced by the growing environments and management practices. Under open-field conditions, it is not possible to control the growing environment of plants thereby affecting the productivity and the quality of the crops.

Sher-e-Kashmir University of Agricultural Sciences and Technology of Kashmir (SKUAST)-Kashmir has standardized construction of low-cost poly houses (10.0 m × 4.0 m × 3.0 m;15.0 m × 4.0 m × 3.0 m; 15.0 m × 3.5 m × 2.2 m; 17.0 m × 4.0 × 3.0 m) using 25 mm and 20 mm GI and MS tubular framework, which can easily last 15–20 years. UV-stabilized 200-micron, 7-m-wide cross-laminated polyethylene film is used as cladding. Sides are also provided with polynet which allows ventilation if the cladding is rolled up to the height of the net. The poly house maintains about 8–10°C higher temperature than ambient, but with sides open only about 23°C higher temperature. The recommendation of the university is that the structural material of the poly houses shall be galvanized iron (GI) instead of mild steel (MS) as the strength of GI pipes is much greater than the mild steel pipes. The covering material in the galvanized pipes also prevents them from rusting and thus enhances the life of these structures. In the case of mild steel pipes, in order to increase the life of the pipes, they are coated with red oxide and along with some suitable paint to increase the life of these pipes. The thickness of the pipes should be of C-class so as to sustain the heavy snowfall during the winter months. The dimensions of the structures and material to be used in the fabrication of these poly houses/greenhouses have been selected so that they can sustain different types of loads that are being thrust on these structures. The telescopic method of inserting the major arcs in the ground also tends to enhance the life of these structures. The dimensions and figures of different types of low-cost poly houses are given in Figures 17.4–17.7.

17.15 GREENHOUSE TECHNOLOGY FOR COLD ARID REGIONS OF LADAKH

Cold arid regions are usually confined to high altitudes and circumpolar regions. Sixteen percent of the total landmass is under cold arid regions. Indian cold arid regions come under the trans-Himalayan zone. Such regions are confined to Ladakh (UT) Lahaul and Spiti (Himachal Pradesh), and small pockets in Uttaranchal. Cold arid regions comprise uneven land, barren mountains, arctic desert, and extremely harsh climate which makes agricultural operations possible only during summer months, i.e., from May to September. In this period, a limited number of vegetables can be grown in open-field conditions. A number of solar-based greenhouses have been designed and

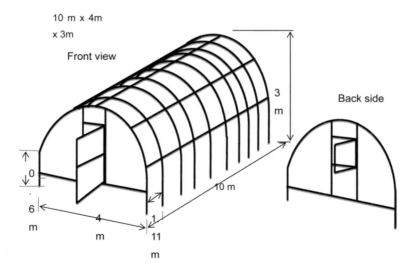

FIGURE 17.4 Quonset-type low-cost poly house

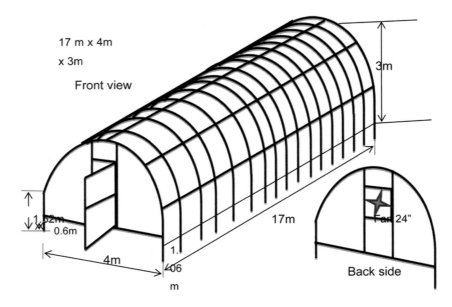

FIGURE 17.5 Tunnel-type low-cost poly house. Ridge-type (120 m²) naturally ventilated poly house

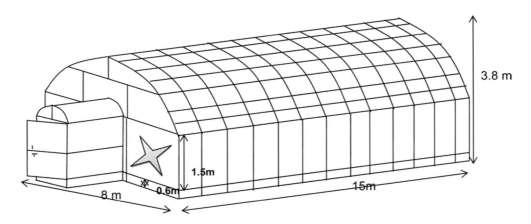

FIGURE 17.6 Gothic-type (120 m²) hi-tech poly house

developed for crop production under cold arid climatic conditions. The various types of greenhouses suitable for these regions are shown in Figures 17.8–17.15.

The Ladakhi greenhouse (Figure 17.8) with its origin from a model similar to the Chinese greenhouse is an indigenous design which has been adopted on a large scale in the Ladakh area. The dimensions of the greenhouse are 30″ × 12″. The back wall height of the greenhouse is 7″ with a front wall height of 1″. The north side of the greenhouse is made up of mud bricks covering the full side. Polythene is on the south side with inclination in such a way that maximum radiations during the winter months are absorbed by the greenhouse. The orientation of the greenhouse lengthwise is preferably east–west. However, the orientation and size of the greenhouse may be modified as per the microclimate of the location and the availability of suitable pieces of land and resources. The polyench greenhouse (Figure 17.9) has standard dimensions of 30″ × 10″ with a center height of 7″. The poly house is made 2.5″ below ground and is suitable for very cold climatic conditions. The temperature inside a polyench greenhouse is 15–25°C higher than the outside conditions and

Design of Protected Structures

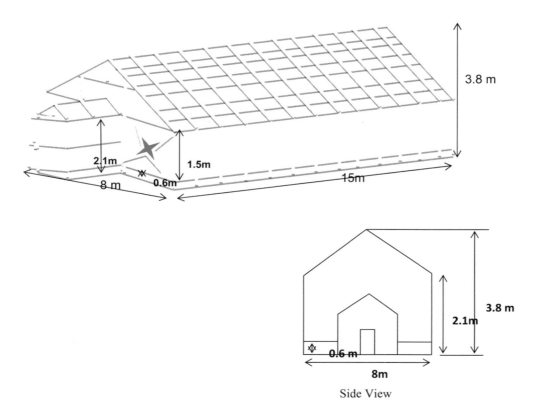

FIGURE 17.7 Ridge-type (120 m²) naturally ventilated poly house

FIGURE 17.8 Ladakhi greenhouse

FIGURE 17.9 Double-wall-type polyench greenhouse

FIGURE 17.10 Polycarbonate-type polyench greenhouse

FIGURE 17.11 Fiber-reinforced polyester (FRP) greenhouse

FIGURE 17.12 Underground greenhouse

is suitable for growing a wide variety of vegetable crops during the winter months. The polyench greenhouse can be constructed with a single wall and double wall depending on the microclimate of the area. The dimensions of the double-wall polyench greenhouse are 60″ × 15″ with a back wall height of 8' and a front wall height of 1″. The orientation of the greenhouse is east–west with a back wall at the northern side of the poly house.

The polycarbonate greenhouse (Figure 17.10) can be fabricated in double layer and triple layer. The standard dimensions of double-layer and triple-layer polycarbonate greenhouses are 55″ × 30″

FIGURE 17.13 Trench technology

FIGURE 17.14 Plastic mulching

and 120″ × 30″, respectively. The center height for both double and triple layers is 11″ and the orientation of the greenhouse is lengthwise east–west. The standard dimensions of fiber-reinforced polyester greenhouse (Figure 17.11) are 100″ × 30″ with a center height of 10″ and a side height of 6″. The orientation of the poly house lengthwise is along the east–west directions. The underground greenhouse (Figure 17.12) has a standard dimension of 30″ × 10″ with a depth of 2.5″. The orientation of the poly house should be along the north–south direction so that maximum radiation enters the poly house through the top of the greenhouse. The orientation and size of the greenhouses

FIGURE 17.15 Concrete trenches

can be modified as per the microclimate of the location, availability of suitable pieces of land and resources. In the case of Ladakhi poly house and polyench greenhouses, the back wall should remain on the northern side so that the front side of the greenhouse can receive sunlight for maximum duration at day time. The greenhouse should be constructed on cultivable fertile land with an irrigation facility. There should not be any obstacles to stop the sunlight of the sight.

17.16 VEGETABLE PRODUCTION

Agro techniques for vegetable production using trench technology (Figure 17.13) are covered by polyethylene sheets to maintain the temperature inside the trench within the optimum range. The trench is a four-unit trench with a size of 6 m × 10 m and a depth of 0.5 m. Spade the soils in the trench till fine-tilled and then level the trench with a six-inch layer of topsoil and 15–20 kg well-rotten farmyard manure (FYM) mixture. Once the trench is ready, one can grow nursery in March and April and it should be covered with a 200-micron polythene sheet. Polythene helps in maintaining temperature and prevention of soil moisture loss by evaporation. During the summer season, farmers can grow tomatoes, brinjal, capsicum, etc. The trenches can also be made in concrete with stone walls on the sides for enhanced life of the structure. This type of trench can also be covered with polythene to maintain the temperature inside the trench.

17.17 COLLECTION OF SOLAR RADIATION

Solar radiation is taken up through a transparent polythene sheet covering the south face of the greenhouse. The angle of the polythene is calculated so that the maximum amount of solar radiation is transmitted into the interior. The angle of the lower section of the polythene is 50° or more (measured from the horizontal) – the best angle to transmit solar radiation in the early morning or late afternoon when the sun is low in the sky. The angle of the upper section is 25° or more (measured from the horizontal) – the best angle to transmit the mid-day solar radiation and allow small amounts of snow to slide off. Moveable insulation is used as a curtain below the polythene after sunset to reduce heat loss; it is removed after sunrise. Moveable insulation can increase the ground and interior temperature at night by up to 5°C. At high altitudes, a double polythene layer can be used to reduce heat loss; it can also increase the interior temperature by up to 4°C at night.

17.17.1 Thermal Storage and Insulation

Several components are used in the design to increase thermal storage and reduce heat loss.

17.17.1.1 Double Wall

The walls are composed of three layers: an outer load-bearing wall built with mud brick, rammed earth, or stone; an inner wall used to store heat during the day and release it at night, also built with mud brick, rammed earth, or stone; and an insulating layer of materials like straw, sawdust, wood shavings, dry leaves, dry grass, or wild bush cuttings pressed between the two.

17.17.1.2 Color

The inner side of the west wall is painted white (whitewash) to reflect the morning solar radiation after the coldness of the night; the inner side of the east wall is painted black to absorb and store the afternoon solar radiation, which is then released at night to heat the interior space; and the bottom two feet of the inner side of the north wall are whitewashed and the upper part painted black for similar reasons.

17.17.1.3 Roof

The fixed roof is sloped (to the north) at an angle of 35°. In winter, when the sun has a low elevation angle, this angle optimizes the solar radiation absorption on the inside surface. During summer, when the sun is high in the sky, the roof partly shades the greenhouse and reduces the risk of overheating. The roof is covered with a layer of insulation; a piece of white cloth or parachute material can be added below it to improve the insulation and reflect solar radiation onto the vegetables. The shape of the roof reduces the interior volume compared to traditional greenhouses, thus increasing the interior temperature.

17.17.1.4 Ground

The greenhouse floor is dug out so that it lies 6″ (15 cm) below the outside surface level. This improves plant growth as the dip acts as a trap for carbon dioxide, as well as provides additional thermal insulation. In extremely cold climates, a 2″ layer of dung should be laid four inches below the surface to insulate the ground and increase the thermal mass efficiency. Horse or donkey dung is the most suitable as they contain straw, but yak or cow dung can also be used.

17.17.1.5 Door

The door is located on the wall opposite to the side from which the prevailing wind blows (the lee side) to reduce the infiltration of cold air.

17.17.1.6 Ventilation

On sunny days, the air in the greenhouse can become very warm. Overheating (over 30°C) can damage the vegetables and encourage diseases and pests. Manually operated openings (ventilators) are provided in the lower part of both sides (door, wall shutter) and in the roof. The warm air rises and leaves the greenhouse through the roof ventilator, drawing in the cooler ambient outer air through the lower ventilators (Figure 17.16).

17.18 LOW-COST POLY HOUSE TECHNOLOGY FOR DRYING

Solar drying is a continuous process where moisture content, air, and product temperature change simultaneously along with the two basic inputs to the system, i.e., the solar radiation and the ambient temperature. The drying rate is affected by ambient climatic conditions which include temperature, relative humidity, sunshine hours, available solar radiation, wind velocity, frequency, and duration of rain showers during the drying period. Poly house technology can be used for drying a wide variety of crops depending on the requirements.

Design of Protected Structures

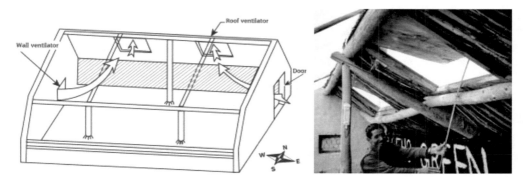

FIGURE 17.16 Ventilation for low-cost greenhouse

FIGURE 17.17 Greenhouse-type solar multitier dryer

The division has standardized the poly house-type solar dryer with the establishment of different models. A poly house-type solar multi-tunnel dryer (Figure 17.17) has a size of 5 m × 4 m and a floor area of 20 m². The front side height of the dryer is 2.25 m and the backside height of the dryer is 3.25 m. The center height of the dryer is 4 m, and the poly house is covered with a transparent UV-stabilized polythene plastic foil of 0.2 m thickness with a transmissivity of 92% for visible radiation which traps the solar radiation in the daytime and enhances the temperature inside to maintain it at the optimum level for drying of fruit and vegetables. The top surfaces of the collector and drying chamber were designed in curved shapes in order to increase the area of radiation. The inclination of the top surface of the dryer was kept 30°. The orientation of the dryer was kept in the north–south direction so that radiation from the sun would not be disturbed. The capacity of the dryer ranged from 100 kg to 150 kg of fresh fruit and vegetables depending upon the material and the thickness of the spreading layer.

The mean ambient temperature and relative humidity with the corresponding temperature and relative humidity inside the loaded and unloaded drier at different hours during the period of experimentation are presented in Figures 17.18–17.19, respectively. The poly house drying was observed to be quicker than in the open condition. Further, the upper tray has more temperature as compared to that of the lower tray. The inside average temperature of the upper and lower tray of loaded drier was constantly observed to be higher than the ambient by a difference of about 20°C and 17.2°C, respectively. The inside temperature of the upper and lower tray of unloaded drier was constantly observed to be higher than the ambient by a difference of about 22°C and 19.3°C, respectively. The maximum temperature inside was observed to be 59.5°C (for upper tray) and 47.6°C (for lower tray) in loaded drier and 61.8°C (for upper tray) and 58.9°C (for lower tray) in unloaded drier. This is because the drier is based on the greenhouse effect and it traps the solar radiation resulting in a subsequent increase in temperature. The temperature range conducive for quick dehydration of fruits and vegetables is recommended as 66–71°C and 60–66°C, respectively (Dauthy, 1995). The chimneys drag the moist air within the chamber and remove excess moisture from the vicinity of the product placed on perforated trays. The relative humidity inside the dryer varied between 21%

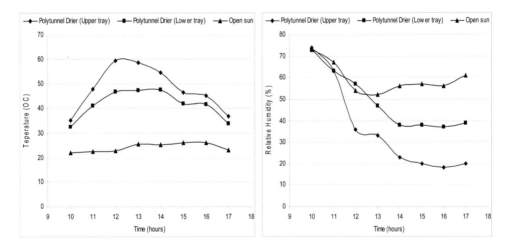

FIGURE 17.18 Average temperature and relative humidity recorded in the open sun and inside poly tunnel dryer when trays are loaded

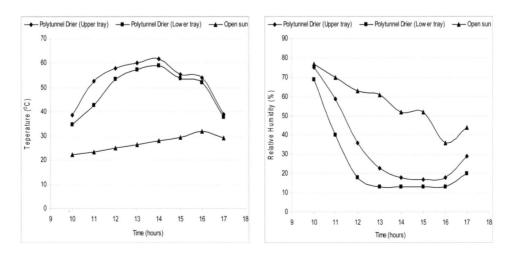

FIGURE 17.19 Average temperature and relative humidity recorded in the open sun and inside poly tunnel dryer when trays are unloaded

and 74% as compared to outside relative humidity which ranged from 40% to 75%. The temperature inside the dryer was 62–76% higher than the ambient conditions. The cost-benefit ratio of the dryer was 1.5 years by adopting solar drying technology.

17.19 LOW-COST POLY TUNNEL DRIER

Poly tunnel drier has dimensions of (1.5 m × 0.67 m) with a height of 0.55 m (Figure 17.20). The drier is provided with two trays one above the other for drying of the produce. Two chimneys of height 0.45 m along with fans inside the chimneys have been provided to maintain the relative humidity inside the dryer. Drier is portable and moveable and could be placed in a sunny location. The drier due to its design was able to significantly reduce the drying time for a wide variety of fruits and vegetables. The drying time was reduced 3–5 times as compared to open-sun drying with the added advantage of protection from rain, insects, and other types of natural calamities. The lower tray of the dryer was found to have greater moisture as compared to the upper tray. The fault in the design was rectified by putting two exhaust fans operated by a 12 V battery for increasing the uniformity of drying inside the poly tunnel-type dryer.

17.20 WALNUT PROPAGATION UNDER POLY HOUSE

The majority of the walnuts growing of seedling origin, however, walnut varieties should be propagated on rootstocks by vegetative methods to maintain their distinctive characteristics and to maintain desirable traits, such as uniform vigor, disease resistance, and quality of nuts and kernels. Scion is brought from a bearing plant of known variety and it is kept intact in the cambium layer of both rootstock and scion. The grafted plant takes around 4 years for fruit while without a grafted plant it takes around 12 years. The walnuts were grafted inside and outside the poly house in the month of February and March at a spacing of 20 cm × 60 cm (Figure 17.21). The effect of environment, mulch, and fertilizer was found to have a profound effect on grafting success and scion girth. Generally, walnuts are not fertilized by the orchardists, but for achieving higher yields of the quality crop, fertilization is very important. During winter, warm temperatures inside the poly house promoted better plant growth, whereas in the open field, plants were dormant. Poly house conditions favored the grafting with maximum grafting success. Walnut plants were ready for transplantation

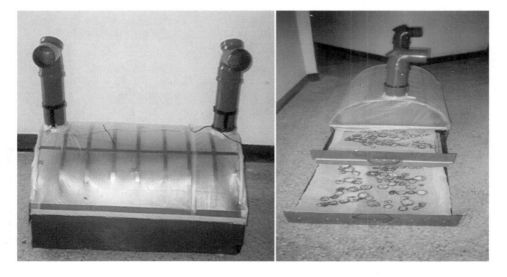

FIGURE 17.20 View of solar poly tunnel vegetable drier

FIGURE 17.21 Walnut propagation inside the poly house

in the field after a gap of one year and were uprooted from the experimental field in mid-February as climatic conditions during this time favored its growth in the field. The poly house plants were found to be consistently better than those of open fields. The observations obtained during the initial stages are as follows:

- Time taken for scion bud sprouting: 25–38 days in the poly house and 42–49 days in the open field
- Grafting success: the grafting success inside the poly house and in the open field was found to be 78% and 37%, respectively
- The walnut plant characteristics like plant height varied from 150 cm to 200 cm inside the poly house and 75–100 cm in outside conditions

The grafting success of walnut plants inside the greenhouse was 80% as compared to only 35% under open-field conditions. The use of greenhouses as solar dryers and animal shelters offers a lot of potential in terms of reducing post-harvest losses of fruits and vegetables. The technical manual offers solutions for the design and development of greenhouses under temperate and cold arid climatic conditions. In temperate regions, special emphasis needs to be given for the design of greenhouses, which can sustain heavy snowfall loads and do not buckle under extreme weather conditions. Similarly, for cold arid regions, design and development parameters are very important which tend to reduce heat losses and absorption of maximum solar radiations during winter months. Since the greenhouses developed are to be utilized throughout the year for economic purposes, it needs to be seen that the greenhouse works equally well during winter as well as summer months. The manual offers details with regard to the successful development of greenhouse technology and large-scale adoption under both temperate regions and cold arid regions. A simple intervention in terms of a pipe bending machine developed by the authors resulted in the development of more than 300 low-cost poly houses, semi-controlled poly houses, and hi-tech poly houses in farmers' fields, government farms, and agricultural entrepreneurs' farms. The department over the past few years

Design of Protected Structures

has fabricated poly houses for more than three crores and resulted in earning considerable revenue for the division. The technical manual offers a simple understanding of various techniques in greenhouse technology and fabrication parts. The results obtained under various experiments conducted by the university under various sponsored schemes are appended for the benefit of the readers and large-scale adoption of this technology.

17.21 DESIGN PARAMETER OF GREENHOUSE

Greenhouse cultivation is the most intensive form of crop production with a yield per cultivated unit area up to ten times superior to that of a field crop. The harsh winter conditions hinder the cultivation of crops in the open field. Off-season and more production (in terms of both quality and quantity) is the obvious goal for creation of said microclimate. It requires high-precision technical knowledge of construction of a greenhouse under different climatic conditions, climate-controlling systems, and the best method of crop cultivation. The technicality of these three aspects and application of the same can create a greenhouse for growing plants. The ultimate design of a greenhouse depends on the following aspects:

- Overall structural design and the properties of the individual structural components
- Specific mechanical and physical properties which determine the structural behavior of the covering materials
- Specific sensitivity of the crop to light and temperature to be grown in the greenhouse
- Specific requirements relevant to the physical properties of the covering material
- Agronomic requirements of the crop

17.21.1 Dead Loads

It is a load that is constant in magnitude and fixed in position throughout the lifetime of the structure. Dead loads are gravity loads that are constant in magnitude throughout the life of the building. They include the walls, roof, glazing, and fixed equipment. Fans, suspended heaters, and overhead piping for water or heat are also considered dead loads, as they are permanently attached to the structure. Long-term crops, such as tomatoes or cucumbers that are suspended from the trusses, are frequently identified as dead load. Short-term crops, such as hanging baskets, are normally considered a live load. Values for wind uplift must not exceed the dead load of the structure; otherwise, the structure could be lifted out of the ground.

$$\text{Total dead load} = W_{ps} + Wp + Wt \text{ kN/m}^2$$

where
W_{ps} = weight of roof covering
W_p = weight of purlins
W_t = self-weight of truss

17.21.1.1 Weight of Roof Covering

$$W_{ps} = D_s \times t_s \times A$$

where
D_s = density of the covering sheet
t_s = thickness of the covering sheet
A = area of the sheet

17.21.1.2 Weight of Purlin

The weight of purlin (W_p) was calculated by multiplying the total length of purlin to weight per unit length of the section used.

17.21.1.3 Weight of Truss

Self-weight of truss is calculated by following the formula used for a truss of span "L":

Total self weight of truss = sum of weight of individual member of truss

i.e., the weight of an individual member of truss includes wt of big arc + wt of small arc + wt of long bracing at big arc + wt of small bracing at big arc + long bracing at small arc + wt of small bracing at small arc + wt of sag tie + wt of bottom chord.

17.21.1.4 Live Loads

Live loads include the weight of moveable equipment and human beings for repair work etc. Live load calculated by following formulae. Live loads are more difficult to calculate, as they can change. The National Greenhouse Manufacturers Association (NGMA)-designed guidelines recommend that purlins, rafters, and trusses be designed to support a minimum concentrated load of 45 kg at midspan. Short-term plant loads are considered live loads.

Up to 10° slope: 0.75 kN/m²
For more than 10° slope: 0.75−0.02 (θ−10)
where θ is the slope of sheeting.

The minimum live loads for the greenhouse structure should be taken as 250 N/m² as per IS 14462-1997.

17.21.1.5 Wind Loads

Loading from the wind can come from any direction, but is usually considered to act in a horizontal direction against the walls. The basic wind speed is adjusted for factors such as site exposure; height and shape of the building; roof slope; and use factor. Greenhouses and sales buildings that are open to the public have a higher use factor than a production greenhouse. The above factors are applied to the wind velocity to get a wind load. Structural member size and resistance to over-turning are then calculated. During heavy wind conditions, doors and vents should be closed to reduce the double effect of external wind pressure and the force of the wind that gets in through the openings. It can be mathematically expressed as follows:

$$V_z = V_b . k_1 . k_2 . k_3$$

where
  V_z = Design wind speed, m/s
  V_b = Basic wind speed, m/s
  k_1 = Risk coefficient
  k_2 = Terrain, height and structure size factor
  k_3 = Topographic factor

Design wind speed up to 10 m heights from mean ground level was considered constant.
The values of risk coefficient (k_1), terrain, height and structure size factor (k_2), and topographic factor (k_3) were found as per standard procedure given in IS:875-1987 (part 3).

17.21.1.6 Design Wind Pressure

The design wind pressure (kN/m²) at any height above mean ground level was obtained by the following relationship between wind pressure and wind velocity.

Design of Protected Structures

$$P_z = 0.0006 V_z^2$$

where
p_z = Design wind pressure, kN/m²
V_z = design wind speed, m/s

17.21.1.7 Wind Load on Individual Members

$$F = (C_{pe} - C_{pi}).A.P_z$$

where F = Wind load, kN
A = Surface area of structural element or cladding unit, m²
C_{pe} = External pressure coefficient
C_{pi} = Internal pressure coefficient
P_z = Design wind pressure, kN/m²

17.21.1.8 Snow Loads

Design snow loads are usually adjusted to consider building exposure, roof slope, heat loss through the roof, and the type of occupancy. As the heat loss through the glazing on a greenhouse is high, the design usually considers that most of the snow will melt or slide off.

Snow can be light and fluffy with a water equivalent of 12 inches equal to 1 inch of rain. It can also be wet and heavy with 3 to 4 inches, equal to the weight of 1 inch of rain. Snow having a 1-inch rainwater equivalent will load a structure with 25 kg/m².

Drifting snow causes unbalanced loads that may collapse greenhouses. Two to four times normal loads may occur where drifts or sliding snow build up. This is especially true in ranges of several hoop houses where they are placed side by side. A space of at least 10 feet should be left between freestanding greenhouses to provide space for the snow that slides off. If there is no adequate space, sidewalls may be crushed in. The external pressure coefficient (C_{pe}) is calculated as shown in Table 17.4.

17.21.1.9 Method of Joint

The method of joints is a way to find unknown forces in a truss structure. The principle behind this method is that all forces acting on a joint must add to zero.

i) If a truss is in equilibrium, then each of its joints must be in equilibrium
ii) The method of joints consists of satisfying the equilibrium equations for forces acting on each joint

$$\sum F_x = 0 \qquad \sum F_y = 0$$

TABLE 17.4
External Pressure Coefficient (C_{pe})

Direction of wind	Windward	Leeward
Wind angle, Θ	0°	0°
C_{pe}	−0.2	−0.5
Wind parallel to the ridge		
Direction of wind	Windward	Leeward
Wind angle, Θ	90°	90°
C_{pe}	−0.8	−0.8

17.21.1.10 Procedure for Analysis
The following is a procedure for analyzing a truss using the method of joints:

i) If possible, determine the support reactions
ii) Draw the free body diagram for each joint. In general, assume all the force member reactions are tension (this is not a rule; however, it is helpful in keeping track of tension and compression members)
iii) Write the equations of equilibrium for each joint
iv) If possible, begin solving the equilibrium equations at a joint where only two unknown reactions exist. Work your way from joint to joint, selecting the new joint using the criterion of two unknown reactions
v) Solve the joint equations of equilibrium simultaneously

17.21.1.11 Load Combinations
As per IS: 875 (part 5), the following load combinations are used during the study.

1) Dead load + Live load
2) Dead load + Wind load
3) Dead load + Live load + Wind load
   - Permissible stresses were increased by 33% when wind load was considered
   - According to IS: 875, out of live load and snow load only one should be considered at a time As snow load is negligible in the Udaipur region only live load is considered

17.21.1.12 Design Strength of Tension Member
To check the strength of truss members, which are under tension and are able to resist the actual load acting on them, the following procedure should be adopted which is given below:
The design strength of a tension member is the lowest of the following:

1. Design strength due to yielding of gross section, Tdg
2. Rupture strength of the critical section, Tdn
3. The block shear, Tdh

Design strength due to yielding of gross section, Tdg
This strength is given by

$$\text{Tdg} = \frac{Ag \times fy}{y_{mo}}$$

Here, fy = Yield of the material
Ag = Gross area of the cross section
Y_{mo} = Partial safety factor for failure in tension by yielding = 1.1

17.21.1.13 Design Strength due to Rupture of Critical Section
This strength for plates is

$$\text{Tdn} = \frac{0.9 \times An \times fu}{y_{ml}}$$

Here, An = Net effective area at critical section
fu = Yield of material
Y_{ml} = Partial safety factor

Design of Protected Structures

17.21.1.14 Design Strength due to Block Shear

IS 800-2007 recommends the following block shear strength Tdb if bolted connections are used.

$$\text{Tdb} = \frac{\text{Avg} \times fy}{\sqrt{3} y_{mo}} + \frac{0.9 \times \text{Atn} \times fu}{y_{ml}} \quad \text{or} \quad \text{Tdb} = \frac{0.9 \times \text{Avn} \times fu}{\sqrt{3} y_{ml}} + \frac{\text{Atg} \times fy}{y_{mo}}$$

Here, Avg and Avn = Minimum gross and net area in shear
Atg and Atn = Minimum gross and net area in tension

17.21.1.15 Design Strength of Compression Member

To check the strength of truss member which are under compression and are able to resist the actual load acting on it, the following procedure has been adopted:

i) First, we have to decide on the buckling class as per IS 800-2007. It divides various cross sections into four buckling classes a, b, c, and d (refer table 10 in IS 800).
ii) Then, find out the slenderness ratio of the member
iii) The slenderness ratio of a column is defined as the ratio of effective length to the corresponding radius of gyration of the section

17.21.1.16 Slenderness Ratio

$$= \frac{le}{r} = \frac{kL}{r}$$

Here, L = actual length of compression member
$le = kL$, effective length
r = appropriate radius of gyration

Then, finding design compressive stress for defined buckling class and calculated slenderness ratio. For finding design compressive stress refer table 9 (a), 9(b), 9(c) and 9 (d) in IS 800-2007.

17.21.1.17 Greenhouse Microclimate

The naturally ventilated greenhouse maintains an atmosphere conditioned considering the temperature and humidity

1. Encourage crop earliness
2. Improve the yield
3. Safeguard the crop
4. Improve water use efficiency

17.21.1.18 Functional Design of Greenhouse

The following points have to be considered for the functional design of any poly house.

- Selection of site
- Orientation
- Maintenance of favorable environment
    i) Light
    ii) Temperature
    iii) Relative humidity
    iv) Air composition
    a) Bottom apron
    b) Gutter height

c) Ridge height
  d) Ventilation area
- **Top Shading**

17.21.1.19 Greenhouse Orientation

Following are the major factors considered for greenhouse orientation.

- Solar radiation and winds
- Single span: east–west orientation
- Multi-span: north–south orientation (gutter)
- Vent opening toward the east side
- The slope along the gutter – 0% to 2%
- The slope along the gable – 1.25%
- Maximum width of greenhouse – 40 m

The relationship between weight, thickness, and area of agri-films is shown in Table 17.5.

SHORT-ANSWER TYPE

1. Define protected cultivation.
2. Differentiate between protected agriculture and greenhouse.
3. What is the greenhouse effect? Explain the significance of greenhouse farming.
4. Define the scope of protected cultivation.
5. What is plastic mulching?
6. Explain the vermi bed.
7. List the application methods of protected cultivation.
8. List the types of structures under protected cultivation. Explain any one of them.
9. Differentiate naturally ventilated and forced ventilated greenhouses.
10. Differentiate conventional agriculture and protected agriculture.
11. Which is the most used irrigation method under protected cultivation?
12. What are the advantages of a micro-irrigation system?
13. What are the disadvantages of greenhouse technology?
14. Enlist the types of containers used in the greenhouse.
15. What are the media used for the plantation of crops?
16. What are the objectives of protected cultivation?
17. List some of the crops used for protected cultivation.

TABLE 17.5
Relationship between Weight, Thickness, and Area of Agri-films

Thickness			Area (m²/kg of film)	Weight (g/m²)
Microns	mm	Gauge		
25	0.025	100	42	23
50	0.05	200	21	46
100	0.100	400	11	93
150	0.150	600	7.16	139
200	0.200	800	5.37	203
250	0.25	1,000	4.29	230

Design of Protected Structures

18. List the effects of cover materials on crop production.
19. What are the problems related to cultivation in agriculture zones?
20. Why is protected cultivation required for crop production?

LONG-ANSWER TYPE

1. What is nutrient film technique in hydroponics? Explain with a diagram.
2. What is protected cultivation? Write in detail the protected cultivation of rose.
3. List different types of greenhouses.
4. Write short notes on Indo Israelis center of excellence of vegetation protected cultivation in khetri.
5. What are the methods used to regulate the climate in the greenhouse?
6. How protected cultivation will help in the growth of vegetable industry?
7. Explain the mechanism of greenhouse ventilation.
8. Differentiate between the covering materials used in the greenhouse.
9. Write short notes on:
    a. Drip irrigation system
    b. Overhead sprinklers
    c. Perimeter watering
10. What is the type of heating system used in the greenhouse under protected conditions of crops?
11. Describe the design criteria for the construction of the greenhouse.
12. Describe insect pest and management of protected crops.
13. What are the disadvantages of protected cultivation in India?
14. What is the status of protected cultivation in India?
15. Enlist the site selection procedure for the greenhouse?
16. What is the effect of greenhouse cultivation on quality parameters?
17. What are the components of a greenhouse?
18. Classify greenhouse on the basis of cost.
19. What are the advantages and disadvantages of hydroponics?
20. Hydroponics – is it as good as growing plants in the soil? What are the plants that can be cultivated under hydroponics?

MULTIPLE-CHOICE QUESTIONS

1. Which of these is not a method adopted for protected agriculture?
    a) Vermi-bed
    b) Greenhouses
    c) Mulching
    d) Tillage
2. Walk-in tunnel structure is covered from which film
    a) UV film
    b) Polyethylene
    c) Net protected
    d) Polycarbonate
3. The greenhouse cover is called the
    a) Bronzing
    b) Glazing
    c) Partitioning
    d) Pumping

4. In India, the greenhouse should be built in such a way that it receives the maximum amount of sunlight throughout the year.
   a) East–west direction
   b) North–east direction
   c) North–south direction
   d) South–east direction
5. Sterilize the soil in the net home using
   a) Ethrel
   b) Formalin (formaldehyde)
   c) Hydrochloric acid
   d) Potassium chloride
6. The relative humidity in a protected structure is calculated using the following:
   a) Psychrometer
   b) pH meter
   c) Lysimeter
   d) Anemometer
7. In protected structures, which of the following growing media is/are employed for cultivation?
   a) Vermiculite
   b) Perlite
   c) Coco peat
   d) All of these
8. Greenhouses are classified according to their structure as
   a) Quonset type
   b) None of these
   c) Gable roof type
   d) Curved roof type
   e) All of these
9. In the construction of shade houses, what colors of the net are used?
   a) White
   b) Black
   c) Green
   d) All of these
   e) None of these
10. Hydroponics refers to
    a) Soil + sand cultivation
    b) Soil + perlite cultivation
    c) Soilless cultivation
    d) None of these
11. What is the most common material used in the construction of a greenhouse?
    a) Stainless steel pipes
    b) Galvanized iron (GI) pipes
    c) Mild steel pipes
    d) None of these
12. The distance between two nearby greenhouses that are naturally ventilated should be –
    a) 300 m
    b) 100 m
    c) 10 m to 15 m
    d) 75 m

Design of Protected Structures 355

13. The greenhouse's primary vertical support structure is known as
    a) Truss
    b) Frame
    c) Binder
    d) None of these
14. What is the best vegetable crop to grow in a greenhouse?
    a) Tomato
    b) Cucumber
    c) Capsicum
    d) All of these
15. Which colored sticky traps are used in greenhouses to control whiteflies?
    a) White
    b) Yellow
    c) Blue
    d) None of these
16. In capsicum, which pollinator agent is the best?
    a) Honey bees
    b) Bumblebees
    c) Sweat bees
    d) Tree bees
17. Plastic mulching conserves
    a) Soil moisture
    b) Runoff
    c) Soil erosion
    d) None of the above
18. The major cost in the greenhouse bedding plant industry comes from
    a) Variable or direct costs
    b) Overhead or fixed costs
    c) Labor costs
    d) Marketing costs
    e) All of these are the same
19. At what percentage should the humidity be kept for new cuttings?
    a) Close to 100
    b) Close to 50
    c) Close to 25
    d) Close to 10
20. The three types of pots used in greenhouse are
    a) Clay, metal, standard
    b) Bulb, geranium, gallon
    c) 2-gallon, azalea, geranium
    d) Standard, azalea, bulb

ANSWERS

1	2	3	4	5	6	7	8	9	10	11
d	a	b	a	b	a	d	e	d	c	b

12	13	14	15	16	17	18	19	20
c	a	d	d	b	a	a	a	d

REFERENCES

Aldrich, R. A. and John, W. B. Jr. 1989. *Greenhouse Engineering, NRAES-33*. Ithaca, NY: Cornell.

Anonymous. 1999. *Drip Irrigation Studies on Cabbage*. Annual report of PDC, Agricultural & Food Engineering Department, IIT Kharagpur, Kharagpur, 14–16.

Bartok, J. W. and Susan, M. 1982. *Solar Greenhouses for the Home*. Cornell University, Northeast Regional Agricultural Engineering Service, Ithaca, NY.

Boodley, J. W. 1996. *The Commercial Green House*. Delwar Publishers, Albany, NY, 612pp.

Chandra, P. and Singh, J. K. 1988. *Instruction Manual for Establishment of a 4 x 24 m Green House*. Division of Agricultural Engineering, Indian Agricultural Research Institute, New Delhi.

Dalrymple, D. G. 1973. *A Global Review of Green House Food Production*. USDA, report no 89.

Dauthy, M. E. 1995. General procedures for fruit and vegetable processing. In: Fruit and vegetable processing. FAO Agricultural Services Bulletin 119, International Book Distribution Co, Lucknow, pp 40–57.

Fisher, R. and Bill, Y. 1976. *The Food and Heat Producing Solar Greenhouse: Design, Construction, Operation*. John Muir Publications, Santa Fe, NM.

Freeman, M. 1997. *Building Your Own Greenhouse*. Stackpole, Mechanicsburg, PA.

Jensen, M. H. and Malter, A. J. 1995. *Protected Agriculture: A Global Review*. World Bank Technical Paper no. 253. The World Bank, Washington, DC.

Nelson, P. V. 2012. *Greenhouse Operation and Management*. Seventh edition. Prentice Hall, Boston, MA, 607p.

Prasad, S. and Kumar, U. 2007. *Greenhouse Management for Horticultural Crops*. Agrobios (India), Jodhpur, 476pp.

Raina, J. N., Thakur, B. C. and Verma, M. L. 1999. Effect of drip irrigation and polyethylene mulch on yield, quality and water use efficiency of tomato. *Indian Journal of Agricultural Science*, 69(6), 430–433.

Shahi, N. C., Khan, J. N. and Dixit, J. 2005. *Annual Project Report on Application of Plastics in Agriculture*. SKUAST, Kashmir, Srinagar.

Stauffer, V. 2004. *Solar Greenhouses for the Trans-Himalayas. A Construction Manual*. International Centre for Integrated Mountain Development, Katmandu, 73pp.

18 Application of Plastic in Farm Machinery

18.1 INTRODUCTION

Agriculture is one of the main components of the economy of India, with about 70% of the population earning livelihood through agriculture. The overall mechanization level is only 40–45% in India in which 90% of the total power is contributed by mechanical and electrical power sources. India is one of the leading countries of the world in the manufacture of tractors and agricultural implements and equipment. Indian agriculture is capable of producing most of the food and horticultural crops of the world in a diverse manner. There is a direct correlation between crop productivity and farm mechanization. It saves inputs like fertilizers and seeds by 15–20%, operational time by 20–30%, labor requirement and crop productivity by 10–15%, and increases cropping intensity by 5–20% (Tiwari et al., 2019). Indian farmers at present are adopting farm mechanization at a faster rate in comparison with the recent past. Farm power availability from tractors has grown from 0.007 kW/ha in 1960–1961 to 1.03 kW/ha in 2013–2014 and it is further estimated to reach 3.74 kW/ha by 2032–2033. It is estimated that the percentage of farmworkers of the total workforce would reduce to 49.9% in 2033 and 25.7% in 2050 from 54.6% in 2011. The share of agricultural workers in total power availability in 1960–1961 was about 16.3%, which is going to reduce to 2.3% in 2032–2033. The overall level of farm mechanization in the country is only 40–45%, and 90% of the total farm power is contributed by mechanical and electrical power sources. To assure timeliness and quality in various field operations, the average farm power availability needs to be increased to a minimum of 2.5 kW/ha by 2020. Farm mechanization has been well-received throughout the world as one of the most important elements of modernizing and sustaining agriculture. The level and appropriate selection of agricultural machinery have a direct impact on land and labor productivity, farm output and income, environmental safety, and the quality of life of farmers in India. Agricultural machines also ensure the timeliness of farm operations and increase work output per unit time. Suitability to small and medium farms needs simple design and technology, versatility for use in several farm operations, affordability in terms of cost and profitability and most importantly, repair and maintenance services are the basic requirements for the expansion of farm mechanization in India. The status of farm mechanization in India is shown in Figure 18.1.

The use of plastics in farm machinery and crop protection has greatly helped the farmers to increase production, improve quality of product, minimize water consumption, and reduce their ecological footprint. It is used to cover crops and silage wrap in irrigation and to transport fertilizer and feed. They suppress the growth of weeds, increase fertilizer uptake, regulate temperature and humidity, and protect plants and soil from bad weather. Agricultural plastic used at the field level are plastic sheets, tubing, and films used in the cultivation; rods, harvesting, and processing of agricultural products; as well as plastics used in equipment and machinery for chutes, bins, liners, and hoppers. Some pieces of equipment used for farm mechanization are very heavy that's why we move towards plastic machinery to enhance working efficiency. Work performance of plastics is better despite their lightweight. In the equipment used for harvesting crops, materials like HDPE, UHMW, ABS, nylon, and polycarbonate add extraordinary value when used to complement or replace metals with plastic. Different ranges of strong plastic are available, which are tough and durable and used in agriculture, snow and ice removal, building construction, road construction and maintenance, towing, drilling, and much more. The agriculture industry has developed a significant amount in the past, and composite plastic parts are being used more and more for agricultural

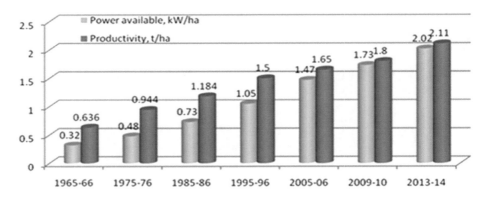

FIGURE 18.1 Status of farm mechanization in India

products. Plastic components are more affordable and lightweight, allowing easy installation and handling. Plastics are often insulated and are good for irrigation equipment, tubing, and more. Proper designing equipment of plastic, its cost, and maintenance and repair are very important for manufacturers. Plastics used in farm mechanization are in

- Prototype development
- Part fabrication and machining
- Component and application design challenges
- Supply, material selection, and inventory management
- Stock shapes cut-to-size based on requirements

With their durability, lightweight, and resistance to impact, wear, heat, chemicals, moisture, and corrosion, plastics improve the performance, efficiency, and longevity of all heavy equipment types. The performances of machinery that can be achieved using plastic are:

- Extend equipment lifetime and reduce downtime
- Minimize fuel consumption
- Reduce noise and improve safety
- Reduce the need for part lubrication
- Optimize equipment performance in extreme conditions and harsh weather
- Wear to protect the more expensive metal of heavy equipment parts

The heavy equipment industry uses plastic materials for different applications which are given as:

- Bumpers
- Bearings and bushings
- Cab interiors, doors, and consoles
- Chute liners
- Crane pads, cribbing pads, and outrigger pads
- Gears
- Hammer cushion pads
- Instrumentation covers
- Pulleys
- Roller components
- Sheaves and pulleys
- Slide pads

Application of Plastic in Farm Machinery

- Snowplow blades
- Telescopic boom components
- Truck and trailer bed linings
- Wear strips and pads
- Windows and protective glazing

Using plastic in farm machinery can save time and money. The most recent fabrication in the plastic and composite sector has resulted in the use of plastics in processing equipment and agricultural products. For instance, plastic films and sheets are now used to cover crops when they need protection from bugs, UV light, or weeds. Sheets can also help conserve water by holding in moisture. There are many applications for custom plastics in agriculture because plastics are lightweight, parts are easy to install, and plastics are easy to handle. They are also simple to be taken care of, which reduces time and money spent on maintenance. There are even insulating properties that make plastic products perfect for irrigation and drainage. In addition, many plastics are degradable, which is better for the environment because of waste reduction.

18.2 MATERIALS USED

The versatility of plastics enables them to be fabricated for use in agricultural equipment and many applications. Many of the applications have a protective element to them, such as the plastic sheets that can be used to protect crops from harmful elements. As for machinery, plastic can be used to fabricate parts used in liners, bins, hoppers, chutes, guards, augers, grills, tanks, and rollers. The materials used in these applications include:

- Polycarbonate
- HDPE
- PVC
- Chlorinated polyvinyl chloride (CPVC)
- PM
- Ultra high molecular weight polyethylene (UHMW)

Energy chains and plastic plain bearings have eight times longer service life for agricultural machinery by the use of metallic plain bearings. Energy chains and plastic bearings have outstanding resilience and long service life characteristics for agricultural applications. These products are used in tractors, disc harrows, fertilizers, hay rakes, seed drills, field sprayers, and other agricultural equipment. Metallic bushings and ball bearing guides can be replaced by lubrication-free polymer plain bearings in many applications of farm operation.

18.2.1 Advantages

The following are advantages of plastic material accessories in heavy and small machinery used in an agricultural operation.

- Dust and dirt resistant
- Extremely durable
- Quiet running
- Compensates misalignment and deflections
- Easy installation
- Longer life

For agriculture equipment different self-lubricating and maintenance-free plastic plain bearings and energy chains are cost-effective, dust and dirt resistant, durable, lubrication and maintenance-free, extremely long service life under extreme conditions (Figure 18.2).

This disc harrow uses a plastic bearing which increases its efficiency and is easily adapted to any width. It can be used in the sliding bearing of the disc harrow and is maintenance free, durable, precise, and very load-bearing, thus guaranteeing a long service life. Plowing by cultivators, which were primarily used earlier for weed control, is mainly used to loosen up the arable land to mix organic matter and prepare the land for the next season. Therefore lubrication-free plastic plain bearings can be used at bearing points (Figure 18.3).

Presently the plastic plain bearings have to be replaced only every few years depending on the load, which accounts for, in addition to the lubricant saved, an additional cost advantage.

FIGURE 18.2 Disc harrow using plastic bearing

FIGURE 18.3 Cultivators used in lubrication-free plastic plain bearings at bearing points

18.3 PLASTIC BEARINGS WITHSTAND HIGH FORCES

In agriculture, high forces are required particularly with the use of heavy equipment, which needs to be very rugged. In earlier times, metallic bearings were the first option. Such types of bearings have a very limited service life, especially with pivoting movements, because lubricants may not be evenly distributed and the same points on the bearing are stressed again and again. Therefore, plastic plain bearings display their advantages, as pivoting movements do not harm them and there is no lubricant to displace (Figure 18.4).

By the use of lubrication-free plastic plain bearings in the overload element, subsequent lubrication can be completely dispensed. Polymer plain bearings are also used in agricultural machinery.

18.4 PLASTIC TYPES

The type of plastic used in agricultural applications varies due to flexibility, strength, and heat resistance. Different types of plastic components are given below.

- Acrylonitrile butadiene styrene (ABS)
- Acrylic (PMMA)
- Nylon (PA)
- High-density polyethylene (HDPE)
- Polycarbonate (PC)
- Polyetherimide (PEI)
- Polyoxymethylene (POM)
- Polypropylene (PP)
- Polystyrene (PS)

18.5 APPLICATION

Plastic parts in any machinery to be made with the utmost precision and care and its application is given in below headings.

18.5.1 Agriculture Injection Molding

Agricultural companies and farms are switching to plastic injection molding as a low-cost alternative to using metal. The high-quality plastic injection molded plays an important role at competitive

FIGURE 18.4 Plastic bearing used in heavy agricultural machinery

prices, without product quality compromise. Injection molding technology advances, so keep pace with the increasing demands of agriculture. In the agriculture sector, sustainability is required. Following are different pieces of machinery used in the farm operation.

a) Chute components
   The plastic chute components for handling crops are shown in Figure. 18.5.
b) Gear components
   The plastic gear components used in different farm machinery are illustrated in Figure 18.6.
c) Bins
   The plastic bins used for agriproduct transportation and handling are given in Figure 18.7.

FIGURE 18.5 Plastic chute components for handling crop

FIGURE 18.6 Plastic gear components used in different farm machinery

Application of Plastic in Farm Machinery

FIGURE 18.7 Plastic bins used for agriproduct transportation

FIGURE 18.8 Plastic auger for digging a hole

d) Augers

The plastic auger for digging holes for transplanting crops, which is given in Figure 18.8.

Polymers and plastics play a significant role in farm machinery and their benefits to agriculture are described below:

- Have high levels of toughness and impact resistance
- Light in weight, easy to clean, and resistant to moisture and dust
- Plastics can withstand adverse temperatures without degrading and are resilient
- Plastics have a hygienic finish and can be safely included in applications and equipment used for watering, feeding, milking, or cleaning animals. It has also valuable reassurance regarding hygiene and health around the harvesting, storing, and handling of crops.

The molding injection process is compatible with both large-scale assemblies and producing tiny components that form crucial parts of the cogs and mechanisms inside agricultural machinery. Identical plastic components can be manufactured in large quantities at one time with accuracy. Accuracy helps farmers enormously during their busier times of harvesting and sowing seasons: Plastic molded (injection) products and their components can be found in a wide range of farming applications.

- Crates, bulk containers, and pallets
- Baskets, bins, and tubs
- Troughs for animals
- Pesticide sprayer
- Spare parts of agricultural machinery
- Irrigation components
- Cleaning brushes and equipment
- Animal-feed sacks and packaging
- Water tanks
- Exterior equipment covers

These different types of machinery are largely attributed to the developing economy of the country where there is a need for modernizing equipment for high productivity and efficiency. India is the largest manufacturer of farm equipment like tillers, tractors, and harvesters. In India, tractor sales are expected to grow by 8.0% from 2018 to 2022. Therefore, modern plastics and polymers offer significant benefits to the agricultural sector. The animal feeder trough is shown in Figure 18.9.

FIGURE 18.9 Animal feeder trough

18.6 OTHER PLASTIC APPLICATIONS IN FARM MACHINERY

Plastics are increasingly used in manufacturing farm machinery equipment and tools. Plastics are applied in different ways including crates for crop collecting boxes (Figure 18.10), transport, and handling; irrigation systems components like fittings and spray cones; tapes that help hold the aerial parts of the plants in the greenhouses, or even nets to shade the interior of the greenhouses or reduce the effects of hail. Some of the examples where plastics are used are given below.

18.6.1 Plastics in Mulching Machines

Plastics are used as mulching material and are very much effective for mulching (Figure 18.11).

FIGURE 18.10 Product store using a plastic crate

FIGURE 18.11 Mulching laying machine

18.6.2 Sprayer Tanks

Sprayer tanks are the most important component to increase agricultural productivity and they should be noncorrosive and should not react with the active ingredients of spraying chemicals. The liquid fertilizer applicator with plastic tank tractor mounted is shown in Figure 18.12. These conditions are fulfilled by plastic tanks. A sprayer is also used in the manufacturing of knapsack sprayers and sprayer nozzles (Figure 18.13).

FIGURE 18.12 Liquid fertilizer applicator with a plastic tank (www.koenigequipment.com)

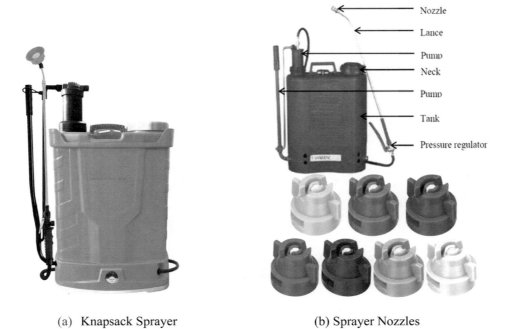

(a) Knapsack Sprayer   (b) Sprayer Nozzles

FIGURE 18.13 (a) Knapsack sprayer (krushikendra.com) and (b) Sprayer nozzles (www.sprayersupplies.com)

Application of Plastic in Farm Machinery

18.6.3 Knapsack Sprayer

A spraying apparatus consisting of a knapsack tank together with pressurizing device, sprayer nozzle, and line are used in spraying fungicides or insecticides.

18.6.4 Poultry Feeders

Poultry feeder trays are also made of plastics as they are durable and cost-effective (Figure 18.14).

18.6.5 Fertilizers Spreader

Plastic also finds its application in fertilizer spreaders as the requirement of fertilizer spreaders is that the material should be noncorrosive and nonreactive (Figure 18.15).

FIGURE 18.14 Poultry feeders (www.huiduobao.en.alibaba.com)

FIGURE 18.15 Agri plastic spreader (Hoare Machinery)

18.6.6 HAND TOOLS

Plastics are increasingly used in making handles for hand tools used in the orchard for farm operation are shown in Figures 18.16–18.18 and are listed below.

1. Garden tool
2. Sickle
3. Direct-seeded rice drum and planters

Direct rice seeders are also used in many parts of the country and these are usually made of plastics. The planter used for the precise sowing of seed is shown in Figure 18.19.

FIGURE 18.16 Plastic handle gardening tools (reysonsindustrial.co.za)

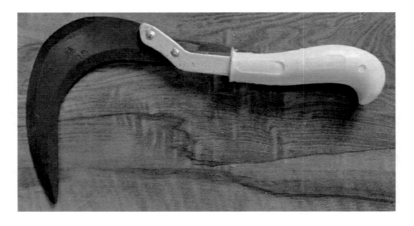

FIGURE 18.17 Sickle with plastic handle (Kissan agro equipments)

FIGURE 18.18 Paddy drum seeder (KSNM Coimbatore, India)

FIGURE 18.19 Planter (www.royermak.com)

18.7 BICYCLE SPRAYER

A bicycle sprayer is an attachment, which essentially uses the circular motions of the sprockets to move the piston of the spray pump which generates pressure and eventually spray pesticides (Figure 18.20). The front and rear sprockets of a thrown away bicycle are interchanged and the small sprocket is connected to the piston pump, so as the circular motion is converted to reciprocating motion of the piston and thus the pressure is generated and thus the pesticides are sprayed on the field. Instead of manual strokes in the case of knapsack sprayers, one can simply drag the bicycle to operate the pump. It is energy and ergonomically efficient, so it increases labor productivity. It has

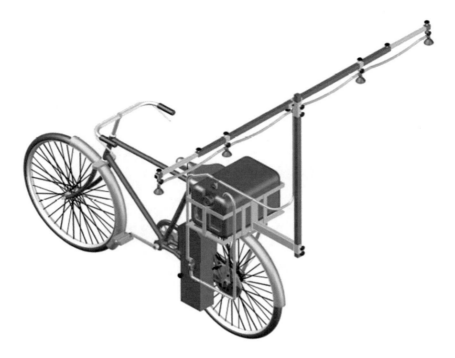

FIGURE 18.20 Bicycle sprayer generating pressure and eventually spraying pesticides

adjustable width and height of the spraying boom. The sprayer can be disengaged and the cycle can be used for regular transportation by interchanging the sprockets.

18.8 SAFFRON CORM GRADER

Saffron is the most valuable crop species in the world. It is used for medicinal purposes, as a spice, and as a condiment for food. Using plastic roller, saffron corms can be reproduced and its production rate depends on corm size thus making corm grading important for saffron production (Figure 18.21).

18.9 SOLAR-OPERATED KNAPSACK SPRAYERS

The use of pesticides is an important part of modern agriculture and contributes to the quality and productivity of the crop grown. For this purpose of pesticide application, a solar-operated knapsack sprayer is available to avoid problems like electricity shortage, fatigue due to continuous operating of a manual knapsack sprayer, and other difficulties in the engine-operated sprayer. This device is developed to reduce the constant application of energy to throttle to regulate air pressure, a problem in conventional knapsack sprayers. The evaluation results showed that the sprayer has a flow rate of 2 l/min to 3 l/min by using different nozzles, the application rate of 850 l/ha to 1,280 l/ha. The sprayer is capable of spraying 850 l/ha to 1,280 l/ha in 7.15 hrs at a walking speed of 0.70 m/s. The 10-W solar panel can produce 0.833 A.

18.10 IMPROVED SICKLE

It consists of a ferrule, serrated blade, and plastic handle. Cutting of crop stalk is being done with the improved (serrated) sickle by sawing action as against by impact or pulling action in case of local (plain) sickle (Figure 18.22). Due to its less weight, i.e., about 180 g, the fatigue coming on the

FIGURE 18.21   Saffron corm grader

FIGURE 18.22   Improved sickle

wrist is less and the drudgery involved in harvesting is reduced as compared to local sickles which are heavier, i.e., weighing about 350 g.

18.11   ZENOAH RECIPROCATOR

With conventional trimmers, be prepared for at least two critical drawbacks. Debris is scattered about by the rotary cutter, and machine kickback when striking stones, trees, or other fixed objects (Figure 18.23). It is used for yard and garden, and the reciprocator is also unsurpassed for trimming around fences, walls, trees, flower beds, golf courses, rivers, or ponds. The engine displacement is 22.5 cc with a power of 0.6 kW (0.8 hp).

18.12 TREE PLANTING AUGER

Tree planting augers make pole and fence installation jobs and tree planting jobs fast and easy. The tree planting auger with different plastic accessories is shown in Figure 18.24.

18.13 BUDDING CUM GRAFTING KNIFE

A budding cum grafting knife is used for the budding. It is a combination of two knives used for carrying out budding and grafting operations with a plastic handle. It consists of two blades: one is used for budding and another for grafting. The length of the knife is around 6.5–7.5 cm and width 1.5 cm (Figure 18.25).

18.14 PRUNING KNIFE

A pruning knife is also known as a slashing knife. It is used for removing the unwanted and dense branches or twigs of plants. The knife tips are slightly hooked or curved for easy cutting of the branches. The knife is made of carbon or alloy steel fitted with a plastic handle (Figure 18.26).

18.15 GRAFTING TOOLS (OMEGA CUT)

There are many machines for making omega grafts, which are hand-operated. It is a combination grafting pliers with both V slot, omega-shaped, and chip cut blade. This tool can perform three

FIGURE 18.23 Zenoah reciprocator (Kisan24.com. Falcon Garden Tools Pvt. Ltd. Alamgir, Malerkotla Road, Ludhiana, Punjab)

FIGURE 18.24 Tree planting digger (Pavan Machine Tools and Services India Private Limited, Peenya, Bengaluru)

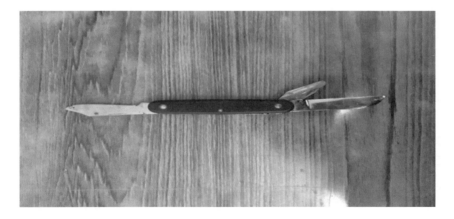

FIGURE 18.25 Budding cum grafting knife (*Source*: agrosiaa.com)

types of cuts, i.e., slot graft (V cut), omega cut, and T cut (chip budding) as in Figure 18.27. This tool cuts through both the rootstock and the scion, one laid on top of the other, making an omega-shaped cut and leaving the two parts interlocked. While these machines work fine for grape grafting, most ornamental nurseries that graft do not use machines. It can cut up to 14 mm diameter rootstock.

18.16 PRUNING SAW

It is a manually operated tool used for the cutting of the thicker branches of the plants. These saws are useful for pruning branches more than a half-inch in diameter. The saw teeth are coarse and

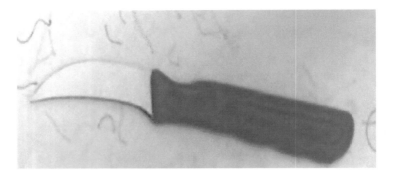

FIGURE 18.26 Pruning knife (*Source*: knifeswork.com)

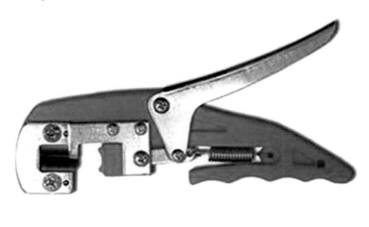

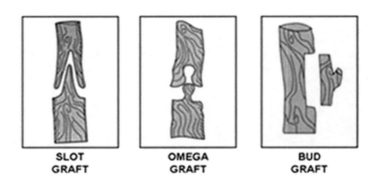

FIGURE 18.27 Omega cut grafting tool with different grafting patterns (*Source*: printerest.com)

wide set, usually cutting on the pull stroke to allow the sawdust to drop out of the cut. It prevents the saw from binding in a green wood cut (Figure 18.28). Pruning saws have sturdier blades and cut faster and straighter through a limb than a carpenter saw. For cutting the blades, it is repeatedly moved over the branch and cutting is done in the forward stroke.

18.17 TREE CLIMBER

It has two frames: upper (40″ × 20″ × 2.5″) and lower frames (30″ × 20″ × 1.5″). The upper frame is operated by hand and another by legs. The upper frame has a seating arrangement for the operator.

Application of Plastic in Farm Machinery

The operator sits on the seat and with up and down movement of the upper and lower frames can climb on the tree (Figure 18.29). It has a four-pin locking system to avoid the possibility of a fall. The locking pins can be fixed at any height to complete the work. The tree circumference preferred is 15″–51″. A person can climb up to 40 ft in 5 minutes.

18.18 TROWELS (PLANTING/DIGGING)

It is a small handheld tool that is used to dig small holes and transplant vegetable seedlings (Figure 18.30).

18.19 RAKE (LEVELING)

A long-handled plastic handle tool is used to create a fine tilt and level the seedbeds. It collects plant debris and stones from the seedbed surface, breaks soil clumps, and spreads fertilizers or compost (Figure 18.31).

FIGURE 18.28 Different types of pruning saws (*Source*: morningchors.com)

FIGURE 18.29 Tree climber (*Source*: nif.org.in)

FIGURE 18.30 Troels (Kisan24.com. Falcon Garden Tools Pvt. Ltd. Alamgir, Malerkotla Road, Ludhiana, Punjab)

FIGURE 18.31 Rake leveling (Unison Engg Industries, An Unison Group of Companies, Jalandhar, Punjab)

FIGURE 18.32 Watering can (Action ware India Pvt. Ltd, Gujarat, India)

FIGURE 18.33 Aerator (Unison Engg Industries, An Unison Group of Companies, Jalandhar, Punjab)

FIGURE 18.34 Wheelbarrow (Shree Vishwakarma Steel Fabrication, Ahmedabad, Gujarat)

18.20 WATERING CAN

A portable plastic water container is used for watering smaller areas and containers (Figure 18.32).

18.21 AERATOR

Effective in reducing soil compaction by removing small plugs in soil and used to aerate the soil (Figure 18.33).

18.22 WHEELBARROW

A good barrow is strong yet light enough to easily transport when full with load (Figure 18.34). A selection of galvanized and hard plastic wheelbarrows is available. Also, as water-carrying bags for wheelbarrows, it is the only way to transport water in a barrow without spilling it.

QUESTIONS

1. Describe the application of plastic in farm machinery.
2. Why do we use plastics in farm machinery?
3. How machinery perform using plastic.
4. Enlist different tools used in horticulture crops.
5. Discuss the mechanization status in India and around the globe.
6. Describe the advantages of plastic machinery used in agriculture.
7. Enlist the names of different grafting tools.
8. Why do we use plastic in heavy machinery for different applications?
9. Describe modified sickle used for drudgery reduction.
10. Explain different types of sprayers.

REFERENCES

Tiwari, P. S., Singh, K. K., Sahni, R. K. and Kumar, V. 2019. Farm mechanization – Trends and policy for its promotion in India. *Indian Journal of Agricultural Sciences*, 89(10), 1555–1562.
www.huiduobao.en.alibaba.com
www.koenigequipment.com
www.royermak.com
www.sprayersupplies.com

19 Smart Farming Using Internet of Things

19.1 INTRODUCTION

Water management practices and nutrient management is essential for increasing the yield of the crop. An automatic irrigation system ensures optimum soil moisture at the root zone of crop and throughout crop growth season for the healthiest growth. This involves a complete understanding of irrigation scheduling, plant soil, and weather parameters. The control systems are designed to accomplish automation using these parameters. The farmers find it a difficult job to irrigate the field twice or thrice daily, as it is time-consuming and involves a lot of drudgeries, along with high labor costs.

The Internet of Things (IoT) is rather an interdisciplinary field that has gained prominence as being one of the primary research areas in both the industrial and academic sectors. The Internet of Things (IoT) corresponds to robust machine-to-machine communications, a key element of the digital market's rapid expansion. IoT has thus been defined with different perceptions, such as

> A dynamic global network infrastructure with self-configuring capabilities based on standard and interoperable communication protocols where physical and virtual "things" have identities, physical attributes, and virtual personalities and use intelligent interfaces, and are seamlessly integrated into the information network, often communicate data associate with users and their environments.
>
> **(Srbinovska et al., 2015)**

In recent years, IoT has begun to play a significant part in our regular activities, expanding our perspectives and the capacity to change the world surrounding us (Puliafito et al., 2019). Nearly every field of today's community has IoT uses (Vermesan and Friess, 2013). IoT delivers effective methods for numerous applications, such as smart health-care systems, smart cities, industrial control, defense, commerce, traffic problems, and agricultural development (Chen et al. 2014). Based on the Juniper Research (Juniper Research, 2015), in 2015, over 13.4 billion systems were connected to the web as a component of IoT, and also a growth of 185% to 38.5 billion systems is anticipated by the year 2020. IoT soon would be one of India's fastest technical realms. As per a recent TechSci Research work, "India Internet of Things (IoT) Business Opportunities and Prediction, 2020," India's IoT market is anticipated to expand more than 28% at a compound annual growth rate (CAGR) during 2015–2020.

IoT integrates sensors and actuators with humans, mechanisms, computers, and technology. This integrated convergence of IoT with human interactions, teamwork, and strategic intelligence allows real-time decision-making to be followed (Giusto et al., 2010). IoT will involve millions of networked embedded smart devices also known as smart things; these smart things will collect knowledge about themselves, their surroundings, and connected smart gadgets and interconnect this knowledge to other devices and systems across all network connections as seen in Figure 19.1 (Yan-E, 2011). IoT has become the new buzzword among next-generation technology that will affect the entire enterprise sector with expanded advantages that integrated end unit, system, and service connectivity. IoT allows devices to be sensed or remotely controlled via existing infrastructure and generates incentives to integrate the universe more directly into computerized systems and enhance functionality (Abbasy and Quesada 2017). It is focused on connectivity among smart sensors, RFID (radio-frequency identification), GPS (Global Positioning Systems), mobile communication, remote

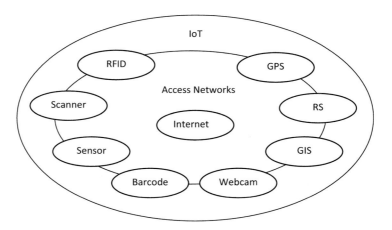

FIGURE 19.1 The IoT concept model

sensing, infrared sensors, and other communications networks (Dlodlo and Kalezhi, 2015). Internet of Things and big data analyses are new developments during the last few years, and applications are also designed in diverse fields using these as core technologies. Big data is a vast volume of data obtained from various outlets, such as sensor data, social networking data, and corporation data, over longer times. Sensor technology is indeed progressing, and several kinds of sensors are designed to be used as required in fields such as environmental sensors and gas sensors. Cloud computing and mobile computing are advanced innovations, and those innovations are found in almost every area of use. Cloud infrastructure includes user experience administration, applications, network node coordination and collaboration, storage, and data processing (TongKe, 2013). The exchange of data at a low cost is enabled by cloud computing. Mobile computing, as a result of its popularity and lower connectivity prices, has changed many aspects of our everyday lives (Channe et al, 2015).

19.2 IOT IN AGRICULTURE

The major backbone of India's economic growth is agriculture. Since the introduction of agriculture, many mechanical and chemical advances have been made, helping farmers to resolve problems such as agriculture and crop diseases. New technologies do have a huge role to play in boosting productivity and reducing the extra workforce. Currently, the Internet of Things (IoT) is turning its attention into an agricultural sector and allowing growers to thrive even in the face of huge challenges. Utilizing IoT, farmers will gain enormous knowledge and information of recent developments in technology. Further, with the IoT revolution, there is an opportunity of building a digital agricultural solution that will enable farmers to make rational decisions on their own farm and help them in advance to handle any unwanted situations which could improve crop quality as well as help farmers. In many respects, IoT can assist farmers. The sensors/actuators in farms and machinery allow farmers to obtain informative data including temperature, fertilizer used, used water, etc. If an IoT framework has been installed, farmers can accurately track and evaluate a range of environmental factors. IoT is a complex technology that gathers, distributes, stores, analyzes, and tests various types of products from numerous firms. In all countries, particularly developing countries, agricultural knowledge has grown rapidly. In Japan, computers are used extensively, for example, in irrigation, seed breeding, crop management, and in the use of forests, insects, fisheries, field activities, agricultural produce processing, etc. Farmers have access to a database of government knowledge centers, universities, research facilities, and libraries in the United States; up-to-date data on market changes, development of plants, new models of farm machinery, the prevention and treatment of plant problems, pesticides, and other pests can be accessed from home. Computers will assist farmers to evaluate the types and methods of planting crops and the farming modes that best help farmers achieve optimum yields and benefits. There are a few significant commercial Financial Management Information Systems (FMIS) suppliers in Finland, namely Wisu and Agrineuvos, both of which have plug-ins or similar applications for different fields of agriculture management (Cheng and Yi, 1999). By 2050, IoT agriculture is expected to increase the yield of foodstuffs by 70%, providing up to 9.6 billion people with food using 2 billion sensors in 525 million farms (Ahmed et al., 2018).

There has been substantial work in agriculture on IoT technologies to promote smart agricultural approaches. IoT innovation has made a big transformation in the agricultural world by exploring various complexities and problems in agriculture (Ray, 2017). Simple IoT technologies such as laser scanners, RFIDs, photoacoustic electromagnetic sensors, etc. can be used to render significant advancements in the field of agricultural knowledge transmitting, precise irrigation, intelligent cultivation monitoring, agricultural product protection, for more (Bo & Wang, 2011) and much more. Software systems are focused on sensors that make it possible to collect more precise information on crops, land, and the environment than those provided through conventional methods. This function aims to significantly boost the consistency of the goods, procedures, and raw materials used in production. As a result of these facts, IoT-based smart farming is much more effective than conventional interventions (Bertino and Choo, 2016). In addition, IoT-based smart farming technologies might also improve organic farming and family farming (Berte, 2018). With the development of technology, IoT-based farmers and technologists are intended to seek a solution to the problems faced by farmers, such as water scarcity, cost management, and robotic production complexities (Kamienski et al., 2019; Elijah et al., 2018). A significant range of innovations has already been used, which have played a critical role in modernizing IoT's agricultural systems, such as cloud and edge computing, big data processing and machine learning, networking, networks and protocols, and robotics (Farooq et al., 2019). Collaborative work of IoT and cloud infrastructure in agriculture offers extensive access to essential resources. A much more efficient collection and storage of information and agricultural functionality has been introduced through cloud computing systems (Botta et al., 2016; Pavón-Pulido et al., 2017). In the IoT edge computing sector, data processing at the center of data generation is regarded as a way to promote sensors, actuators, and many other connected devices. Big data is made up of a large number of essentials produced by agricultural sensors. Big data processing provides diverse and effective approaches for tracking crops at various times (Gill and Chana, 2017). Neural networks are really well known for providing high-speed, optimized solutions. An IoT-based hydroponic system was developed through the use of deep neural networks (Mehra et al., 2018). IoT agricultural network infrastructure and applications include networking protocols as the backbone (Al-Sarawi et al., 2017). It is used for network sharing of all agrarian data or knowledge. Various IoT application networks help to develop sensors and software for crop or field tracking (Navulur and Prasad, 2017). Numerous different agribots have also been designed to cut down on the number of farmworkers by increasing the effectiveness of work with technological advancement to the goal of intelligent farming. For characterization and also ground mapping, a multi-sensor robotics approach has been suggested (Milella et al., 2019). Agribots execute essential tasks such as weeding, sprinkling, etc. All these robots are managed by IoT to boost crop production and resource consumption.

For any IoT program, four main components – IoT devices, communication, the Internet, and data storage and processing are necessary. Under the IoT framework, low-cost computer technologies can enhance human contact with the real environment, and useful information can be generated by the processing resources and tools accessible on the Internet. The IoT equipment consists of embedded systems that communicate with sensors, devices, and wireless technology. The sensors are being used to track and quantify various farm variables such as soil nutrients, weather data, and production factors. Communication technology plays an important role in implementing IoT programs effectively. Connecting IoT computers to the Internet makes it easy to have data accessible wherever and whenever required. The development of wireless networks, handheld devices, and digital broadband has opened the way for vast Internet access. Muangprathub *et al.* (2019) developed and deployed systems for optimally watering crops through crop field sensors and data processing using a web-based smartphone that included three components – hardware, web-based application, and mobile device, as can be seen in Figure 19.2. The amount of linked agricultural devices is projected to rise from 13 million by the end of 2014 to 225 million by 2024. The Internet has led to the creation of cloud computing, which stores massive data for storing and processing. Data-driven agriculture requires the gathering of large, diverse, complex, and spatial data that require processing and storage (Yan-E, 2011). The data can vary from relating to historical data to sensor data, live streaming data, and industry and consumer data. Using cloud IoT systems allows for the collection of large data from sensors into the cloud. A variety of agricultural knowledge management technologies have been established to handle different types of data (Yan-E, 2011). Examples of some of the commercially accessible solutions include On-Farm Network, Farmobile, the silent herdsman network, Cropx, Farmx, Easyfarm, KAA, and Farmlogs.

19.3 IOT SENSORS FOR AGRICULTURE

Smart farming is an emerging concept considering the fact that IoT sensors are designed to provide data on agricultural areas and to respond to user-dependent input. The remarkable development of nanotechnology during the past few years has allowed the production of smaller and affordable sensors. Some major IoT sensors are also available, such as motion detectors, passive infrared (PIR) sensor, soil moisture, temperature, humidity, barometric pressure, ultraviolet, pH, and gas sensor. Various agricultural implementations based on the sensors have been addressed in Table 19.1. The number of different commercial agricultural sensor systems and their descriptions are discussed in Table 19.2 (Ray, 2017).

The wireless sensors network (WSN) is extensively used nowadays to establish decision-making support systems to solve several real-world problems. Precision agriculture is one of the most important areas in which decision support systems are increasingly needed. Kassim *et al.* (2014) designed a low-cost and environmentally friendly intelligent greenhouse monitoring system

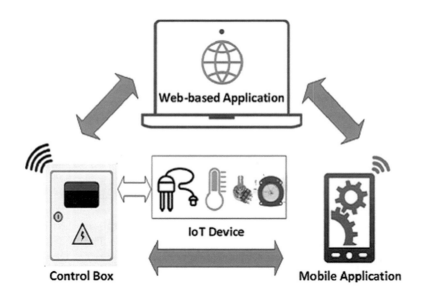

FIGURE 19.2 A summary of the smartphone-connected device

TABLE 19.1
Different Sensors Based on Farm Implementations

Sensors	IoT Roles
Temperature	The temperature of the soil plays a major role in agricultural production. Changes in the soil temperature specifically influence soil moisture and the uptake of soil nutrients. A new sensing system (Alahi et al., 2017), designed to quantify nutrient balance in groundwater and surface water, has been developed. Electrochemical impedance is used for tracking and measuring soil nitrate levels. The test samples were measured with the LCR (**Inductance (l)**, Capacitance (C), and Resistance (R)) meter. In this study, standard library tests were used to determine the concentration of nutrients in water
Humidity sensor	This system is used for air humidity monitoring and measurement. The average ambient temperature and humidity level of air are measured. The primary and indirect growth of plants is affected by humidity. Leaf formation, for example, depends not only on photosynthesis but also on physical cell growth processes, which can be controlled by the moisture sensor (Balaji et al., 2019)
pH sensor	A pH sensor is used to track the precise amount of soil nutrients required for irrigation. The quantity of nutrients required by controlling the value of pH for plants or crops is supplied for safe production (Garcia-Sanchez et al., 2011)
Soil moisture	It calculates the amount of moisture and water in the soil. Soil moisture sensor operates on the conductivity element. The soil's humidity sensor resistance is inversely proportional to the humidity content, which defines plant growth. This sensor is used in the entire field to control the volume of water and other required automation (Brewster et al., 2017). A wireless moisture-sensing system was used by Sicari et al., (2015) to boost the greenhouse irrigation system
Barometric pressure sensor	Pressure sensors monitor air pressure while the rainfall is low; so if the pressure is high, it is estimated that the odds of precipitation are decreased. Often used for regulating water flow is a barometric pressure sensor. If the pressure is less than the threshold value, water flow is regulated by stopping the water source. To calculate the average pressure magnitude, multiple small size sensors are mounted at various points in the field (Yang et al., 2017)
Gas sensor	Gas sensor tests the level of poisonous gases in the greenhouses and animal shelters by measuring the amount of infrared radiation absorbed. It is comprised of two low-range and high-range variables. The low range usually begins from 0 to 10,000, and the high range from 0 to 100,000 (Bapat et al., 2017)
Ultraviolet sensor	UV sensors are used for converting photocurrent to voltage to track ultraviolet rays. They are fitted with an external circuit, which is analog to digital signal converter signal. This sensor is most successful in detecting light rays for effective crop growth (Patil et al., 2017)
Motion detector	This sensor is extremely useful during the night for the detection of animals and field theft. If an unexpected activity happens in the land, the farmer gets a warning call. Farmers may, therefore, take corrective steps by identifying the passage across the field of an unacceptable entity or an animal (Mat et al., 2016).

(IGMS) relying on WSN technology to evaluate vital environmental parameters such as temperature, humidity, and soil moisture, confirming that automatic irrigation optimized the use of water and fertilizer and further maintained the plant's level of moisture and salinity. Commercial sensors for agricultural systems and their irrigation are quite costly, and hence there is a growing interest in relatively inexpensive sensors for agriculture and water surveillance. New low-cost sensors are established for investigation such as a leaf water stress monitoring sensor (Daskalakis et al., 2018), a multilevel soil moisture sensor consisting of copper rings mounted around a PVC pipe (Parra et al., 2013), a water salinity monitoring sensor made from copper coils (Guruprasadh et al., 2017), or a water turbidity sensor made from colored and infrared LED emitters and receptors (Sendra et al., 2013). Watthanawisuth et al. (2009) identified an agricultural IoT approach that can be characterized in the subdomain of air monitoring. The authors proposed a microclimate

TABLE 19.2
IoT Sensor System Comparison

Sensor Systems	Sensors	Cloud Supported	Application Type
Bitponics	Water/Air temperature, humidity, pH, brightness	Yes	Irrigation
Botanicalls	Air temperature, humidity, brightness, soil moisture	No	Garden
Edyn	Temperature, light, pH, soil moisture, soil humidity	Yes	Irrigation
Parrot	Temperature, light, pH, soil moisture, soil salinity	No	Garden
PlankLink	Soil moisture	Yes	Irrigation
HarvestGeek	Temperature, light, pH, soil moisture, CO_2	Yes	Garden, irrigation
Iro	-	No	Irrigation
Spruce	Temperature, soil moisture	No	Irrigation
Open Garden	Temperature, humidity, light, pH, soil moisture, conductivity	Yes	Garden, irrigation, aquaponics
Koubachi	Soil moisture, air temperature, soil temperature, ambient light	Yes	Irrigation
Niwa	Temperature, humidity, light	Yes	Aquaponics

real-time monitoring device based on a WSN that included temperature and relative humidity sensors (SHT15) driven by solar panels and assisted by communication technology ZigBee. Postolache et al. (2013) suggested an IoT approach for water quality evaluation by calculating conductivity, temperature, and turbidity based on a WSN architecture that incorporates low-cost sensing instruments and control of several water quality parameters of surface waterways in metropolitan areas. Likewise, Fourati et al. (2014) suggested a web-based decision support system to communicate with a WSN for irrigation scheduling in olive fields utilizing sensors to monitor humidity, solar radiation, temperature, and weather. An illustration of animal tracking was presented by Jain et al. (2008), where an IoT system was responsible for tracking Swamp Deer's activity and movement patterns, collecting, at the same time, knowledge about the animal's location and environment.

A novel form of communication was created to establish a two-way information channel in which instructions could be transmitted back to the field referred to as wireless sensor and actuator network (WSAN). In this scenario, server or data center information traveled to a wireless sensor and actuator network (WSAN) to monitor a series of actuator devices to change process or system status. Instructions were sent via a human–computer interface or as a function of an analytic modules-supported decision algorithm. Among others, the actuators featured switches, pumps, humidifiers, and alarms. Many of these programs aimed to optimize the use of water, fertilizers, and pesticides depending on the climatic prediction model and on-site WSN details. Kanoun et al. (2014) suggested a precision irrigation approach based on a wireless sensor network to develop an integrated irrigation system that might minimize water consumption, conserve electricity, resources, and time. Shuwen and Changli (2015) identified a Zig Bee-based remote management system for irrigation on agricultural land. Pahuja et al. (2013) created an innovative online greenhouse microclimate monitoring and control device funded by a WSN to capture and interpret plant-related sensor data to generate climate management, fertilization, irrigation, and pest control behavior. Yoo et al. (2007) identified a WSN-based automated farming system for controlling the greenhouses that are used to grow melons and cabbages. The framework supervised the crop-growing process and regulated the atmosphere of the greenhouse. Some of the observed factors included ambient light, temperature, and humidity. The mechanism could monitor the lighting for the greenhouse with melons by adjusting the luminous status through a relay.

19.4 IOT SOFTWARE FOR AGRICULTURE

Many organizations invest in IoT-driven agricultural software creation, and several software solutions are available to support the various agricultural processes. AG-IoT (Uddin *et al.*, 2017) is an unmanned aircraft that locates and supports on-the-ground IoT-based devices for groups of data transmission. Agro 4.0 (Fonseca *et al.*, 2016), on the other hand, incorporates high-level computing procedures, the sensor network, device access, cloud, and predictive approaches for managing large data volumes and delivering decision support systems. The data obtained from different sensors available in a given crop area are collected, processed, and updated by Agro-Tech (Pandithurai, *et al.* 2017). Farmers can also access this information to monitor their crops with this software. A virtual end-to-end system for smart agriculture, AgroTick (Roy *et al.* 2017), operates on multiple levels, such as adaptive water management through WSN, numerical optimization through data science, and robust cloud computing connectivity. Malthouse is an Artificial Intelligence framework (Dolci, 2017) that enables specific farming and food-processing setups and schedules to be controlled. Agricultural surveillance systems are available that transmit live video to remotely carry out these processes through IoT-based devices incorporating cameras and Raspberry Pi (Shete & Agrawal, 2016; Anvekar et al., 2017). Cropx, on the other hand, is a method for sustainable irrigation, which helps farmers to maximize crop yields and conserve water and energy at the lowest possible prices. Farmlogs is a program for farm management that enables to record crop conservation activities that photographs. Last, the program Mbegu Option helps farmers to pick the suppliers of the right seeds that are resistant to drought. The IoT solutions for agriculture are listed in Table 19.3.

As an open-source EU-funded approach composed of a collection of program modules known as generic engineers (GE) that execute functions needed in the smart IoT applications, FIWARE is a significant source of interest. For several IoT-based technologies for smart farming, FIWARE has been used as a computing interface. The FIWARE platform architecture set up for the agricultural precision domain was defined in depth by Martinez *et al.* (2016). The implementation of FIWARE for a particular scenario of precise agriculture irrigation was proposed in Southern Spain by Lopez-Riquelme *et al.* (2017).

19.5 SOME APPLICATIONS OF IOT IN AGRICULTURE

In the agro-industrial supply chain, IoT can be used at various stages (Medela *et al.*, 2013). It may help to determine farm variables including soil levels, environmental parameters, and crop or livestock production. It is also useful in determining and monitoring product transportation factors, e.g.,

TABLE 19.3
Examples of IoT Solutions for Agriculture

IoT Solutions	Services
On-Farm	A farm administration toll displaying and reviewing data from several different outlets
Phytech	Provides IoT forum for direct sensing, data processing, plant activity, and suggestions
Semios	Focuses on network access, pests, frost, diseases, and orchids irrigation
EZfarm	An IBM project emphasizes water management, soil monitoring, and plant protection
KAA	Provides remote crop surveillance, resource mapping, predictive crop and livestock analysis, livestock monitoring, feed and livestock data, smart logistics, and storage
MbeguChoice	Is an application that allows farmers to have access to high-quality seeds, drought-resistant seeds as well as a range of seeds from different suppliers
Farmlogs	Farm-monitoring tools for automated activity-tracking and processing of crops' health images
Cropx	Provides adaptive irrigation management systems that produce improvements in crop yields, reduction of water consumption, and energy costs

temperature, humidity, vibration, and shock (Pang etc. 2015). IoT has become a promising medium for the goals of self-organization, decision-making, and automation in agriculture, due to the self-contained structure of its processes, along with its flexible and cost-efficient hardware systems. In this context, precision farming (Barcelo-Ordinas *et al.*, 2013; Díaz, *et al.*, 2011; Cambra *et al.*, 2014), automatic irrigation scheduling (Reche *et al.*, 2014), plant development optimization (Hwang *et al.*, 2010), farmland tracking (Corke *et al.*, 2010), greenhouse control (Mao *et al.*, 2012; Shifeng *et al.*, 2011), and seed farming process management (Dong *et al.*, 2013) are among a few main applications in the management of crop production. The major applications of IoT technologies in agriculture are seen in precision agriculture (Mekala and Viswanathan 2017; Rajeswari *et al.*, 2017), whose framework involves urban agriculture IoT techniques and agronomy precision in smart cities. Other IoT applications are agricultural drones (Uddin *et al.*, 2017), which are relatively cheap drones with sophisticated sensors that offer farmers new approaches to improve yields and, among many other things, minimize crop damage. Another field where IoT technologies are implemented is vertical agriculture (Bin Ismail and Thamrin, 2017), which enables soil moisture and water content to be monitored by computers or mobile devices such as tablets and smartphones. At last, there are systems that integrate IoT technology with Artificial Intelligence, such as Malthouse (Dolci *et al.*, 2017), which is an Artificial Intelligence device that enables modifications and schedules to be implemented in the fields of precision farming and food processing. Figure 19.3 shows the categorization of IoT and sensor-dependent applications as provided by Farooq *et al.* (2019). A few of IoT's agricultural applications are reviewed herein.

19.5.1 Irrigation

IoT is becoming increasingly popular around the world in irrigation management-related systems. Modern agriculture needs an advanced method of irrigation management to maximize water use in agriculture and associated operations (Vijay, *et al.*, 2010; Shimizu *et al.*, 2015). Depending on the satellite positioning network (Li *et al.*, 2016) and "shallow wells underground cables + field + automated irrigation system pipe" (Li *et al.*, 2016) technology, these can acquire irrigation water, electricity, and time data to achieve farmland irrigation automation through a thorough application of irrigation-monitoring systems in information technology. It's necessary to determine the right requirements for crop water. Field data obtained from sensors can help achieve a high degree of precision in measuring the water requirement. A conceptual framework for precision farming using a wireless sensor network with an IoT cloud is suggested in this study (Karim *et al.*, 2017). In this study, an alarm system was provided for the management of plant water stress using IoT technology. For irrigation control, the farmer uses a dashboard program in the form of a graph to track changes in soil conditions in real time, and, on the other hand, an SMS-warning mechanism will be delivered via the application software when a critical threshold is achieved to prevent water stress. Luan *et al.* (2015) planned and developed a computational framework that combined the tracking and forecasting of drought and the volume of irrigation into an IoT, hybrid programming, and parallel computing network. Kanoun *et al.* (2014), Kaewmard and Saiyod (2014) focused on irrigation systems using WSN for mobile data collection and irrigation system monitoring of the climate. Chen *et al.* (2014) proposed systems for tracking multilayer soil temperature and humidity in a cultivation region using WSN to enhance water usage and gather specific data for research on variability in soil water infiltration and smart irrigation accuracy. When such requirements are met, automatic decisions will be taken by the IoT device and may require less to no human interference. Such an automatic decision could vary from temperature management to water source control from an irrigation system. Sushanth and Sujatha (2018) suggested a framework using IoT, WSN, and cloud computing to help farmers prepare their farm's irrigation schedule from an agricultural profile that can be updated as per their specifications. The device consisted of a distributed wireless network with sensors of soil moisture and temperature located in the plant's root zone (Figure 19.4). This system produces an irrigation schedule based on field and weather databases and sensed real-time

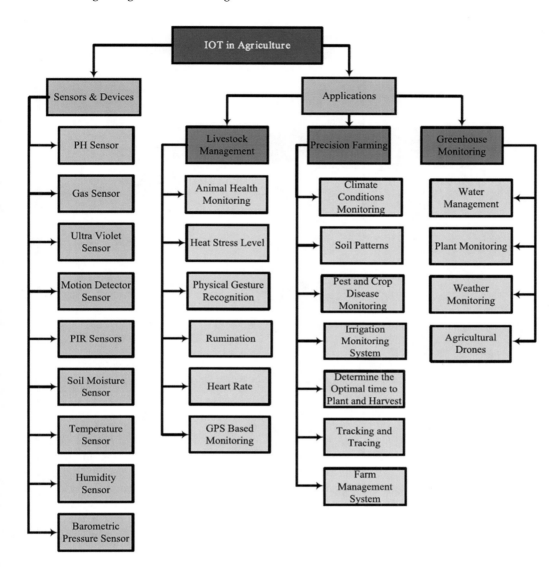

FIGURE 19.3 IoT structure in agriculture

data. Owing to its energy efficiency and low cost, the device has the ability to be effective in geographically remote regions which are confined in water resources.

19.5.2 Water Quality Monitoring

Water quality is tracked via wireless communications-enabled sensor nodes. Regulation of water quality can be achieved by positioning sensor nodes that are improved by wireless communications. IoT technology evaluates pH, dissolved solids, temperature, conductivity, and oxygen physical and chemical constraints (Bin Ismail and Thamrin, 2017). A new study explores water quality in real time using IoT. Physical and chemical water parameters including temperature, pH, turbidity, strength, and dissolving oxygen are measured through the device. Cloud platforms are used to display sensor data on the Internet (Paventhan et al., 2012). A solution for IoT water quality assessment by conductivity, temperature, and turbidity assessment was suggested by Postolache and others

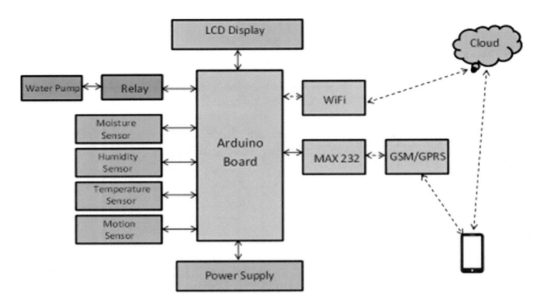

FIGURE 19.4 Function block diagram given by Sushanth and Sujatha (2018)

(2013). The approach relies on a WSN design, which integrates cheap detectors with several water quality criteria for surface waters in urban areas (lakes, estuaries, rivers).

19.5.3 Soil Monitoring

Soil monitoring both for industries and farmers is becoming one of the most difficult activities for agriculture. The soil variations examined are soil moisture, humidity, fertilization, and temperature. To regulate the soil moisture content, moisture sensors and temperature sensors are used (de Morais *et al.*, 2019). Optic sensors to quantify surface properties; photodiodes and photodetectors for the soil, organic matter, and soil moisture; and humidity sensors to determine the amount of water in the soil are some of the most common IoT-based instruments used in farming. In addition, the detection of contaminated soils using IoT technology prevents the land from over-fertilization and crop losses. Several steps have indeed been introduced by the Government of India to provide online and mobile messaging services to farmers on agricultural questions, to provide information to farmers on agro-vendors, and to provide static data on soil quality in each area. Real-time control of these properties helps to keep soil quality intact by adding only the necessary quantity of fertilizers. Sakthipriya (2014) recommended the implementation of a real-time monitoring system for rice cultivation to increase rice yield. This device includes the use of external sensors for leaf wetness, soil moisture, soil pH, and air pressure sensors mounted on it. The amount of fertilizer to be used can be determined by the farmer using the pH values.

19.5.4 Greenhouse Condition Monitoring

Greenhouse production is more complex, which implies that it requires high accuracy in terms of control and surveillance. Findings have suggested how greenhouse IoT can be implemented to minimize human costs, save electricity, make tracking greenhouse sites more effective, and link greenhouse farmers directly to clients (Zhao *et al.*, 2010; Dan *et al.*, 2015; Kodali *et al.*, 2016). A smart greenhouse architecture removes handheld operations and tests various climatic parameters according to plant specifications with intelligent IoT devices and sensors (Ma, *et al.*, 2015;

Katsoulas *et al.*, 2017; Ibrahim *et al.*, 2019). The use of machine learning in greenhouses will help to identify ideal conditions for a crop by analyzing the data acquired from nutrient, yield, growth, transpiration, color, taste, and retransplantation sensor data as well as the level of light, pest, temperature, and air quality. Zhao *et al.* (2010) suggested a "greenhouse control system" with a wireless communications and Internet implementation. The "greenhouse surveillance system," built with IoT, is accurate and regulated, operates very simply, is user-friendly, and allows for real-time management of the environmental factors within the greenhouse. This system also features high performance, stable working, and simple upgrade features. Li *et al.* (2012) suggested an IoT-based agricultural greenhouse environmental surveillance system that incorporates the Internet, wireless network, mobile network, and centralized real-time tracking of greenhouse environmental information. The online microclimate monitoring and control system for greenhouses has been developed by Pahuja *et al.* (2013). The device was assisted by a WSN to capture and analyze sensor data relating to plants for climate control, fertilization, irrigation, and pests. A significant issue is calculating the exact quantity of water in greenhouses. The use of automatic drip irrigation in greenhouses is carried out using the soil moisture level, which has to be calculated accordingly (Windsperger, *et al.*, 2019).

A variety of experiments have been performed on the use of WSNs in the greenhouse to track environmental or meteorological conditions. Yoo *et al.* (2007) has identified an integrated WSN farming system for controlling greenhouses for the cultivation of melons and cabbages. The machine regulated production and managed the climate of the greenhouse. A few of the measured variables included external illumination, humidity, and temperature. The lighting could be regulated by adjusting the light status via a relay for the greenhouse with melons. The intelligent greenhouses (Kim *et al.*, 2008;Wortman *et al.*, 2015) that have hydroponic and small-scale aquaponic systems (Montoya *et al.*, 2017; Atmadja *et al.*, 2017) are another area for IoT applications. Smart greenhouses are becoming much more common in urban areas, as these allow the monitoring of many nutrient solution parameters (Barbosa *et al.*, 2015) and the improvement of plant growth, yield, and quality.

19.5.5 Pest and Disease Control

Prediction of early-stage crop diseases allows farmers to produce more profits by protecting crops from pest attacks. Regulated use of pesticides and fertilizers tends to boost crop production and reduce agriculture costs. However, in order to manage the use of pesticides, we need to monitor the risk and incidence of pests in crops. In order to foresee this, we will need to gather disease and insect pest data using sensor nodes, data processing, mining, etc. with the aid of the IoT infrastructure (Wenshun *et al.*, 2013). IoT defends crops in many forms by identifying various pathogens and avoiding crop attacks. The IoT-based control system for the management of wheat pathogens, pests, and weeds has been developed by Zhang *et al.*, (2014). A surveillance and repulsion framework for the protection of crops against wild animal attacks has been proposed by Giordano *et al.*, (2018). Identification of the early stages of crop disease is a big problem in the field of agriculture since a team of experts is called to diagnose seed or leaf disease, which is costly and time-consuming. Whereas the automated diagnosis of diseases is very useful, reliable, and cheaper for farmers compared to manual inspection by specialists. Image recognition technology also plays a crucial role in the early identification of plant disease (Barbedo *et al.*, 2018).

19.6 BENEFITS OF IOT IN AGRICULTURE

The implementation of Information and Communication Technology (ICT) has proved to expand prospects in developed countries for fostering agriculture in different areas and realms. IoT can also be integrated into environmental applications to create detailed and real-time maps of air and water contamination, noise pollution (Torres-Ruiz *et al.*, 2016; Hachem *et al.*, 2015), temperature, and hazardous radiation among many others. The key advantages and benefits of IoT for improving farming in agriculture are as follows:

i. Without much wastage of water using IoT sensors, water control can be successfully achieved. Sensors are used to obtain automatic irrigation systems (Kaewmard and Saiyod, 2014) which operate as per temperature, humidity, and soil moisture parameters
ii. Crop tracking can be performed with ease to detect crop growth, thus lowering costs as well as equipment loss
iii. IoT aims to analyze the land constantly so that steps could be taken early. Soil conditions such as pH level, moisture content, etc. can be easily monitored and hence the farmworkers can sow the seeds based on soil quality
iv. Automatic gathering of environmental parameters for further processing and analysis via sensor networks
v. Sensors and RFID chips help in the detection of pathogens in plants and crops. RFID tags give the electronic product code EPC (information) to the reader, and they are exchanged online. From a remote area, the farmer or scientist may access this knowledge and take appropriate action, automatically protecting crops against future diseases (Jain *et al*. 2012)
vi. It increases efficiency, decreases manual labor, cuts time, and makes agriculture more productive
vii. In urban and rural areas, community agriculture takes advantage of the hardware and technological tools and vast volumes of data. Decision support systems analyze large volumes of data in order to enhance operational productivity and efficiency (Nandyala and Kim, 2016; Cambra *et al*., 2017)
viii. Logistic and quality traceability of food production reduces costs and loss of resources by using real-time decision-making data
ix. Business model generation (Dolci, 2017) permits direct customer relationships in the agricultural context. In the global market, crop sales are to be increased. Without restricting geographical area, farmers can easily bind to the global market

19.7 AUTOMATION IN WATER MANAGEMENT

Automation is controlling things without being in direct physical contact with the system and/or controlling the things without touching them. Automation of irrigation systems mainly used drip irrigation with IoT as a novel concept which helps us to apply water precisely to the field, which, in turn, will increase the net water use efficiency by plant, resulting in high productivity of the crop.

This concept will work eco-friendly by converting naturally available solar energy into electrical energy, which is used to run the Arduino and other gadgets. It will work on the principle that whenever moisture content in the soil goes below the threshold, it turns the motor on by Arduino programming (predetermined) with the help of a soil moisture sensor based on capacitance/resistance. The process can be visualized on the display unit or with the help of a cell phone. The whole process will be monitored by our fingertips cautiously without involving the drudgery and labor cost. It includes a global system for mobile (GSM) module facility which is responsible for connecting the system with the smartphone (Figure 19.5).

19.8 CONCLUSION

In the years to come, IoT will be an important platform for people working within an agro-industrial system: manufacturers, producers, designers, dealers, businesses, customers, and members of governments. The introduction of IoT is meant to be useful for the development of food and the industry by adding new dimensions. Real-time collection of relevant data is feasible using IoT-based technologies. Some of the most excellent innovations integrated with IoT to develop agricultural

Smart Farming Using Internet of Things

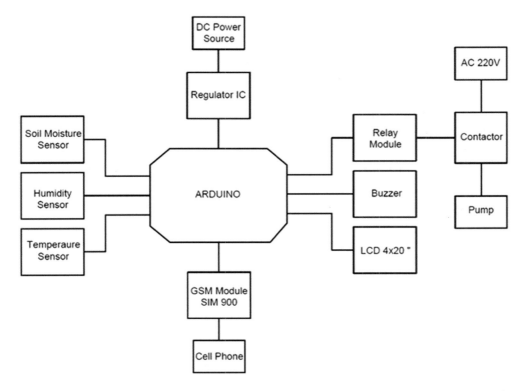

FIGURE 19.5 Automation for water management

solutions include wireless sensor networks, cloud computing, software systems, and smartphone apps. The productivity of IoT-based agriculture is high when compared with traditional techniques. By using IoT in agricultural fields, farmers can understand the current status regarding agricultural soil selection through means of intelligent analysis and better management, the crops that are more suitable for agriculture presently, and other environmental information about the farmland. Agricultural cloud and IT advantage provide farmers with strategic management knowledge in the areas of the yield of crops, field assessment, manures, diseases, and their comprehensive cure strategies. Excessive water use issues and inappropriate use of fertilizers could also be reduced so as to minimize water consumption and reach maximum production. The irrigation schedule established is dynamic and versatile, since it enables the farmers to prepare irrigation as per their requirements. It also becomes apparent that a number of major companies have begun to invest and create new farm management system strategies using IoT. Continued and accelerated growth of microelectronic technologies and network infrastructure offers an incentive for experts to systematically pursue the technical advancement of modern agriculture.

SHORT ANSWER-TYPE-QUESTIONS

1. Define precision agriculture.
2. Outline the major components of precision agriculture.
3. What are soil moisture sensors?
4. How do soil moisture sensors help with water management?
5. What are the benefits of smart farming for the farmers?
6. List the challenges of smart farming.
7. What is the role of IoT (Internet of Things) in agriculture?

8. What is smart irrigation?
9. What is irrigation scheduling?
10. What is farm automation?
11. What are the differences between traditional and smart farming?
12. What are the objectives of smart water management?
13. What are the system requirements for a typical automated irrigation system?
14. Explain the working principle of the Internet of Things.
15. Explain the working principle of continuous monitoring of soil moisture.
16. What is a GSM-based automatic irrigation system?
17. What are the disadvantages of AI in agriculture?
18. How can different diseases be detected using automation?
19. Which software or programming language is utilized to control the Arduino?
20. What are climate-controlled greenhouses?

LONG ANSWER-TYPE QUESTIONS

1. Write the framework towards technology adaptation of smart farming.
2. What are the ways to incorporate automation in agriculture?
3. Enlist the benefits of agriculture automation.
4. Enlist IoT based agriculture applications.
5. Applications of Artificial Intelligence in agriculture.
6. How can Artificial Intelligence contribute considerably to the agriculture sector in India?
7. How is Artificial Intelligence solution presently exist for the agriculture and horticulture sector?
8. What are the advantages of implementing Artificial Intelligence in agriculture?
9. What is the scope of Artificial Intelligence in the agriculture sector?
10. What are the main challenges for the adoption of agricultural robotics?
11. What are the approaches to agriculture Polyhouse Monitoring Solution?
12. Describe the architectural system of automation in the agriculture system.
13. What are ways to plan automatic irrigation systems using a microcontroller?
14. How does AI enable the right decision-making process in farms?
15. Why Artificial Intelligence is playing an important role in Indian agriculture?
16. What is the state of automation in the horticulture sector?
17. What are the climate control systems for greenhouse technology?
18. What are the environmental entities that can be measured in a greenhouse?
19. Explain the system design and working of greenhouse environment monitoring and control systems.
20. Write the difference between IoT and Artificial Intelligence.

MULTIPLE-CHOICE QUESTIONS

1. Smart agriculture includes
    a) Alert generation in case of above-threshold pollutants in the air
    b) Scheduling harvesting and arranging a proper transfer of harvests to warehouses or markets
    c) Both the above
    d) None of these
2. The navigation system which is based on a satellite network that allows users to record positional data is
    a) GPS
    b) GIS

c) Remote sensing
d) All of the above

3. The concept of SSNM in rice was created in collaboration with researchers from
   a) Africa
   b) South America
   c) Asia
   d) Europe

4. _____ can be defined as a system for collecting, storing, verifying, and displaying data that are spatially referenced to the earth.
   a) GIS
   b) GPS
   c) Variable-rate applicator
   d) Remote sensing

5. A precision instrument for determining chlorophyll concentration is
   a) SPAD meter
   b) LCC
   c) Green seeker
   d) Quantum sensor

6. The Coordinate Reference System (CRS) is a system for determining coordinates of GPS referred to as
   a) WGS 45
   b) WGS 84
   c) WGS 89
   d) WGS 88

7. Which software or programming language is utilized to control the Arduino?
   a) Assembly language
   b) C language
   c) Java
   d) Any language

8. Does Arduino provide an integrated development environment (IDE) environment?
   a) True
   b) False
   c) None of the above

9. DHT11 is a _____ sensor.
   a) Proximity sensor
   b) Humidity sensor
   c) Touch sensor
   d) Pressure sensor

10. Between field capacity and permanent wilting point, the soil moisture is
    a) Free moisture
    b) Bound moisture
    c) Available moisture
    d) None of these

11. A sensor is a _____.
    a) Subsystem
    b) Machine
    c) Module
    d) All the above

12. A sensor's function is to
    a) Detect events within a specified environment
    b) Separate physical parameters

c) Track and transfer data to a computer processor
d) Both a and b
13. The effectiveness of the sensor is determined by _____ parameter.
    a) Sensitivity
    b) Radiation
    c) Restively
    d) All the above
14. Which of the following is considered as input to a sensor?
    a) Microcontroller
    b) Processor
    c) Both a and b
    d) Drivers
15. Sensors convert analog signals to _____ domain.
    a) Digital
    b) Electrical
    c) Mechanical
    d) Both a and b
16. The abbreviation for LDR sensor is
    a) Light-Dependent Resistor
    b) Light determinant resistor
    c) Luminous duplicated resistor
    d) None of the above
17. GSM module is used for
    a) Remote access
    b) Bluetooth
    c) Color sensing
    d) None of the above
18. Which of these cannot be determined using automation?
    a) pH
    b) Moisture content
    c) Nutrient condition
    d) Soil type
19. What is the role of GPS in creating useful GIS?
    a) The GPS can obtain the soil profile information without having to dig holes
    b) GPS generates a map that combines soil information with other land properties
    c) GPS enables soil data to be accurately associated with a specific location as it is collected in the field
    d) All of the above
20. Map made for site-specific or precision management of agricultural field easily involve map layers showing the spatial distribution of which type of information
    a) Soil series
    b) Previous crop yields
    c) Soil organic matter
    d) Soil nutrient level
    e) All of the above

ANSWERS

1	2	3	4	5	6	7	8	9	10	11	12	13	14
b	a	c	a	a	b	d	a	b	c	d	d	a	c

15	16	17	18	19	20
d	a	a	d	c	d

REFERENCES

Abbasy, M. B. and Quesada, E. V. 2017. Predictable influence of IoT (Internet of Things) in the higher education. *International Journal of Information and Education Technology*, 7(12), 914–920.

Ahmed, N., De, D. and Hussain, I. 2018. Internet of Things (IoT) for smart precision agriculture and farming in rural areas. *IEEE Internet of Things Journal*, 5(6), 4890–4899.

Alahi, M. E. E., Xie, L., Mukhopadhyay, S. and Burkitt, L. 2017. A temperature compensated smart nitrate-sensor for agricultural industry. *IEEE Transactions on Industrial Electronics*, 64(9), 7333–7341.

Al-Sarawi, S., Anbar, M., Alieyan, K. and Alzubaidi, M. 2017, May. Internet of Things (IoT) communication protocols. In *2017 8th International Conference on Information Technology (ICIT)*. IEEE, 685–690.

Anvekar, R. G., Banakar, R. M. and Bhat, R. R. 2017, April. Design alternatives for end user communication in IoT based system model. In *2017 IEEE Technological Innovations in ICT for Agriculture and Rural Development (TIAR)*. IEEE, 121–125.

Atmadja, W., Liawatimena, S., Lukas, J., Nata, E. P. L. and Alexander, I. 2017, December. Hydroponic system design with real time OS based on ARM cortex-M microcontroller. *IOP Conference Series: Earth and Environmental Science. IOP Publishing*, 109(1), 012017.

Balaji, S., Nathani, K. and Santhakumar, R. 2019. IoT technology, applications and challenges: A contemporary survey. *Wireless Personal Communications*, 108(1), 363–388.

Bapat, V., Kale, P., Shinde, V., Deshpande, N. and Shaligram, A. 2017. WSN application for crop protection to divert animal intrusions in the agricultural land. *Computers and Electronics in Agriculture*, 133, 88–96.

Barbedo, J. G. A., Koenigkan, L. V., Halfeld-Vieira, B. A., Costa, R. V., Nechet, K. L., Godoy, C. V., Junior, M. L., Patricio, F. R. A., Talamini, V., Chitarra, L. G. and Oliveira, S. A. S. 2018. Annotated plant pathology databases for image-based detection and recognition of diseases. *IEEE Latin America Transactions*, 16(6), 1749–1757.

Barbosa, G. L., Gadelha, F. D. A., Kublik, N., Proctor, A., Reichelm, L., Weissinger, E., Wohlleb, G. M. and Halden, R. U. 2015. Comparison of land, water, and energy requirements of lettuce grown using hydroponic vs. conventional agricultural methods. *International Journal of Environmental Research and Public Health*, 12(6), 6879–6891.

Barcelo-Ordinas, J. M., Chanet, J. P., Hou, K. M. and García-Vidal, J. 2013. A survey of wireless sensor technologies applied to precision agriculture. In *Precision Agriculture'13*. Wageningen Academic Publishers, Wageningen, 801–808.

Berte, D. R. 2018. Defining the IoT. *Proceedings of the International Conference on Business Excellence, Sciendo*, 12(1), 118–128.

Bertino, E., Choo, K. K. R., Georgakopolous, D. and Nepal, S. 2016. Internet of Things (IoT) smart and secure service delivery. *ACM Transactions on Internet Technology*, 16, 1–7.

Bin Ismail, M. I. H. and Thamrin, N. M. 2017. IoT implementation for indoor vertical farming watering system. In *2017 International Conference on Electrical, Electronics and System Engineering (ICEESE)*. IEEE, 89–94.

Bo, Y. and Wang, H. 2011. The application of cloud computing and the internet of things in agriculture and forestry. In *2011 International Joint Conference on Service Sciences*. IEEE, 168–172.

Botta, A., De Donato, W., Persico, V. and Pescapé, A. 2016. Integration of cloud computing and internet of things: A survey. *Future Generation Computer Systems*, 56, 684–700.

Brewster, C., Roussaki, I., Kalatzis, N., Doolin, K. and Ellis, K. 2017. IoT in agriculture: Designing a Europe-wide large-scale pilot. *IEEE Communications Magazine*, 55(9), 26–33.

Cambra, C., Díaz, J. R. and Lloret, J. 2014. Deployment and performance study of an ad hoc network protocol for intelligent video sensing in precision agriculture. In *International Conference on Ad-Hoc Networks and Wireless*. Springer, Berlin, Heidelberg, 165–175.

Cambra, C., Sendra, S., Lloret, J. and Garcia, L. 2017. An IoT service-oriented system for agriculture monitoring. In *2017 IEEE International Conference on Communications (ICC)*. IEEE, 1–6.

Channe, H., Kothari, S. and Kadam, D. 2015. Multidisciplinary model for smart agriculture using Internet-of-Things (IoT), sensors, cloud-computing, mobile-computing & big-data analysis. *International Journal of Computer Technology & Applications*, 6(3), 374–382.

Chen, K. T., Zhang, H. H., Wu, T. T., Hu, J., Zhai, C. Y. and Wang, D. 2014. Design of monitoring system for multilayer soil temperature and moisture based on WSN. In *2014 International Conference on Wireless Communication and Sensor Network*. IEEE, 425–430.

Chen, S., Xu, H., Liu, D., Hu, B. and Wang, H. 2014. A vision of IoT: Applications, challenges, and opportunities with china perspective. *IEEE Internet of Things Journal*, 1(4), 349–359.

Cheng, J. and Yi, S. 1999. Digital agriculture–One of application domain of digital earth. In *Proceedings of the International Symposium on Digital Earth*. Science Press.

Corke, P., Wark, T., Jurdak, R., Hu, W., Valencia, P. and Moore, D. 2010. Environmental wireless sensor networks. *Proceedings of the IEEE*, 98(11), 1903–1917.

Dan, L. I. U., Xin, C., Chongwei, H. and Liangliang, J. 2015. Intelligent agriculture greenhouse environment monitoring system based on IOT technology. In *2015 International Conference on Intelligent Transportation, Big Data and Smart City*. IEEE, 487–490.

Daskalakis, S. N., Goussetis, G., Assimonis, S. D., Tentzeris, M. M. and Georgiadis, A. 2018. A uW backscatter-morse-leaf sensor for low-power agricultural wireless sensor networks. *IEEE Sensors Journal*, 18(19), 7889–7898.

de Morais, C. M., Sadok, D. and Kelner, J. 2019. An IoT sensor and scenario survey for data researchers. *Journal of the Brazilian Computer Society*, 25(1), 4.

Díaz, S. E., Pérez, J. C., Mateos, A. C., Marinescu, M. C. and Guerra, B. B. 2011. A novel methodology for the monitoring of the agricultural production process based on wireless sensor networks. *Computers and Electronics in Agriculture*, 76(2), 252–265.

Dlodlo, N. and Kalezhi, J. 2015. The internet of things in agriculture for sustainable rural development. In *2015 International Conference on Emerging Trends in Networks and Computer Communications (ETNCC)*. IEEE, 13–18.

Dolci, R. 2017. IoT solutions for precision farming and food manufacturing: Artificial intelligence applications in digital food. In *2017 IEEE 41st Annual Computer Software and Applications Conference (COMPSAC)*. IEEE, 2, 384–385.

Dong, X., Vuran, M. C. and Irmak, S. 2013. Autonomous precision agriculture through integration of wireless underground sensor networks with center pivot irrigation systems. *Ad Hoc Networks*, 11(7), 1975–1987.

Elijah, O., Rahman, T. A., Orikumhi, I., Leow, C. Y. and Hindia, M. N. 2018. An overview of Internet of Things (IoT) and data analytics in agriculture: Benefits and challenges. *IEEE Internet of Things Journal*, 5(5), 3758–3773.

Farooq, M. S., Riaz, S., Abid, A., Abid, K. and Naeem, M. A. 2019. A survey on the role of IoT in agriculture for the implementation of smart farming. *IEEE Access*, 7, 156237–156271.

Fonseca, S. M., Massruhá, S. and Angelica De Andrade Leite, M. 2016. Agro 4.0 – Rumo À Agricultura Digital, 28–35.

Fourati, M. A., Chebbi, W. and Kamoun, A. 2014. Development of a web-based weather station for irrigation scheduling. In *2014 Third IEEE International Colloquium in Information Science and Technology (CIST)*. IEEE, 37–42.

Garcia-Sanchez, A. J., Garcia-Sanchez, F. and Garcia-Haro, J. 2011. Wireless sensor network deployment for integrating video-surveillance and data-monitoring in precision agriculture over distributed crops. *Computers and Electronics in Agriculture*, 75(2), 288–303.

Gill, S. S., Chana, I. and Buyya, R. 2017. IoT based agriculture as a cloud and big data service: The beginning of digital India. *Journal of Organizational and End User Computing (JOEUC)*, 29(4), 1–23.

Giordano, S., Seitanidis, I., Ojo, M., Adami, D. and Vignoli, F. 2018. IoT solutions for crop protection against wild animal attacks. In *2018 IEEE International Conference on Environmental Engineering (EE)*. IEEE, 1–5.

Giusto, D., Iera, A., Morabito, G. and Atzori, L. (eds.). 2010. The internet of things. In *20th Tyrrhenian Workshop on Digital Communications*. Springer Science & Business Media.

Guruprasadh, J. P., Harshananda, A., Keerthana, I. K., Krishnan, K. Y., Rangarajan, M. and Sathyadevan, S. 2017, September. Intelligent soil quality monitoring system for judicious irrigation. In *2017 International Conference on Advances in Computing, Communications and Informatics (ICACCI)*. IEEE, 443–448.

Hachem, S., Mallet, V., Ventura, R., Pathak, A., Issarny, V., Raverdy, P. G. and Bhatia, R. 2015, March. Monitoring noise pollution using the urban civics middleware. In *2015 IEEE First International Conference on Big Data Computing Service and Applications*. IEEE, 52–61.

Hwang, J., Shin, C. and Yoe, H. 2010. A wireless sensor network-based ubiquitous paprika growth management system. *Sensors*, 10(12), 11566–11589.

Ibrahim, H., Mostafa, N., Halawa, H., Elsalamouny, M., Daoud, R., Amer, H., Adel, Y., Shaarawi, A., Khattab, A. and ElSayed, H. 2019. A layered IoT architecture for greenhouse monitoring and remote control. *SN Applied Sciences*, 1(3), 1–12.

Jain, D., Krishna, P. V. and Saritha, V. 2012. A study on Internet of Things based applications. *arXiv preprint arXiv:1206.3891*.

Jain, V. R., Bagree, R., Kumar, A. and Ranjan, P. 2008. wildCENSE: GPS based animal tracking system. In *2008 International Conference on Intelligent Sensors, Sensor Networks and Information Processing*. IEEE, 617–622.

Juniper Research. 2015. *Internet of Things' Connected Devices to Almost Triple to Over 38 Billion Units by 2020 Juniper Research* [Online]. Available at http://www.juniperresearch.com/press/pressreleases IoT connected-devices-to-triple-to-38-bn-by-2020

Kaewmard, N. and Saiyod, S. 2014. Sensor data collection and irrigation control on vegetable crop using smart phone and wireless sensor networks for smart farm. In *2014 IEEE Conference on Wireless Sensors (ICWiSE)*. IEEE, 106–112.

Kamienski, C., Soininen, J. P., Taumberger, M., Dantas, R., Toscano, A., Salmon Cinotti, T., Filev Maia, R. and Torre Neto, A. 2019. Smart water management platform: IoT-based precision irrigation for agriculture. *Sensors*, 19(2), 276.

Kanoun, O., Khriji, S., El Houssaini, D., Viehweger, C., Jmal, M. W. and Abid, M. 2014. Precision irrigation based on wireless sensor network. *IET Science, Measurement & Technology*, 8, 98–106.

Karim, F. and Karim, F. 2017. Monitoring system using web of things in precision agriculture. *Procedia Computer Science*, 110, 402–409.

Kassim, M. R. M., Mat, I. and Harun, A. N. 2014. Wireless sensor network in precision agriculture application. In *2014 International Conference on Computer, Information and Telecommunication Systems (CITS)*. IEEE, 1–5.

Katsoulas, N., Bartzanas, T. and Kittas, C. 2017. Online professional irrigation scheduling system for greenhouse crops. *Acta Horticulturae*, 1154, 221–228.

Kim, Y., Evans, R. G. and Iversen, W. M. 2008. Remote sensing and control of an irrigation system using a distributed wireless sensor network. *IEEE Transactions on Instrumentation and Measurement*, 57(7), 1379–1387.

Kodali, R. K., Jain, V. and Karagwal, S. 2016. IoT based smart greenhouse. In *2016 IEEE Region 10 Humanitarian Technology Conference (R10-HTC)*. IEEE, 1–6.

Li, J., Gu, W. and Yuan, H. 2016. Research on IOT technology applied to intelligent agriculture. In *Proceedings of the 5th International Conference on Electrical Engineering and Automatic Control*. Springer, Berlin, Heidelberg, 1217–1224.

Li, S. L., Han, Y., Li, G., Zhang, M., Zhang, L. and Ma, Q. 2012. Design and implementation of agricultural greenhouse environmental monitoring system based on Internet of Things. In *Applied Mechanics and Materials*. Trans Tech Publications Ltd. 121, 2624–2629.

López-Riquelme, J. A., Pavón-Pulido, N., Navarro-Hellín, H., Soto-Valles, F. and Torres-Sánchez, R. 2017. A software architecture based on FIWARE cloud for precision agriculture. *Agricultural Water Management*, 183, 123–135.

Luan, Q., Fang, X., Ye, C. and Liu, Y. 2015. An integrated service system for agricultural drought monitoring and forecasting and irrigation amount forecasting. In *2015 23rd International Conference on Geoinformatics*. IEEE, 1–7.

Ma, J., Li, X., Wen, H., Fu, Z. and Zhang, L. 2015. A key frame extraction method for processing greenhouse vegetables production monitoring video. *Computers and Electronics in Agriculture*, 111, 92–102.

Mao, X., Miao, X., He, Y., Li, X. Y. and Liu, Y. 2012. CitySee: Urban CO_2 monitoring with sensors. In *2012 Proceedings IEEE INFOCOM*. IEEE, 1611–1619.

Martínez, R., Pastor, J. Á., Álvarez, B. and Iborra, A. 2016. A testbed to evaluate the fiware-based IoT platform in the domain of precision agriculture. *Sensors*, 16(11), 1979.

Mat, I., Kassim, M. R. M., Harun, A. N. and Yusoff, I. M. 2016. IoT in precision agriculture applications using wireless moisture sensor network. In *2016 IEEE Conference on Open Systems (ICOS)*. IEEE, 24–29.

Medela, A., Cendón, B., González, L., Crespo, R. and Nevares, I. 2013. IoT multiplatform networking to monitor and control wineries and vineyards. In *2013 Future Network & Mobile Summit*. IEEE, 1–10.

Mehra, M., Saxena, S., Sankaranarayanan, S., Tom, R. J. and Veeramanikandan, M. 2018. IoT based hydroponics system using deep neural networks. *Computers and Electronics in Agriculture*, 155, 473–486.

Mekala, M. S. and Viswanathan, P. 2017. A survey: Smart agriculture IoT with cloud computing. In *2017 International Conference on Microelectronic Devices, Circuits and Systems (ICMDCS)*. IEEE, 1–7.

Milella, A., Reina, G. and Nielsen, M. 2019. A multi-sensor robotic platform for ground mapping and estimation beyond the visible spectrum. *Precision Agriculture*, 20(2), 423–444.

Montoya, A. P., Obando, F. A., Morales, J. G. and Vargas, G. 2017. Automatic aeroponic irrigation system based on Arduino's platform. *Journal of Physics: Conference Series*, 850(1), 1.

Muangprathub, J., Boonnam, N., Kajornkasirat, S., Lekbangpong, N., Wanichsombat, A. and Nillaor, P. 2019. IoT and agriculture data analysis for smart farm. *Computers and Electronics in Agriculture*, 156, 467–474.

Nandyala, C. S. and Kim, H. K. 2016. Green IoT agriculture and healthcare application (GAHA). *International Journal of Smart Home*, 10(4), 289–300.

Navulur, S. and Prasad, M. G. 2017. Agricultural management through wireless sensors and internet of things. *International Journal of Electrical and Computer Engineering*, 7(6), 3492.

Pahuja, R., Verma, H. K. and Uddin, M. 2013. A wireless sensor network for greenhouse climate control. *IEEE Pervasive Computing*, 12(2), 49–58.

Pandithurai, O., Aishwarya, S., Aparna, B. and Kavitha, K. 2017. Agro-tech: A digital model for monitoring soil and crops using internet of things (IOT). In *2017 Third International Conference on Science Technology Engineering & Management (ICONSTEM)*. IEEE, 342–346.

Pang, Z., Chen, Q., Han, W. and Zheng, L. 2015. Value-centric design of the internet-of-things solution for food supply chain: Value creation, sensor portfolio and information fusion. *Information Systems Frontiers*, 17(2), 289–319.

Parra, L. O. R. E. N. A., Ortuño, V., Sendra, S. and Lloret, J. A. I. M. E. 2013. Low-cost conductivity sensor based on two coils. In *Proceedings of the First International Conference on Computational Science and Engineering (CSE'13)*, Valencia, Spain. 68, 107112.

Patil, G. L., Gawande, P. S. and Bag, R. V. 2017. Smart agriculture system based on IoT and its social impact. *International Journal of Computer Applications*, 176(1), 0975-8887.

Paventhan, A., Allu, S. K., Barve, S., Gayathri, V. and Ram, N. M. 2012. Soil property monitoring using 6lowpan-enabled wireless sensor networks. In *Proceedings of the Agro-Informatics and Precision Agriculture*, Hyderabad, India, 1–3.

Pavón-Pulido, N., López-Riquelme, J. A., Torres, R., Morais, R. and Pastor, J. A. 2017. New trends in precision agriculture: A novel cloud-based system for enabling data storage and agricultural task planning and automation. *Precision Agriculture*, 18(6), 1038–1068.

Postolache, O., Pereira, M. and Girão, P. 2013. 'Sensor network for environment monitoring: Water quality case study. In *4th Symposium on Environmental Instrumentation and Measurements*, Lecce, Italy, 30–34.

Puliafito, C., Mingozzi, E., Longo, F., Puliafito, A. and Rana, O. 2019. Fog computing for the internet of things: A survey. *ACM Transactions on Internet Technology (TOIT)*, 19(2), 1–41.

Rajeswari, S., Suthendran, K. and Rajakumar, K. 2017. A smart agricultural model by integrating IoT, mobile and cloud-based big data analytics. In *2017 International Conference on Intelligent Computing and Control (I2C2)*. IEEE, 1–5.

Ray, P. P. 2017. Internet of things for smart agriculture: Technologies, practices and future direction. *Journal of Ambient Intelligence and Smart Environments*, 9(4), 395–420.

Reche, A., Sendra, S., Díaz, J. R. and Lloret, J. 2014. A smart M2M deployment to control the agriculture irrigation. In *International Conference on Ad-Hoc Networks and Wireless*. Springer, Berlin, Heidelberg, 139–151.

Roy, S., Ray, R., Roy, A., Sinha, S., Mukherjee, G., Pyne, S., Mitra, S., Basu, S. and Hazra, S. 2017. IoT, big data science & analytics, cloud computing and mobile app based hybrid system for smart agriculture. In *2017 8th Annual Industrial Automation and Electromechanical Engineering Conference (IEMECON)*. IEEE, 303–304.

Sakthipriya, N. 2014. An effective method for crop monitoring using wireless sensor network. *Middle-East Journal of Scientific Research*, 20(9), 1127–1132.

Sendra, S., Parra, L., Ortuño, V., Lloret, J. and De Valencia, U. P. 2013. A low cost turbidity sensor development. In *Proceedings of the Seventh International Conference on Sensor Technologies and Applications (SENSORCOMM)*, Barcelona, Spain, 25–31.

Shete, R. and Agrawal, S. 2016. IoT based urban climate monitoring using Raspberry Pi. In *2016 International Conference on Communication and Signal Processing (ICCSP)*. IEEE, 2008–2012.

Shifeng, Y., Chungui, F., Yuanyuan, H. and Shiping, Z. 2011. Application of IOT in agriculture. *Journal of Agricultural Mechanization Research*, 7, 190–193.

Shimizu, S., Sugihara, N., Wakizaka, N., Oe, K. and Katsuta, M. 2015. Cloud services supporting plant factory production for the next generation of agricultural businesses. *Hitachi Review* 64(1), 63–68.

Shuwen, W. and Changli, Z. 2015. Study on farmland irrigation remote monitoring system based on ZigBee. In *2015 International Conference on Computer and Computational Sciences (ICCCS)*. IEEE, 193–197.

Sicari, S., Rizzardi, A., Grieco, L. A. and Coen-Porisini, A. 2015. Security, privacy and trust in Internet of Things: The road ahead. *Computer Networks*, 76, 146–164.

Srbinovska, M., Gavrovski, C., Dimcev, V., Krkoleva, A. and Borozan, V. 2015. Environmental parameters monitoring in precision agriculture using wireless sensor networks. *Journal of Cleaner Production*, 88, 297–307.

Sushanth, G. and Sujatha, S. 2018. IOT based smart agriculture system. In *2018 International Conference on Wireless Communications, Signal Processing and Networking (WiSPNET)*. IEEE, 1–4.

TongKe, F. 2013. Smart agriculture based on cloud computing and IOT. *Journal of Convergence Information Technology*, 8(2), 210–216.

Torres-Ruiz, M., Juárez-Hipólito, J. H., Lytras, M. D. and Moreno-Ibarra, M. 2016. Environmental noise sensing approach based on volunteered geographic information and spatio-temporal analysis with machine learning. In *International Conference on Computational Science and Its Applications*. Springer, Cham, 95–110.

Uddin, M. A., Mansour, A., Le Jeune, D. and Aggoune, E. H. M. 2017. Agriculture internet of things: AG-IoT. In *2017 27th International Telecommunication Networks and Applications Conference (ITNAC)*. IEEE, 1–6.

Vermesan, O. and Friess, P. (eds.). 2013. *Internet of Things: Converging Technologies for Smart Environments and Integrated Ecosystems*. River Publishers.

Vijay, G., Bdira, E. B. A. and Ibnkahla, M. 2010. Cognition in wireless sensor networks: A perspective. *IEEE Sensors Journal*, 11(3), 582–592.

Watthanawisuth, N., Tuantranont, A. and Kerdcharoen, T. 2009. Microclimate real-time monitoring based on ZigBee sensor network. In *SENSORS, 2009*. IEEE, 1814–1818.

Wenshun, C., Shuo, C., Lizhe, Y. and Jiancheng, S. 2013. Design and implementation of sunlight greenhouse service platform based on IOT and cloud computing. In *Proceedings of 2013 2nd International Conference on Measurement, Information and Control*. IEEE. 1, 141–144.

Windsperger, B., Windsperger, A., Bird, D. N., Schwaiger, H., Jungmeier, G., Nathani, C. and Frischknecht, R. 2019. Greenhouse gas emissions due to national product consumption: From demand and research gaps to addressing key challenges. *International Journal of Environmental Science and Technology*, 16(2), 1025–1038.

Wortman, S. E. 2015. Crop physiological response to nutrient solution electrical conductivity and pH in an ebb-and-flow hydroponic system. *Scientia Horticulturae*, 194, 34–42.

Yan-e, D. 2011. Design of intelligent agriculture management information system based on IoT. In *2011 Fourth International Conference on Intelligent Computation Technology and Automation*. IEEE. 1, 1045–1049.

Yang, Y., Wu, L., Yin, G., Li, L. and Zhao, H. 2017. A survey on security and privacy issues in Internet-of-Things. *IEEE Internet of Things Journal*, 4(5), 1250–1258.

Yoo, S. E., Kim, J. E., Kim, T., Ahn, S., Sung, J. and Kim, D. 2007. A 2 S: Automated agriculture system based on WSN. In *2007 IEEE International Symposium on Consumer Electronics*. IEEE, 1–5.

Zhang, S., Chen, X. and Wang, S. 2014. Research on the monitoring system of wheat diseases, pests and weeds based on IOT. In *2014 9th International Conference on Computer Science & Education*. IEEE, 981–985.

Zhao, J. C., Zhang, J. F., Feng, Y. and Guo, J. X. 2010. The study and application of the IOT technology in agriculture. In *2010 3rd International Conference on Computer Science and Information Technology*. IEEE, 2, 462–465.

Sample Question Papers

Model Question Paper 1

PLASTICULTURE ENGINEERING AND TECHNOLOGY

1. a) Discuss the types of plastics used in soil and water conservation. Also, explain the future prospect of plasticulture in India.
   b) Describe the design steps of poly-lined farm pond
2. Write short note on
   i) Use of plastics in in situ moisture conservation
   ii) Plastic in postharvest management
   iii) Application of plastic in rainwater harvesting
3. a) What is plastic mulching? Describe the role of plastic mulching to enhance water use efficiency.
   b) Describe design steps of the drip irrigation system. Also, explain its advantages and disadvantages.
4. Write short note on
   i) Use of plastic in the subsurface drainage system
   ii) Plastic to prevent water losses from canals, ponds, and reservoirs
   iii) Soil conditioning and soil solarization
5. a) Describe the application of plastic in protected cultivation.
   b) How is plastic used for nursery production?
6. Write short note on
   i) Poly house drying
   ii) Role of plastics in drying and storage
   iii) Plastic crop cover
7. a) Discuss the application of plastics in aquaculture engineering. Also, enlist agencies involved in the promotion of plastics in agriculture.
   b) Describe the application of plastic for inland fisheries.
8. Write short note on
   i) Silage bag technique and its design
   ii) Vermi bed technology
   iii) Status and scope of plasticulture in India and the world

Model Question Paper 2

PLASTICULTURE ENGINEERING AND TECHNOLOGY

1. Discuss the present status of plastic materials and the future prospect of plasticulture in India.
2. Write short note on
   a) BIS standard in plastic materials
   b) Government policies to promote plastic in agriculture
   c) Describe the role of plastic for water management in rainfed area

3. Write short note on
   a) Use of plastic to enhance water use efficiency
   b) Plastic to prevent water losses from canals, ponds, and reservoirs
4. Describe the application of plastic in protected cultivation.
5. Write short note on
   a) Plastic in nursery raising
   b) Plastic packaging and use of plastic in food grain processing
6. Discuss the application of plastics in farm equipment and machinery. Also, enlist the agencies involved in the promotion of plastics in agriculture.
7. Write short note on
   a) Silage bag
   b) Role of plastics to reduce drudgery in agriculture
   c) Status and scope of plasticulture in the world

Model Question Paper 3

PLASTICULTURE ENGINEERING AND TECHNOLOGY

1. Discuss the application of plastics in agriculture and the future prospect of plasticulture in India.
2. Write short note on
   a) Types and quality of plastics used in agriculture
   b) Enlist the agencies involved in the promotion of plastics in agriculture at the national level
3. Describe the role of plastics in increasing crop productivity.
4. Explain the design steps of the LDPE farm pond.
5. Write short note on
   a) Use of plastic in moisture conservation
   b) Plastics for losses of water from canals, ponds, and reservoirs
6. Describe the application of plastics in the food grain processing and packaging sector.
7. Write short note on
   a) Plastic in protected cultivation
   b) Plastic packaging and use of plastic in food grain processing
8. Discuss the application of biodegradable plastics for crop production. Also, describe the application of plastics in farm equipment and machinery.
9. Write short note on
   a) Plastics vermin beds and leno bags
   b) Role of plastics to reduce drudgery in agriculture
   c) Use of plastics in farm sheds and vacuum packaging

Index

A

Abandoned, 219
Active Heating, 249
Active Refrigeration, 249
Adaptability, 95
Advantages, 41, 71, 96, 185, 191, 239, 315
Aerator, 377
Aeroponic, 320, 322, 323
Agencies, 16
Agricultural, 188
Agriculture Drainage, 129
Agriculture Need, 311
Air Permeability, 188
Air-Pruning Pots, 291
Air Release, 79
Analysis, 350
Animal Production, 227
Animal Shelter, 238
Anionic Polymers, 138
Anti-Hail, 189
Anti-hail Net, 189
Anti-Insect, 191
Application, 23, 361
Application of Plastics, 226
Approximate Irrigation, 160
Aquaculture, 214, 218, 220
Areas of Application, 300
Arid Regions, 335
Assessment, 164
Atmospheric Packaging, 206, 277
Auger, 372
Automation, 390

B

Benefits, 330, 389
Benefits of Irrigation, 149
Bicycle Sprayer, 369
Biochar, 137
Biodegradable Films, 206, 276
Biodegradable Polymers, 29
Biological Mechanism, 139
Bioplastic, 293
Black Mulch, 301
Block Shear, 351
Blue Mulch, 302
Budding, 372
Buildings, 250
Bulk Bins, 267
Bunker Silo, 230
Bunker Walls, 234

C

Cages, 213
Calculation, 160
Canal Lining, 112
Canal Lining, 114
Canopy Temperature, 159
Capacity, 106
Capacity, 94
Catfish Hatchery, 214
Cationic Polymers, 138
Centrifugal, 88
Chain Polymers, 26
Channel Capacity, 114
Chemical, 141
Chemical Application, 43
Chemical Mechanism, 139
Chemical Properties, 32
Chemical Use, 313
Clamp Shells, 271
Classification, 20, 205
Climate, 131, 140, 164, 333
Climatological Approach, 158
Clogging, 45
Cold Arid, 335
Collection, 341
Color, 187, 342
Color and Thickness, 234
Color Mulching, 299, 301
Color of Film, 299
Commanded Area, 114
Components, 49
Components, 50, 97
Composts, 136
Compression Member, 351
Condensation Polymers, 27
Conditioners, 136
Condition Monitoring, 388
Conservation, 227
Constraints, 331
Construction, 230
Consumer Packs, 276
Containers, 269
Contribution, 218
Control of Pests, 141
Conveying, 204
Cost of Installation, 252
Covering Materials, 251
Cover Silage, 234
Creek-Fed, 117
Criteria, 163
Critical Section, 350
Crop Coefficient, 172
Crop Coefficients, 173
Crop Residues, 136
Crops, 203, 229
Crop Yield, 44
Cultivation, 314
Current Usage, 143

D

Dead Loads, 347
Deficit Irrigation, 148

403

Definition, 1
Design, 63, 75, 102, 333
Design and Components, 59
Design of Pump, 64
Design of Sub-main, 64
Design Strength, 350
Die and Screw, 35
Disadvantages, 45, 319
Discharge, 75
Door, 342
Double Wall, 342
Drawbacks, 323
Drier, 345
Drip Irrigation, 48, 49, 70, 75, 94, 165
Drippers, 75
Dripper System, 47
Drip Tap, 47
Drying, 197, 199, 203, 342
Drying of Crops, 197
Dugout Farm Ponds, 116

E

Eco-Friendly, 93
Effect, 301
Efficacy, 142
Elasticity, 31
Elastomers, 27
Electrical Properties, 32
Embankment, 117
Emitting Devices, 63
Environmental Impacts, 312
EPS, 276
Equipment, 87, 320
Estimation Methodologies, 167
Evaluation, 217
Even-Span, 246
Extrusion, 23

F

Factors Affecting, 139, 149
Farm Machinery, 365
Farm Ponds, 116, 117, 125
Feasibility, 324
Feed Dispenser, 214
Fenced Silo, 232
Fertilizer Dissolver, 88
Fertilizer Injection, 87, 88
Fertilizers Spreader, 367
Fertilizer Tank, 87
Fibers, 27
Field Curing, 203
Field Handling, 203, 262
Field Heat, 203, 263
Field Heat, 203
Film Blowing, 23
Film Manufacturing, 3
Film Properties, 19
Films Used for Packing, 213
Filtration Systems, 81
Fish Feeder, 214
Fishnet, 212
Fish Pickle, 218

Fish Retailing, 217
Floods, 114
Flow Control, 81
Fly Ash, 138
Fodder, 227
Food Grains, 276
Formation, 16
Framed Timber, 250
Fresh Fish, 216
Fruit, 10
Full Irrigation, 148
Functional Design, 351
Fungi and Bacteria, 141
Furrow, 248
FYM, 136

G

General Principles, 41
Glass Greenhouses, 252
Grader, 370
Grading and Sorting, 204
Grafting Knife, 372
Grafting Tools, 372
Granular Matrix Blocks, 157
Gravel, 83
Gravimetric Method, 157
Greenhouse, 244, 245, 248, 249, 252, 254, 335, 351
Greenhouse Microclimate, 351
Greenhouse Orientation, 352
Greenhouse Technology, 330
Green Manure, 136
Green Mulch, 302
Ground, 342
Growing Containers, 93
Growing Media, 286
Grown, 322
Guidelines, 160
Gypsum, 137
Gypsum Blocks, 156

H

Hail Net, 189
Hand Tools, 368
Heteropolymer, 27
High-Density, 23
High-Pressure, 24
High-Temperature, 29
Hilly and Mountainous, 331
History, 40, 283, 314
Hi-Tech Greenhouse, 253
Homomer, 27
Hydrocyclones, 85
Hydroponics, 315, 319, 329
Hydroponic Setups, 316

I

Importance, 12
Improved Fertilizer, 43
Improved Sickle, 370
Improved Soil, 141
Infrared Thermometer, 158

Index

Initial Assessment, 60
Injection Molding, 361
Inline Drip, 47
Inorganic Polymers, 28
Insects, 188
Insert Trays, 269
Installation, 50
Insulation, 342
IoT, 380, 385, 389
IoT Sensors, 382
IoT Software, 385
Irrigation, 93, 147, 386
Irrigation Depth, 160
Irrigation Interval, 148
Irrigation Interval, 149
Irrigation Scheduling, 147, 149, 151, 165, 166, 174
Irrigation Techniques, 304

J

J-Turbo, 47

K

Kinds of Silos, 230
Knapsack Sprayer, 367, 370

L

Ladakh, 335
Land Use, 312
Laterals, 63, 77
Layout, 64
LDPE, 23, 24
Lean, 246
Leno Bag, 276
Less Maintenance, 114
Light Transmittance, 189
Lime, 137
Limitations, 97, 300
Linear-Chain, 26
Lining, 125, 234
Lining Material, 125
Live Loads, 348
Livestock, 239
Livestock Shelter, 237
Load Combinations, 350
Low-Cost, 197, 252, 332, 342, 345
Low Tunnels, 183, 254
Lysimeter, 128, 165

M

Main and Submain, 76
Maintenance, 51
Major Requirements, 278
Manifold, 78
MAP, 278
Marine Litterm, 218
Materials, 186
Materials Used., 359
Measuring Devices, 154
Mechanical Properties, 31
Mechanical Properties, 187

Medium-Cost, 253
Mesh Size, 187
Meteorological Hazards, 188
Method of Joint, 349
Methods, 154
Micro-Irrigation, 5, 40, 41, 46, 52, 54, 59, 111
Micro Jets, 48
Miniature Forms, 254
Mini Sprinklers, 48
Modeling Approach, 166
Modification, 173, 206, 277
Moisture Conservation, 112
Moisture Distribution, 45
Molecular Mass, 29
Monitoring, 154
Mulches, 7, 184, 300, 304
Mulching Machines, 365
Mulch Laying, 303

N

Natural Polymers, 26
NCPAH, 16
Nematodes, 141
Net, 189
Net houses, 185
Nethouses, 185, 285
Netting, 273
Net Types, 186
Neutron Probe, 15
Nonpeak, 160
Nonreturn Valves, 81
Notation, 168
Nursery Containers, 289
Nursery Management, 285

O

Observation, 151
Off-Stream, 117
Online Emitter, 47
Open Sun, 203
Optical Properties, 32
Organic Polymers, 28
Organic Soil, 136
Oxygen Permeability, 235

P

Packages, 206, 267
Packaging, 10, 205, 216, 218, 264
Packaging Systems, 205, 265
Parameter, 303, 347
Peat, 137
Pens, 213
Periphery, 45
Pest, 389
PET Bottles, 218
Petroleum, 290
Physical, 29, 141, 291
Pipeline, 78
Pipes, 76
Pit, 230
Planning Irrigation, 160

Plant, 164
Plant Growth, 142
Plants and Soils, 151
Plastic, 19, 125, 215, 227, 231, 237, 254, 272, 361, 365
Plastic Applications, 365
Plastic Bags, 205, 266
Plastic Containers, 290
Plastic Debris, 220
Plastic Film, 140, 234, 235, 252, 278
Plastic Mulch, 131, 298
Plastic Packaging, 216
Plastic Pouch, 273
Plastic Sacks, 272
Plastic Tension, 273
Plastic Types, 361
Plasticulture, 11
Plasticulture Development, 1
Plasticulture in India, 14
Polyethylene, 21, 23, 26, 217
Poly House, 197, 199, 333, 342
Poly House Ponds, 214
Polymer Extrusion, 32
Polymerization, 24, 35
Polymers, 19, 20, 26–29, 185, 225
Polypropylene Packaging, 217
Poly Tunnel, 345
Ponds, 116
Postharvest, 10, 225, 261
Postharvest Management, 9
Potential, 54
Pot-In-Pot, 291
Pots, 85
Poultry Feeders, 367
Preharvest, 225
Pressure-Regulating, 81
Pressure Relief, 81
Preventions, 113, 304
Principles, 329, 333
Procedure, 350
Production Agriculture, 9
Products, 225
Properties, 29, 291
Protected, 331
Protected Cultivation, 5, 51, 243, 329, 332
Protection, 188
Pruning, 373
Pruning Knife, 372
Purlin, 348

Q

Quality of Plastics, 4
Quonset Greenhouse, 249

R

Radiation, 188
Rake, 375
Recycling, 292
Red Mulch, 301
Reducing Water, 8
Reduction, 188

Reference Evapotranspiration, 166, 167
Reflectivity, 187
Refractive, 201
Removal, 203
Resistance Devices, 156
Restricted Root, 46
Reusable, 292
Reusable Plastic, 269
Ridge, 248
Rigid Panel, 252
Rigid Plastic, 206
Rigid Plastic, 267
Roof, 342
Roof Covering, 347
Root Zone, 45
Rupture, 350

S

Safety, 114
Saffron Corm, 370
Salinity, 44
Salt Accumulation, 45
Sand Separator, 88
Saw, 373
Sawtooth, 248
Schedules, 160
Scheduling, 164, 165
Scheduling Irrigation, 163
Screen Filters, 84
Sealing Methods, 232
Sealing Methods, 232
Seed-Rearing, 213
Seepage Reduction, 113
Selection, 229
Selection of Mulch, 03
Semisynthetic Polymers, 26
Sensors, 155
Settling Basins, 82
Sewage Sludge, 136
Shading Factor, 187
Shading Nets, 257
Shading Nets, 257
Shrink-Wrap, 205, 266
Silage, 228
Silage Bags, 228
Silage Cover, 11
Silage Making, 229
Silo, 229, 230
Site Characteristics, 333
Site Selection, 332
Sleeve Packs, 272
Slenderness Ratio, 35
Snow, 333
Snow Loads, 349
Soil, 160
Soil Conditioner, 135
Soilless Cultivation, 313, 315
Soilless Peat, 288
Soil Moisture, 140, 154, 155, 163
Soil Monitoring, 388
Soil Solarization, 138

Index

Soil Temperature, 139, 140
Solarization, 139, 142
Solarization Results, 140
Solar-Operated, 370
Solar Radiation, 188, 341
Sources of Water, 67
Specifications, 302
Specific Gravity, 216
Sprayer Tanks, 366
Spring, 117
Sprinkler Irrigation, 49, 50, 95, 97, 106, 165
Sprinkler System, 102
Stack Silo, 230
Status of Plastics, 15
Stomata Resistance, 158
Storage, 204, 278
Storage Management, 10
Storage Ponds, 117
Strength, 350
Stretch, 276
Structural Design, 330
Structure, 28, 185
Structures Framed, 250
Subsurface Drainage, 131
Suitability, 324
Surface Irrigation, 16, 164
Synthetic Polymers, 26
System Capacity, 63

T

Techniques, 303
Technology, 335
Tensile Strength, 31
Tensiometer, 154
Tension, 350
Texture, 186
Thermal Mechanism, 139
Thermal Storage, 342
Thermocouple Psychrometry, 158
Thermoplastic, 12, 20
Thermosetting Plastics, 12
Thermosetting Polymers, 13
Thermosetting Polymers, 28
Threads and Texture, 186
Tower Silo, 232
Transmissivity, 187
Transportation, 205, 279
Transport Properties, 31
Trays, 213
Tree Climber, 374
Tree Planting, 372
Troubleshooting, 51
Trowels, 375
Truss, 250
Twin-Wall, 47

U

Underground Pipeline, 127
Uneven-Span, 247
Uniformity Coefficient, 102
Unit Operation, 203
Unsealed Silos, 233
Usage, 8, 13
Use of Plastics, 212

V

Valves, 78, 81
Vegetable, 10
Vegetable Production, 341
Ventilation, 342

W

Walk-in Tunnels, 256
Walnut Propagation, 345
Water, 216
Water Application, 42
Water Balance, 162
Water Bodies, 125
Water Budget, 159
Water Distribution, 76
Water Harvesting, 115
Watering Can, 377
Waterlogging, 113
Water Management, 390
Water Quality, 87
Water Requirement, 93
Water Saving, 41
Water Sources, 44
Water Use, 312
Weather, 140
Weeds, 141
Weight, 347, 348
Weight of Truss, 348
Wetting Patterns, 41
Wheel barrow, 378
White Mulch, 301
Wind, 330, 333
Wind Loads, 348, 349
Window Drying, 201
Wind Pressure, 348
World Scenario, 52, 315, 328
Wrapped, 276

Y

Young's Modulus, 31

Z

Zenoah Reciprocator, 371